AN INTRODUCTION TO ORNITHOLOGY
AND
BIOLOGY OF THE BLUE ROCK PIGEON

AN INTRODUCTION TO ORNITHOLOGY
AND
BIOLOGY OF THE BLUE ROCK PIGEON

B. N. Bhattacharyya M.Sc., Ph.D., F.Z.S. (Cal.)
Former Reader and Head of the Department of Zoology, Bangabasi College, Kolkata
and Professor of Zoology, St. Edmund's College, Shillong (Meghalaya)
Member, International Committee on Avian Anatomical Nomenclature (ICAAN)

New Central Book Agency (P) Ltd
8/1 Chintamoni Das Lane, Kolkata 700 009
INDIA

An Introduction to Ornithology
and
Biology of the Blue Rock Pigeon

First Published : April 2008

ISBN : 81-7381-561-5

Publisher : New Central Book Agency (P) Ltd
8/1 Chintamoni Das Lane, Kolkata 700 009

Typesetter : Compu House
Ashirbad, 3rd Floor
Flat 2C
172 Regent Estate, Kolkata 700 092

Printer : New Central Book Agency (P) Ltd
Web-Offset Division, Dhulagarh, Sankrail, Howrah

Cover Designer : Soumen Paul

Cover Printer : Liba Graphics
90C Christopher Road, Kolkata 700 046

Proof Checking : Subas Maitra and Author

Project Team : Prabhat Jas, Somenath Ganguly

Price : **Rs 495.00** [Rupees Four Hundred Ninety-five only]

Tearful dedication to

Late Dr. Biswamoy Biswas

who first inspired me in the study of birds

and

Late Shefali Bhattacharyya

my wife who could not see the publication come true

CONTENTS

Part II

BIOLOGY OF THE BLUE ROCK PIGEON : *COLUMBA LIVIA* GMELIN

Centre for Environment & Development

Dr. A. K. Ghosh
Director

329 Jodhpur Park
Kolkata 700 068
Phone: (91) (033) 2424 9801
E-mail: cedkolkata@sify.com
cedkolkata@yahoo.com

Date : 08-10-2007

FOREWORD

"Ornithology" is one of the most studied areas in Indian Biodiversity. Contributions from both professionals and amateur ornithologists have enriched the subject over a century. The present contribution is an outcome of lifetime research of the author Prof. Dr. Biren Bhattacharyya. Dr. Bhattacharyya had his initiation in ornithology under noted expert Dr. Biswamoy Biswas in the Bird section of Zoological Survey of India at the Indian Museum. He has assiduously followed it up over decades and the present contribution is a logical conclusion of his lifelong passion. I am sure students and researchers will find the publication most useful. I extend my heartfelt congratulations to Prof. Dr. Bhattacharyya for his excellent contribution to the science of ornithology in India.

A. K. Ghosh, Ph.D.
Member, Biodiversity Authority
Government of India
Former Director
Zoological Survey of India

PREFACE

The book depicts glimpses in the life of birds in all aspects, with particular attention to Indian birds.

The book has been written in two parts in a simple language with recent information as far as available. It is intended for both the undergraduate (Pass and Honours courses) and Post-graduate students studying Zoology. A number of chapters have been included in the book covering syllabi for these courses. Alongside, the book may be useful to the young researchers who are keen to learn the fundamentals of Ornithology and the Biology of the Blue Rock Pigeon. Any naturalist or a bird-watcher—or even a common man interested in the study of the structure and function of birds in more details—will find this book immensely helpful and knowledgeable. As for the students of colleges and universities, a pigeon or a fowl is considered as a model textbook specimen for the purpose of dissection, demonstration, description and explanation of the intricate life processes of a flying bird. The author has described the bodily organisation of a pigeon with that very view in mind, although a reference to the fowl (*Gallus* sp.) has also been cited here and there.

In Part I, the diversity of habit and habitats of birds, their classification, origin, evolution, anatomy, feeding and drinking, endocrine glands, reproduction, nest building, songs and song dialects, flight, migration and other behavioural aspects of bird-life have all been described briefly and lucidly to fit into the requirements of the beginners as well as those interested in higher studies on birds. In this part, for an easy reference of the readers, chapters like (i) A profile of some common Indian Birds, (ii) The birds that lost their flight, (iii) Endangered and threatened birds of India and conservation strategies to protect them, (iv) Economic importance of birds, and (v) Agricultural Ornithology have also been included as added attraction.

Part II of the book deals with the biology of the Blue Rock Pigeon, *Columba livia* Gmelin, as an illustrative model of a flying bird that is included in the curricula of many colleges and universities. An uptodate knowledge of the structures and functions of various organs and organ-systems of the Blue Rock Pigeon has been provided in this part of the book.

The author has taken a rather long time in the preparation of the manuscript due to the unfortunate demise of his wife and other predicaments which delayed publication of the book.

This book is an humble endeavour of the author to create interest among the students in the study of the fascinating world of birds. Any omission in the text or mistake that might have crept in inadvertently is regretted.

The author will gratefully acknowledge comments and suggestions from ardent readers towards further improvement through revision and corrections wherever found necessary.

2nd October 2007 **B. N. BHATTACHARYYA**

ACKNOWLEDGEMENTS

The list of my colleagues, friends and well-wishers to whom I am obliged is so long that I must apologise for my inadvertent omission of any of their names.

I will remain ever—grateful to Late Dr. Biswamoy Biswas (of the Zoological Survey of India, Kolkata) who was not only my research-guide but also took pains to look into most of my publications. I am indebted to Professor W. J. Bock of Columbia University, New York, and Professor D. G. Homberger of Louisiana State University, USA, for their help and valuable suggestions in shaping my works in avian morphology. I sincerely thank Professor Julien J. Baumel and Dr. J. C. Vanden Berge of USA, Dr. Gart Zweers and Dr. H. Berkhoudt of Leiden, Netherlands, and Dr. Mikio Yasuda of Japan for helping me with relevant literature.

I express my boundless gratitude to my elder-brotherly Late Professor Asok Ghosh of the Department of Zoology, Calcutta University, for his valuable suggestions in arranging the text of the book. From this Department, I must specially thank Professor B. Manna for introducing me to the publisher; Professors Naresh Datta, Pulak Lahiri, Samir Banerjee and Dr. Gautam Saha for their help and assistance in the publication. I thank Dr. Dipankar Ghosh of WWF for helping me with literature. I also thank the President and the Hony. General Secretary of the Zoological Society, Kolkata, for their best wishes in the publication of the book.

I sincerely thank my ex-colleague and friend, Professor B. Ray (Retd.) of the Department of Zoology, Bangabasi College, Kolkata, and his co-authors (Prof. B. Basu, Prof. S. Guha and Prof. P. Ray Basu) for my reference to their earlier textbooks in Zoology (in Bengali). I am indebted to Prof. Ray for his valuable advice in the publication of the book. I also thank the Head of the Department and all the teachers of the Department of Zoology, Bangabasi College, for their eagerness to see my venture materialised.

I thank Dr. N. C. Ghorai and Prof. Arup Sinha (Retd.) of Presidency College and Prof. Dipankar Sengupta of Moulana Azad College for their initiative and interest in the publication of the book. I thank Mr. Ranjan Basu and Dr. Amalesh Misra of Golf Green as wellwishers of this publication. Mr. Basu helped me with some literature on the protection of birds in West Bengal. I thank Dr. Sudhin Sengupta who helped me with references of his works, and Prof. P. Sensarma, my ex-colleague, for his best wishes for the publication of this book.

I sincerely thank Dr. Asish Kumar Ghosh, Ex-Director of the Zoological Survey of India, Kolkata, for his best wishes to see the publication come to light and for his writing the Foreword of this book.

I owe apology for not mentioning the names of many of my close friends and relatives whose eagerness for the publication must be thankfully acknowledged. Among my close family kins,

I specially thank Mrs. Patralekha Chatterjee, Miss Tanuka Banerjee and Mr. Basab Bhattacharyya for their help and assistance in multifarious ways.

I will remain grateful to the publishers of New Central Book Agency (P) Ltd., Mr. Amitabha Sen, Director, and to his competent associates: Mr. Siladitya Sinha Roy, the Chief Executive Officer; Mr. Soumen Paul, the Chief Producer and Designer of the book; Mr. Prabhat Jas, in-charge of Production; Mr. Subas Maitra, the Senior Proof-reader; Mrs. Sushmita Das, the composer; Mr. Aparup Mukherjee, the artist; Mr. Asish Ghosal—for typing a portion of my manuscript, and all others who are in some way or other connected with the publication of this book.

B. N. BHATTACHARYYA

PART I

GENERAL ORNITHOLOGY

1

Introduction

If 'A thing of beauty is a joy for ever', then the birds must be classed as a symbol of beauty and aesthetic pleasure for us. The beautiful colour combinations of their plumage, the melodious notes of some songbirds, their spectacular flight on wings and dive in water, and their artistic skill in nest building evoke joy and admiration of mankind since time immemorial.

Through millions of years, the origin and evolution of the plants and animals on the surface of the earth have been the most spectacular phenomenon in the organic evolution. Among the vertebrate animals, the birds and mammals have acquired the greatest adaptational diversity, but, because the birds have conquered the air, they have occupied a very distinctive position in the animal kingdom. Even a few hundred years before the birth of Christ, Aristotle (384-322 BC), the great Greek philosopher and naturalist, described some aspects of bird life such as volant adaptation, feeding habit, nest building, migration. But to naturalists of those days and to many others in the later centuries, many aspects of bird life still remained mysterious. During the last few hundred years, improved researches and scientific study of birds have unveiled most of the mysteries of bird life—but not all.

Originated from a distinct and divergent group of dinosaurian reptile, either in Triassic or in the Jurassic period of Mesozoic era, the modern birds are presumed to have evolved through several transitional stages in the remote past. Prof. T. H. Huxley (1867), being impressed by the pronounced reptilian affinities of birds, designated them as 'Glorified Reptiles'. On the other hand, the flight adaptations in birds have been acquired and improvised at the level of such an ingenious mathematical precision that many authors have compared the birds to man-made "heavier than air machine"—an automatic aeroplane gifted by the nature.

Ornithology (*Ornithos*: beaked animals; *logos*: knowledge) is a specialised discipline in Vertebrate Zoology. It deals with the study of birds in all scientific aspects—relating to their structure and classification and their habits, song, flight and value to agriculture as destroyer of insect pests (Hutchinson Dictionary of Science). Integration of the vast array of knowledge gathered form the study of birds in the last few decades through application of modern techniques and technologies forms the prime objective in the study of Ornithology. With this objective in view, the present text has been described in two parts.

In Part I, the diversity of the habits and habitats of birds, their classification, origin and evolution, anatomy, feeding and drinking, songs and song dialects, endocrine glands, reproduction, nest building, flight, migration etc. have been described in brief. In the origin and evolution of birds, a description of the legendary reptile-bird, the *Archaeopteryx,* has been given with the analyses and opinions by different scientists on the impact of the discovery of *Archaeopteryx*. Also in this part chapters have been included on (i) A profile of some common birds of India (both resident and migratory), (ii) The birds that lost flight, (iii) Endangered species of birds and their conservation, (iv) Economic importance of birds, and (v) Agricultural Ornithology. In Part II, the biology of our common pigeon, the Blue Rock Pigeon (*Columba livia* Gmelin) has been described in some details as a model representative of a carinate (keeled/ridged) bird. The pigeon is considered as easily available specimen for both theoretical and practical studies in the colleges and universities. That is the main reason for describing a pigeon in totality of its structures and functions.

From the long list of Indian birds, the author has selected nearly three dozens of common species that have diverse habits, habitats and feeding adaptations. The interesting behavioural patterns in avian feeding, drinking, courtship, reproduction, defending territories and the seasonal migratory movements in birds need to be discussed for a complete understanding of the structural adaptations. But that has not been possible within the limited scope of this text that envisages to convey only an elementary knowledge in Ornithology to the students who may pick up herein fundamentals of the subject and be interested to learn more.

A chapter has been included to highlight the endangered and threatened birds of India and possible measures for their protection and conservation. It is a global problem that has been a matter of great concern for the conservationists since the past few decades.

The nomenclature used in the text for the skeletal parts, joints, ligaments, muscles and in the anatomy is based on the "Nomina Anatomica Avium" (NAA, J. J. Baumel et al eds, 1993) as well as on the conventional terminology used in the current textbooks.

2

Birds From Antiquity

It is on record that mankind, even in the prehistoric days, was in the habit of bird watching, more so with amazing curiosity than being urged by any scientific inquiry. In the human habitations of the Neolithic Age, even six to fifteen thousand years before the birth of Christ, the wall-drawings of birds bear testimony about the eagerness and curiosity of the people of that age about birds. The earliest cave painting of a bird—probably a totem, may be traced back to 15,000 years (Lascaux cave, France). In ancient scripts of early human civilisation, in the Bible and in ancient Sanskrit literature, mention has been made of birds like Cuckoo, Myna, Duck, Parrot and Peacock. In the 'Puranas' ,'Ramayana' and 'Mahabharata' quite a number of bird species and their beneficial and harmful activities have been mentioned. In ancient Greece, it was Aristotle (384-322 BC) who first of all recorded many facts about the morphology, internal anatomy and even embryology of many birds. It is remarkable indeed that he devised a system of classification of the birds known to him. In the 1st century A.D., Pliny the Elder made a detailed description about birds as a part of his voluminous works on Natural History. Again, from the time of Renaissance—that commenced from the 12th century onward–Fredrick II, Lamarck, Darwin and many others had a truly scientific approach in describing the life of birds. Thus they made valuable contribution in the field of Ornithology.

3

Definition—Number—Distribution and Habitat of Birds

Some authors have defined a bird by a single unique feature as a "feathered biped'. This definition is too precise, and, to make it more convincing and broad-based, Hans Gadow (1889) and Gadow and Selenka (1891) defined as "A bird is an oviparous, warm-blooded and amniotic vertebrate whose fore-limbs have been modified as wings, the metacarpus and fingers bear feathers and whose first of the four toes in each foot is called Hallux".

This definition is brief and comprehensive, and it clearly isolates the birds from the reptiles or the mammals within the amniotic vertebrates.

The number of bird-species is much greater than that of any other vertebrate group, except the number in fishes. Brodkorb (1971) estimated that the Pleistocene Avifauna contained 4,806 non-passerines and 5,847 passerines, or a total of 10,653 species. According to his estimate, at the beginning of 1970s, the number of living species of birds was about 8,600. Brodkorb (1971) also estimated that the total number of birds, past and present, was approximately 154,000 species. King and McLelland (1984) estimated the total number of bird-species in the world to be 8,948, although the number may not be absolutely rigid. And, taking into account the number of sub-species, the total comes to about 30,000. In Indian sub-region, the species-subspecies number comes to about 2,500, of which the number of winter-visitors is about 500. However, more recently, Monroe and Sibley (1993) published "A world Checklist of Birds" (University Press, New Haven and London) containing 9,702 species of birds based on their "Taxonomy of Birds of the World" (1990) and its "Supplement" (1993). Such an estimate is also subject to revision since (i) more than a decade has elapsed when the estimate was made, and (ii) different opinions held by avian systematists to ascertain the status of species, subspecies and geographical races. So it would not be incorrect to make a statement that at present the number of species of birds in the world stands below 10,000. As regards total number of birds now prevailing on earth, Young (1981) has given an approximate low estimate of about 100 billion birds (10,000 crores) which number—estimated about 26 years ago—is also subject to a revision at present.

The birds have a worldwide distribution, except in some perpetually snow-clad south polar regions. Because of the worldwide distribution of birds, P. L. Sclater and W. L.

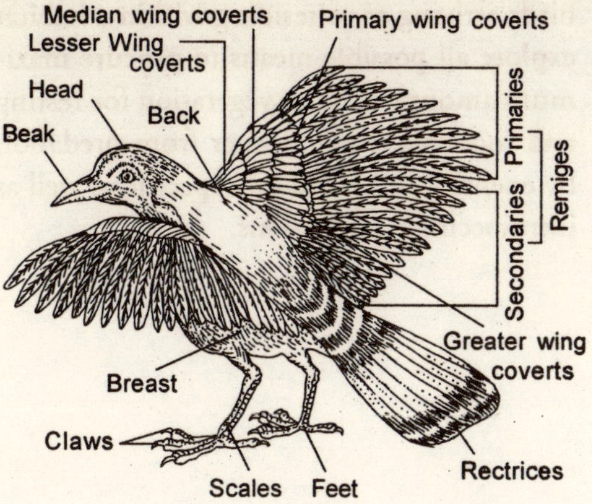

Fig. 3.1: Generalised parts of a bird

Sclater (1858) considered avifauna as an important criterion in their classification of the "Zoogeographical regions". However, the number of bird species and subspecies varies in different continents of which South America or the Neotropical region possesses the largest number of unique endemic forms of birds and mammals (Keast, 1969). Evolving from the middle or late Jurassic period, diversifying throughout the Cretaceous period and then crossing and surviving through the Cretaceous-Tertiary boundary (KT boundary), the present-day geographical distribution of the surviving avifauna is the outcome of multiple factors which changed and shaped the surface of the earth through long geologic time scale to the present state. Such factors broadly include geophysical conditions of the earth, climate, topography, vegetation, barriers, glaciations and interglacial periods, ocean bottom spreading and the large bodies of water isolating land masses. The existing avifauna had to adapt themselves to the changing environmental conditions. This led to diverse faunal characteristics in different zoogeographical regions.

With the acquisition of volant life, the birds enjoy the greatest potentiality in geographical distribution, as the usual geographical barriers are no limiting factors to them (expect vast stretches of the seas and oceans for the terrestrial birds). The birds are able to adapt themselves to almost all types of habitats and food-niches. Thus they inhabit altitudinous mountains, plateau, open field, forest, shrubs, savanna, deserts and semi-deserts, marshes and wetlands, as also small to large bodies of water—the ponds, rivers and the seas. However, selection of the habitat is an important phenomenon in the life of the birds that are remarkably adapted to live in contrasting habitats. The ecosystems of habitats also change with time, and very often the population of allied species adjust themselves suitably to some amount of overlapping of habitats and the food-niches. The birds are a very specialised group of vertebrates, and, as such, they have been able to occupy every possible habitat and niche on the earth, ranging from the Arctic belt to the Antarctica. The biotic and abiotic factors of the environment forming complex ecosytems determine the characteristics of different habitats. If the habitat refers to the ecological unit of space occupied by one or more than one species of birds, the ecological niche, on the other hand, refers to the functional role or the 'profession' of an individual of a species in relation to individuals of other species in the community, and so it is closely interrelated to the habitat.

In bird community it has been observed from both the migrant and resident birds that the most successful ecological niche provides a balanced combination of the habitat occupied by the individuals and their behavioural responses to the habitat. The factors contributing to the productivity of the habitat-niche complex must be sustainable for the individuals living within the habitat, but not completely excluding factors of the surrounding habitats. Thus the individuals of a species of birds arriving at or residing within a habitat explore all possible means to procure maximum amount of food, vegetation for resting and roosting, locate danger from predation by enemies and avoid interspecific as well as intraspecific competitions.

4

Shape & Size in Birds: Uniformity in Diversity

Although there is a wide range of variation in shape and size, colouration, arrangements of feathers, bill-shape, feeding habit, habitat etc. among different species of birds, an overall uniformity is observed in the spindle-shaped body form in all birds. Comparable to other vertebrates, much more similarities are observed in the general characteristics of the body in all birds. The birds possess a remarkable trend of uniformity among much diversity. The uniformity with respect to body shape has been attained by the birds mainly due to their flight adaptation through millions of years.

Spectacular diversity is observed as regards size of the birds. The African Ostrich (*Struthio* sp.) is the largest living bird, measuring about 8 feet (2.4 meters) in standing height and about 140 kilograms in weight. The extinct flightless birds were, however, much more gigantic than the modern ratite birds (e.g. ***Aepyornis***, the elephant bird). Young (1981)states that the egg of ***Aepyornis*** weighed more than 10 kg, which is a unique example of the largest single cell that ever existed. Among the Indian birds, the Sarus Crane (*Grus antigone*) and the Himalayan Bearded vulture (*Gypaetus barbatus aureus*) are the largest. The former has a length of about 5 feet (1.5 meters) and the standing height of 4-5 feet (1.2-1.5 meters) and the latter has a wing spread between 8.5 and 10.5 feet (2.6-3.2 meters). Tickell's Flower pecker, *Dicaeum erythrorhynchos,* is the smallest among the flying Indian bids, whereas the Cuban Humming Bird (*Mellisuga helenae*, the Bee Humming Bird) is the smallest in the world.

Sexual Dimorphism: Size differences in birds are very common in sexual dimorphism. Mayr (1988) comments: "Orinthologists showed, earlier than students of other groups, that sexual dimorphism and age difference often result in greater difference between male and female than are found between closely related species. This fooled Linnaeus into describing male and female Mallard (*Anas platyrhynchos*) as well as adult and immature Northern Goshawks (*Accipiter gentilis*) as different species. There is perhaps no other group of organisms in which there are so many families with strong sexual dimorphism as we find in ducks, the gallinaceous birds, birds of paradise, humming birds, and many groups of songbirds."

Fig. 4.1: A. *Grus antigone* (Sarus Crane); largest flying bird in India; standing height 156 cm; B. *Dicaeum erythrorhynchos* (Tickell's Flowerpecker); smallest bird in India; 8 cm long

5

Plumage, Plumage Colouration and Diversity in Colour

Among various epidermal structures in birds the feathers of different sizes and types form the plumage which—besides being the most vital organs in flight—offers a wide range of colour patterns. The colouration in feathers results from combination of various factors of which the production of pigments such as melanin and carotin depends on the extent of assimilation through food. These two pigments produce red, yellow and orange colours in the plumage of birds, together with dark brown, black and various other intermediate shades. Bologna (1995) reports that a particular structure of the birds—combined with a pigment that absorbs rays with a longer wave length–determines blue and white (except the white of albino birds). Birds' inability to synthesise specific pigments has given rise to several pigment aberrations such as albinism, isabellism ad acyanism.

Turacin is an interesting pigment which may fade or turn pale on the plumage of a living bird. Turacin is present in the dark-red markings on the wings of a bird called Turaco (e.g., *Turaco macrorynchus*; 13 species; Order Cuculiformes; Fam. Musophagidae) distributed in tropical Africa. If this bird is held by hand, the hand is smeared with the reddish colour of turacin, although the bird is not affected by this loss of pigment. Bologna (1995) reports that the green colour of Turaco is due to the presence of another pigment called **Turacoverdin**. This pigment, slightly differment from turacin,is reponsible for the rich green clour on the plumage. However, some modifications in plumage coluration may also occur in the worn out feathers or due to the effects of dust and dirt, environmental pollution, reflection and diffractions of light.

DIVERSITY IN COLOUR AS ADAPTIVE DEVICE

One of the greatest attributes of birds in general is the remarkable diversity in their body colour, particularly among the birds in the tropics. Although quite a few species of birds exhibit uniform body colour (Black Crow, White Egret, Roseringed Parakeet, etc.), most birds exhibit beautiful colour variation in their plumage. Some birds resemble the colour of their environment either to protect themselves (Protective coluration) or to outwit the insects, arachnids and other animals in their attempt for predation. Such colouration may thus be an adaptive device for the Willow Ptarmigan *(Lagopus* sp. Order Galliformes*)* which changes the brown colour of its plumage to white during the winter when areas around this bird remain completely covered with snow. The characteristic markings on the eggs of many birds are also adaptive, either matching with the colours of the environment for protection or for recognition by the particular species of birds. The resemblance of colouration in birds—either bright or dull—to the colour of the environment is mostly adaptive and has a greater survival value.

Many birds exhibit sexual dimporphism–not only based on size but also on plumage coluration. During breeding season, the beautiful coluration of feathers in birds (more often in males) is very helpful to attract the opposite sex. The males of Large Indian Parakeet, the Fairy Bluebird, the Purple Sunbird and the Koel are brightly and differently coloured from the females during the breeding season. Young (1981) comments that in most species of birds, colouration is a compromise between concealment and display, and the cryptic colour is most common in birds. Like an insect using warning coluration to avoid predation by birds or any other animal, the birds themselves sometimes use the same strategy of warning colouration by possessing some distinct colour patterns.

The external colouration of the birds' body is biologically highly significant.

6

Temperature Regulations

J. Z. Young (1981) comments:"It is likely that in evolutionary history the function of heat regulation came first. The two main functions, of heat conservations and flight, are indeed today performed by feathers of different types."

Birds and mammals—being homoiothermic and endothermic animals—possess unique devise of regulating their body temperature within the optimal range between 90°F (32°C) and 106°F (42°C). In extreme case the body temperature may even reach 44°C in some passeriform birds. In birds this has been achieved due to their higher metabolic rate, absence of sweat glands in the skin, and feathers acting as a perfect insulating covering. However, endothermy has also been achieved by some of the extinct reptiles (e.g. some thecodonts, dinosaurs, pterosaurs).

Adaptation to temperature regulation in birds has been exhaustively studied by Midtgard (1980, 1981, 1984 and 1988) and Baumel et al (1983). In homoiothermic animals such as bird, the circulatory system is very important in thermoregulation. According to Midtgård (1988), three different types of specialisation of the circulatory system concerned with thermoregulation are recognised in birds. These are (a) **arteriovenous anastomoses (AVAS)**, (b) **Venous plexuses**, and (c) **Arteriovenous (AV) heat exchange systems** in which warm arterial blood is cooled by counter-current heat exchange with venous blood. AVAS are particularly abundant in the feet and incubation patch and the venous plexuses in the mucosa of the mouth and upper respiratory tract. However, Baumel et al (1983) report that a dense venous plexus located subcutaneously in the neck appears to be unique to pigeons.

The AV heat exchange systems are particularly well-developed in the hind limb, wing and head, either in the form of a '**rete mirabile**' which is a vascular bundle consisting of numerous arteries and veins or a '**renae comitantes**' system in which a single artery is surrounded by a variable number of veins. Midtgard (1988) has tabulated both the forms of AV heat exchange systems in the hind limbs of some short-legged and long-legged birds from different orders.

If the internal device of heat regulation in birds is mainly performed by the circulatory system, the external device for heat conservation of several types, the contour feathers and the underlying down feathers are very useful exoskeletal structures conserving body heat. The feathers thus form a very efficient insulating device.

7

Moulting

Moulting of feathers—like other epidermal structures of the body—is an usual phenomenon in birds. Alongside, the other keratinised coverings of the bill and claws are also moulted. Generally, the adult birds moult their feathers at least once in a year. The old feathers are replaced by the new ones which may look brighter. Moulting may not be always seasonal in all the birds as different birds moult differently. The hatchlings of many birds usually moult 3 or 4 times within the first year; after which their moulting may be seasonal. The nestling down may give rise to contour feathers. Thus moulting does not follow a constant trend in all the species of birds. Some birds, both males and females, become brightly-coloured with nuptial plumage in their breeding season through moulting of feathers.

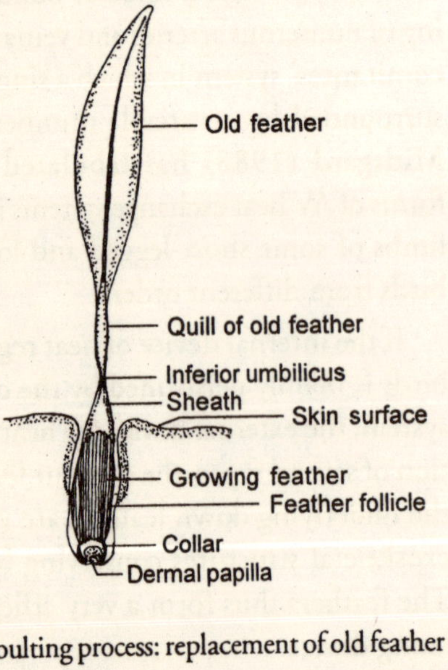

Fig. 7.1: Moulting process: replacement of old feather

In most species, however, the birds moult twice a year—once before and once after reproduction. Bologna (1995) reports that the prenuptial or winter moulting does not affect the wing and tail quills, as these are vitally concerned with flight. But unlike this incomplete winter moulting, the post-nuptial or the autumnal moulting is complete, affecting the entire plumage in the body.

As regards growth of plumage in bird chicks, all birds have been broadly classified into two categories : (i) **nidicolous** (altricial) which are born almost naked (without feathers) and have to live in the nest for certain period during which they develop feathers and then leave the nest; and (ii) **nidifugous** (precocial) which are born with their down feathers and are able to move out of their nest soon after their hatching (may be reared by the parents for a few hours or days).

The moulting period varies in different species of birds, sometimes fairly long as in albatrosses, pelicans, gannets and birds of prey. The conspicuous changes of coluration in the plumage during spring-summer and autumn-winter phases caused due to moulting provide some important clues in the identification of adult birds. The adult colouration of plumage may appear even after several years in some species of birds like large gulls, gannets, birds of prey, or it may appear within a few months in others.

According to Watson (1963), moulting is a single growth process producing new generation of feathers which push the old feathers to fall off passively. This has been termed as '**active moult**' which may not be followed in all the birds.

The keratinised coverings of the beaks and claws are also moulted likewise. According to Van Tyne and Berger (1976) and Young (1981), the large flight feathers are not normally moulted all at a time, as these are concerned with flight. Ducks and Geese are exceptions, as these birds at the time of moulting assemble near the water-point and throw off their old flight feathers all at a time. These birds are unable to fly during such period.

8

Habit and Intelligence

The birds—being endowed with the marvellous power of flight and an exceedingly well-developed nervous system compatible with their freedom of movement and tireless activity—are far advanced than their poikilothermous predecessors and comparable only to the mammals. Their social habits and family organisations give an impression of their better intellect and a more advanced parental care than exhibited by their reptilian ancestors. The faculties of vision (microscopic as ell as telescopic) and hearing, the steering, balancing and speed in flight, the overall sense perception, recognition of prey, predators and other birds, foraging strategies, the language and communication and the remarkable migratory movements in birds–all combine to give the impression of a superb neuro-behavioural coordination. Thus the birds are no doubt a very intelligent group of homoiothermic vertebrates with countless adaptive specializations. However, from experimental evidences gathered so far, it appears that not all their activities are guided by intellect, but some also by their instinct.

Arora (1992; In: Animal Behaviour) has described with illustrations many aspects of the life of birds which show intelligent behaviour–either individually or in a colony. Arora has cited many notable examples of bird behaviour such as, aggression, submission, motivation, feeding in groups, defending territory, learning and communication, in which birds exhibit a very high degree of intelligence.

Bird's songs as a means of communication or signals in various ways or their superb skill in nest building (e.g., hornbills, tailor birds, weaver birds, colonial nests of flamingos) or cooperative feeding by the pelicans are some of the examples showing their instinct as well as intelligence. Not all movements and activities of birds are guided by their instincts, but quite a number of behavioural patterns of birds can be explained by their intelligence. For example, their selection of stop-over sites during migratory flight or diversion of route (if the conventional route is not amenable) in long distance migration or diversion in feeding adaptations may be cited as intelligent manoeuvres accountable by superior brain function.

Of different parts of the brain, the cerebrum and the cerebellum are very well-developed in birds. The better development of the cerebrum may be directly correlated to greater intelligence in birds and that of the cerebellum is indicative of a perfect sense of orientation in flight and movements. Young (1981) states that greater development of the forebrain in crows and parrots is explainable by their outstanding intelligence.

9

Food, Feeding and Drinking

A. FOOD AND FEEDING

Flight adaptations in birds have brought in multiple changes in every system of their bodily organisation. Flight in air demands enormous conservation of energy derived from a steady and abundant source of food. The form-function complexes of the digestive system in birds have undergone adaptational changes in course of evolution to meet the demands of nutritional specialization. Various experiments and field observations have revealed that while some birds have a rather "exclusive specialization" for a diet, the vast majority of birds have a "diversification" in their choice of food; i.e. these birds, although having preference for a particular diet, may switch over to other food-niches, should their preferential food be not amenable. Thus our common Blue Rock Pigeon (*Columba livia*), although preferentially grain-eater by habit, may also feed on small sized fruits such as berries and even insect-eggs and larvae (Bhattacharyya 1980). Likewise, many insectivorous birds have been occasionally found to be sap-suckers or nectar-feeders.

Study of the food and feeding habits of birds form a very interesting subject in Ornithology. It is observed in the nature that most birds adapt themselves to a wide range of diversity in selecting their food; very few species stick to their exclusive diet. While the vast majority of birds collect their food from both the plant and animal sources, their selection of food is not random, but, even up to the species level, most birds exhibit distinct preference for some items of food and less so for the others.

The following is the list of some of our representative flying birds with their food items—"exclusive", "preferential" or "diversified": The interesting feeding habits of some birds have also been described:

i. The Grey Pelican (*Pelicanus philippensis*) and the Little Cormorant (*Phalacrocorax niger*) are exclusively fish-eaters. The Pelicans swim gregariously in a semicircle, drive the fishes towards shore or shallow waters and then scoop them up individually in their mandibular pouch. The Cormorant also occasionally (though not habitually) hunt fishes like the Pelicans.

ii. The Snake-bird (*Anhinga rufa*), which is a black cormorant-like water bird, has its staple food fish, which is chased with great speed and caught below the surface of water.

iii. The Spoonbill (*Platalea leucorodia*) feeds on tadpoles, frogs, molluscs, insects and vegetable matter.

iv. The Egrets and Herons (*Bubulcus, Ardea*), in general, feed mainly on crustaceans, molluscs, insects and fishes, but also sometimes on frogs and even small reptiles. The White Stork (*Ciconia ciconia*), a spectacular migratory bird, feeds on frogs, lizards and large insects like locusts.

v. The shape of the bill in the Open-billed Stork (*Anastomus oscitans*) helps the bird in opening the thick molluscan shell to feed on its viscera and soft parts. It also feeds on frogs, crabs and large insects.

vi. The large, tall [about 4 feet (1.2 meters) in standing height] Adjutant Stork (*Leptoptilos dubius*) feeds on fishes, frogs, reptiles and large insects.

vii. The Flamingo (*Phoenicopterus roseus*) prefers to feed on algae, fungi and seeds of marshy plants. The rostral half of the upper beak of the bird being sharply bent down, the culmen of the upper beak in this region scoops a hollow in the mud from which minute organic particles are collected as food. The bird also feeds on crustaceans, worms and insect larvae.

viii. The Bar-headed Goose (*Anser indicus*) feeds mainly on green shoots of winter crops. The other ducks and geese are also chiefly vegetarians, but they occasionally feed on crustaceans, molluscs and insects.

ix. The White-backed or Bengal Vulture (*Gyps bengalensis*) is exclusively flesh-eater. Its beaks and claws are suitably adapted for tearing and piercing flesh.

x. Most Kites (e.g., *Milvus, Haliastur*) feed on large insects, fish, frogs, lizards, small snakes and nestling birds.

xi. Most Eagles and Falcons (*Spizaetus, Haliaeetus*), besides sharing same items as the Kites, have a greater preference for Lizards, Snakes, Rats, Crabs and pigeon-sized birds.

xii. The Red Jungle Fowl (*Gallus gallus*), the common Peafowl (*Pavo cristatus*) and the Cranes (e.g. *Grus antigone*) mostly feed on grains, vegetable shoots, insects, lizards etc.

xiii. The Shore Birds like Gulls and Terns (e.g. *Larus, Sterna*) feed on fishes, curstaceans, insects, slugs and shoots of various crops.

xiv. While the Green Imperial Pigeon, Maroon-backed Imperial Pigeon (both *Ducula* sp.) and the Common Green Pigeon (*Treron* sp.) are almost exclusively frugivorous, most Pigeons and Doves mainly feed on seeds and grains, cereals, pulses and also fruits. Zweers (1982) has worked extensively on the feeding system of the Common Pigeon–*Columba livia*.

xv. The Koel (*Eudynamys* sp.), the Cuckoos (*Cuculus* sp.) and the Barbets (*Megalaima* sp.) mainly feed on fruits, berries and wild figs, but they also feed on hairy caterpillars and various insects.

xvi. The Barn Owl (*Tyto alba*) feeds exclusively on rats and mice; most other owls feed on crabs, large insects, fish, frogs, reptiles, birds and small mammals.

xvii. Among the Coraciid birds, the Small Green Bee-eater (*Merops orientalis*) mainly prefers the Dipteran and Hymenopteran insects, their larvae and pupae, the King Fishers (*Alcedo* sp., *Hal-*

cyon sp.), though mainly feed on fishes, also feed on frogs, tadpoles and aquatic insects. The Hornbills (*Buceros* sp), though mainly frugivorous, also feed on lizards, mice and small birds.

xviii. The Old World Insect-eaters like the Flycatchers (*Muscicapa* sp.) the Tailor Bird (*Orthotomus sp.*), the Jungle Babbler (*Turdoides* sp.), the Magpie-Robin (*Copsychus* sp., etc.) are primarily insect-eaters but they also feed on wild figs, berries, grains and flower-nectar.

xix. The Black Drongos (*Dicrurus* sp.), the Golden Orioles (*Oriolus* sp.), the Pied Myna (*Sturnus* sp.) the Bulbul (*Pycnonotus* sp.), etc., have almost equal preference for fruits, berries, figs, flower-nectar and insects.

xx. While the Flower-peckers (*Dicaeum* sp.) mainly feed on fruits and flower-nectar, the Sunbirds (*Nectarinia* sp.), though largely feeding on flower nectar, also feed on insects and spiders. The tubular tongue in these birds is suitably adapted for flower probing.

xxi. The Baya Weaver Birds (*Ploceus* sp.) mainly prefer to feed on ripening crops and cereals, but they also feed on insects.

Thus, the diversity of food and feeding habits of a good number of birds belonging to different orders reveal that very few species of birds stick to 'exclusive' diet, while the great majority of birds are adapted to 'diversified' food-niches.

For a more detailed study of the food, feeding behaviour and functional morphology of the feeding apparatus of different birds, references may be made to the notable works of the following authors, out of an exhaustive list :

	Author	Subject
i)	Goodman & Fisher (1962)	Functional anatomy of the feeding apparatus of Waterfowl.
ii)	Bock (1966)	An approach to the functional analysis of the bill shape.
iii)	Bock (1972)	Morphology of the tongue apparatus of *Ciridops anna* (*Drepanididae*).
iv)	Burton (1974)	Feeding and feeding apparatus of Waders.
v)	Zweers (1974)	Structure, movement and myography of the feeding apparatus of the mallard (*Anas platyrhynchos L*).
vi)	Zeigler, Levitt and Levine (1980)	Eating in the Pigeon.
vii)	Bühler (1981)	Functional anatomy of the avain jaw apparatus.
viii)	Zweers (1982 a)	Pecking of the Pigeon (*Columba livia L*).
ix)	Zweers (1982 b)	The feeding system of the Pigeon (*Columba livia L*).
x)	Gerritsen and Sevenster	Foraging behaviour and bill anatomy in sand pipers.

xi)	Homberger (1986)	The lingual apparatus of the African grey parrot (*Psittacus erithacus L*), (Aves: *Psittacidae*).
xii)	Homberger and Meyers (1989)	Morphology of the lingual apparatus of the domestic chicken, *Gallus gallus*, with special attention to the structure of fasciae.
xiii)	Kooloos & Zweers (1992)	Different integrations of pecking, drinking and filter feeding mechanisms in Waterfowl.
xiv)	Zweers, Berkhoudt and Vanden Berge (1994)	Behavioural mechanisms of avian feeding

Regarding pecking in the pigeon, Berkhoudt (1985), quoting Zweers (1982 a) and Zeigler (1976) has described eight phases:

i) **Preparatory phase** for searching and approaching followed by fixation of the seed.

ii) **Final approach or Peck phase** by lowering the head and closing the eyes.

iii) **Grasp phase** in which the seed is seized between the beak-tips.

iv) **Repositioning or stationing phase** in which a large seed may be required to be repositioned before transportation to the gullet.

v) **Transport** phase in which small grains and seeds are transported from the mouth through 'slide-and-glue' mechanism down the oropharynx. This movement is mainly aided by the protraction and retraction of the tongue. Large seeds are usually transported by 'catch-and-throw' method in which the tongue has no direct role.

vi) **Collection Phase** I, where the foodgrains collect at the base of the tongue.

vii) **Swallowing Phase** in which the grains are swallowed.

viii) **Collection Phase** II in which the foodgrains reach the rostral oesophagus and collect there.

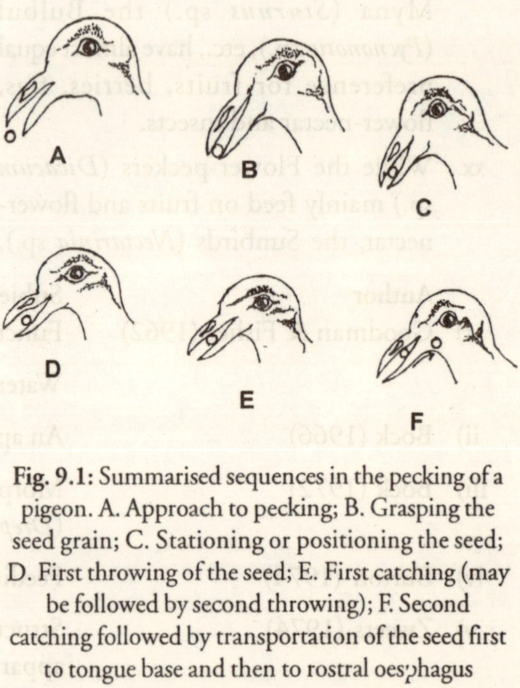

Fig. 9.1: Summarised sequences in the pecking of a pigeon. A. Approach to pecking; B. Grasping the seed grain; C. Stationing or positioning the seed; D. First throwing of the seed; E. First catching (may be followed by second throwing); F. Second catching followed by transportation of the seed first to tongue base and then to rostral oesophagus

Zweers, Berkhoudt and Vanden Berge (1994) have worked on the diversity of feeding of quite a good number of bird-families, including foraging, food acquisition (which includes ingestion and deglutition), food-handling in the oropharynx, digestion, etc. of an appreciable number of bird-families. In food acquisition, they have described ground pecking, filter feeding, husking seeds, extracting snails, biting fruit, nut-cracking, catching prey by striking as well as by aquatic and aerial sweeping, probing sand-holes (by sandpipers) and flowers (by nectar-feeders) and various other feeding adaptations by birds. These authors have clearly stated one thing– All neognathous birds show pecking mechanism, irrespective of specialist food acquisition behaviour. Zweers (1991 a, b) assumes that pecking is the ancestral feeding design from which wide range of trophic diversities have evolved through modifications and specialisations.

The author of this text, while working on the structural adaptations of the bill, skull elements, tongue and hyoid of some Indian insect-eating birds (Bhattacharyya, 1987) and on the diversity of feeding adaptations of some columbid birds (Bhattacharyya, 1992) has observed that most birds are adapted to diverse food and feeding behaviour, in spite of their preference for some items. And the birds that are perfectly adapted for aerial sweeping of insects are also not unaccustomed for pecking non-insect food from the ground. Gianfranco Bologna (1995,In: Macdonald Encyclopedia of Birds of the World) reports "Members of large flocks feed better and more plentifully than isolated individuals of the same species".

B. DRINKING

Just as food and feeding habits in birds present wide range of diversities among different avian groups, so also are the diverse ways of drinking in birds. Avian drinking behaviour occurs in two general ways: **Tip-up drinking and Suction drinking,** although there may be several modifications in both.

Extensive works on the drinking of several groups of birds by Zweers (1982), Homberger (1986), Kooloos and Zweers (1989), Heidweiller and Zweers (1992) and Zweers (1992) reveal that birds use several kinematic and behavioural models for drinking water. These methods have been described as 'tipping-up', 'sucking', 'scooping', 'lapping', etc. Zweers (1982), quoting White (1970) about drinking of water, states that most fowls and Waterfowls (e.g. *Gallus*) drink water by 'tipping-up' method in which the bird bends the head down, opens and closes the beaks several times, takes some water in the mouth and then tips up the head so that the water runs down the pharynx by gravity. Cameron (1938) mentioned that Swallows fly back and forth over water, scooping up water with the help of their lower mandible. Homberger's (1980) exhaustive study on the drinking of psittaciform birds confirms that among this group of birds, at least four different methods of drinking are observed: (i) scooping water with the help of lower mandible, (ii) lapping water with the brush-tipped tongue, (iii) ladling water with the tongue tip, and (iv) pressing the tongue against the palate—thereby causing suction force with the help of only the upper beak and tongue. The columbid birds drink water by sucking up with the help of the tongue but in a different way from that of the psittaciform birds.

Heidwiller and Zweers (1992) have lucidly described the development of drinking mechanisms in the chicken (*Gallus gallus*). According to them–"hatchlings display pecking behaviour towards water. Week-old chicks protrude the tongue far out and the ingested water is sucked up by capillary action between the tongue, the floor and the roof of the oropharynx. With increased age, laryngeal suction develops. At the age of four weeks and after, the chickens use the adult mechanism to drink. Water transport by capillarity action is not sufficient in the older age group; therefore, additional mechanisms such as pressure development by closing the bill and laryngeal suction must develop". Heidweiller, Van Loon and Zweers (1992b) elsewhere have also discussed the flexibility of the drinking mechanism in adult chickens.

The drinking behaviours and mechanics of the pigeons have been elaborately described by Zweers (1982) with the help of high speed films and X-ray motion pictures. He has analysed the movement cycle of the swallowing of water into five phases: (i) capillary action of the beak-tips, (ii) lingual suction, (iii) pharyngeal preparation, (iv) pharyngeal suction; and (v) oesophageal collection.

The author has formulated a double suction model for the mechanics of drinking. The retraction of the tongue causes a low air pressure area in the buccal cavity and the water is sucked into the mouth. Secondly, the depression of the pharyngeal floor causes a lower air pressure area in the pharynx. As a result, the water in the mouth is forced backward over the larynx into the oesophagus. Hence, according to Zweers (1982), the earlier view of Lorenz (1939) that the pigeons pump the drinking water by peristalsis of the oesophagus is not tenable. Stresemann (1927-1934) reports that the Sandgrouses (Pteroclidae) too show similar drinking behaviours as the pigeons.

Kooloos and Zweers (1989), while discussing the mechanics of drinking in the mallard (*Anas platyrhynchos*), mention that a special type of 'tip-up' drinking occurs in Anseriformes, different from the normal 'tip-up' drinking as observed among the Passeriformes and Galliformes. The authors conclude that, in the former, water transport is primarily caused by pressure-changes and capillarity, whereas, in the latter groups, by grasping and holding actions, and gravity. Besides the above examples, authors like Wickler (1961), Ziswiler and Farner (1972), Immelman (1967), Cameron (1938), Hunt and Smith (1967), McLelland (1979), Gibson and Bacon (1983) and Kuenzel (1983) have described the behaviours and mechanics of drinking in different groups of birds.

10

Selection of Nesting Site and Territoriality

The habitat selection by birds is done by keeping their eyes open to look out for the foraging ground which should be within a reasonable distance from the habitat. This is somewhat different from the selection of a mate and a nesting site, usually by the male bird, at a convenient location having a rather restricted boundary. This location is referred to as the '**territory**' of a pair of birds belonging to a species. Such territory becomes obligatory to the bird pair at the onset of the reproductive season and the area with its boundary is very faithfully and valiantly guarded by the male from possible intrusion by other males of the species. This phenomenon is known as '**mate guarding**'.

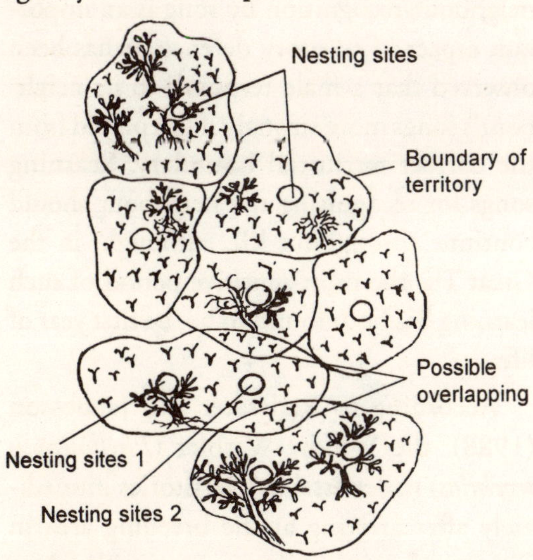

Fig. 10.1: Diagrammatic representation of bird territories with nesting site usually at the centre of each territory. Selection of nest-sites depends on several factors, and hence, within the territory of a pair of individuals, the nest-site may be shifted from location 1 to 2

10.1. RANGE OF TERRITORY

The territorial range in different species of birds is different and variable, depending on the nature of habitat, available resources of food, suitable roosting as well as nesting sites, interspecific and intraspecific competitions, etc.

Arora (1992) reports that the feeding site of bird territory varies from 0.07 hectare of a flycatcher to 9,300 hectare of the Golden Eagles. Herbivores have usually smaller territories than carnivores. Schoener (1968) pointed out, as quoted by Arora (1992),that there is an approximately linear relationship between the logarithm of body weight and that of the range of territory. A bird may cover a wide range of territory where the food supply is scarce or unpredictable, and, conversely, the range of the territory may be limited within one to a few square miles (kilometres) where the food supply is abundant. Arora (1992), while discussing territorial behaviour in birds, states that the cost and benefits in the territorial behaviour of birds are ultimately determined by their fitness where selection plays an important role. Here cost means the time and energy spent and the competitions faced in search of food or in defence of the territory. Hence the cost increases with the increasing area of the territory.

Although birds usually do not capture extra space of the territory held by other individuals, some overlapping of territorial boundaries or extension of the territorial space in the area previously occupied by another species and at present left vacant by them (due to environmental changes or other factors) may always occur.

10.2. DEFENCE OF THE TERRITORY

Van Tyne and Berger (1976) comment that a territory may be defended against intruders not only by physical threat or attack, but often merely by song, or even by the presence of the defending bird. But in territorial behaviour, one behaviour is almost certain—a bird defending a territory is more aggressive than the intruder, and it becomes successful in driving the latter away. If, however, the defending bird—in course of fighting with the aggressor—enters the area of the latter, the role of the defender and aggressor is reversed!

McKinney (1988) reports that in many species of basically monogamous birds, extrapair copulations (EPCS) occur. In such species, 'mate-guarding' is a widespread 'anticuckoldry adaptation'. The males guard their mates by remaining close to them and watching vigilantly, warning them through their calls and dialects of the approaching danger and defending them from any possible threats or assaults by other males. According to Baker et al (1986), 'the songs of male oscine birds are thought to serve at least two functions—mate attraction and territorial defence.' Exhaustive studies on the behaviour and recording of songs by song birds—by Krebs et al (1978) and McGregor and Avery (1986)—reveal that many species of songbirds use song in defence of territories. They have shown in the Great Tit (*Parus major*) through playback of song in the absence of a male that it is effective in deterring settlement of an area for a number of days.

10.3. SONGS IN GUARDING TERRITORY

The significance of bird song in establishing and guarding territory has been amply illustrated. Falls (1982) reports that males of many species holding territories can well-recognise their neighbours' songs. And such a close neighbour—usually possessing a territory, a nest-site and mate—is unlikely to cause ejectment of another territory-holder. Hence, neighbour-recognition by song is an important aspect of territory defence. It has been observed that a male responds to its neighbour's songs more strongly if it is played from the correct territorial boundary. Learning songs for recognising the neighbour should continue throughout life, although, in the Great Tit, the most sensitive period of such learning has been found to be the first year of life.

According to Radesater and Jakobsson (1988), the Willow Warbler (*Phylloscopus trochilus*) males establish territories immediately after arriving at the breeding area in spring and start 'singing contest' with others before the territorial borders are permanently fixed. In this 'singing contest', song-rate matching is commonly observed, particularly during initial phase of territorial establishment. These authors have observed that males in some territories spend more time singing and less time in eating than others.

Some studies by Krebs et al (1978) and Krebs and Kroodsma (1980) have revealed that large song repertoires are important in territorial defence and mate attraction. Catchpole (1980, 1983) reports that the territorial males of Great Reed Warblers (*Acrocephalus arundinaceus*) sing shorter and sterotyped songs during encounter with rival males, but longer and more varied songs during spontaneous singing by unpaired males. As soon as the paired bond is established, the practice of long songs is discontinued. Thus it is commonly observed that during initial phase of interactions between two males, singing becomes intensive for a quite a long duration after which such interaction stops as one of the contestants withdraws.

10.4. SIZE OF THE TERRITORY

The size of the territory is species-specific, although, within a species, variation in the size of the range may occur. The nesting territories of the colonial sea-birds lie close to one another, whereas those of the falconid birds may extend over a few square miles (kilometres):

TERRITORIES OF SOME BIRD SPECIES

BIRD	TERRITORY (ACRES)
Robin	: 0.11 to 0.6
Red-eyed Vireo	: 1.4 to 2.1
Black-capped Vireo	: 2.5 to 4.6
Prothonotary Warbler (*Protonotaria citrea*)	: 1.9 to 6.4
Carolina Chikadee	: 10 to 15
Black-capped Chikadee	: 8.4 to 17.1

11

Birds Songs and Song-Dialects

11.1. SONG DIALECTS

Do birds have their own language? If language is meant for intercommunication of behavioural responses, it has been well-established in Bird Acoustics that the various sounds, signals and musical vocalisation used among individuals of the same species, or among neighbours and population of different species, are the language and songs of birds. Through these languages and songs, the birds express their joy, anger, fear, choice of mates and territorial security.

Since early 1970s, many researches have been carried out from various aspects–ecological, behavioural and structural patterns–in the field of bird-songs and song-dialects. Different workers have interpreted song dialects in different ways–according to their study models and parameters. Of many authors working in this field, particular mention may be made of Payne (1973, 1981a, b, 1982, 1983a,), Kroodsma (1974, 1979, 1982), Baker (1975, 1982a, b, 1983), Baptista (1975), King et al (1980), Slater et al (1980, 1984), Slater and Ince (1982), King and West (1983b), Baker and Cunningham (1985), Baptista and Petrinovich (1986) and Payne et al (1987).

According to Payne (1981, 1982, 1983) and Baker (1982), in dialect species, birds sing a song more similar to that of their neighbours than to that of birds in other populations. Baker and Cunningham (1985) mention that the regional song differences might be involved in the maintenance and origin of genetic races or even in speciation. According to Kroodsma (1974) and Payne (1983, 1987), in some species, fathers and sons have different songs, and sons learn the songs of their new neighbours after dispersing from natal territory. Many workers consider song as a social signal and the range of variation of song is also limited within a species. Even among the songbirds having considerable variations within individual populations, Marler (1984) and Marler and Pickert (1984) observe that the species information may be present in acoustic uniformity, note structure, pattern, or the tonal quantity.

Adret-Hausberger (1988), in his comparative study of the European Starling whistles (*sturnus vulgaris*) with other distant populations, suggests that within a range of local variations, dialects in Starling whistles are based on a balance between species-specificity and individual learning. Kroodsma (1974), Payne et al (1981) and Jenkins (1978) found that the song traditions of neighbours were copied after the young dispersed and established their own breeding territories. McGregor and Krebs (1982), recording song-dialects of Great Tits (*Parus major*), observed that there was no difference between the proportions of a son's repertoire, shared with his father, with other males born in the same year, male present in the year of breeding or with brothers. Millington and Price (1985) found that male Darwin's finch (*Geospiza* sp.) and his brothers always sang the father's song-types. But Payne et al (1987), working with a large number of female Indigo Buntings (*Passerina cyanea*), observed that they mated with males having songs unlike their fathers' songs and success in breeding was not related to matching or mismatching of song-types.

From the above evidences in song-dialects and extensive literature on the subject, it appears that, in general, songs can be transmitted from father to son, but not necessarily.

Sorjonen (1983-1986), working with the Thrush Nightingale of Europe, observes that the male of this species has no clear cut dialect boundaries between populations and that the local dialects change continuously from year to year.

11.2. SONG CHARACTERISTICS OF SOME COMMON BIRDS

	BIRD		CHARACTERISTICS
1.	Blue headed Rock Thrush (*Monticola cinclorhynchus*)	:	Loud, clear but monotonous whistling song.
2.	Blue Rock Thrush (*Monticola solitarius*)	:	Less vocal, but the males have sweet whistling songs, particularly at the advent of breeding season.
3.	Himalayan Whistling Thrush	:	During breeding season, male has a very rich and remarkable human-like whistling song.
	(*Myiophonus caeruleus*)		
4.	White-throated Ground Thrush (*Zoothera citrine*)	:	Males have a rollicking whistling song of several rich notes, often repeated and often birds' calls within the song.
5.	Pied Bushchat (*Saxicola caprata*)	:	In the breeding season, male has a sweet whistling song, resembling that of Indian Robins.
6.	Pied-crested Cuckoo (*Clamator jacobinus*)	:	Commonly known as '*Papiya,*' this bird has a characteristic metallic call that sings 'piu-piu-pee-pee-piu'.
7.	Koel (*Eudynamys scolopacea*)	:	In the early morning, the call begins with a low 'kuoo' which rises in scale with each successive 'kuoo' until it reaches fever-pitch at the seventh or eighth and breaks off abruptly.
8.	Magpie-Robin* (*Copsychus saularis*)	:	The male bird sings in the early morning and late afternoon; songs punctuated by upward jerks of the tail. This bird is also a very good mimic of other birds' calls.
9.	Shama (*Copsychus malabaricus*)	:	This bird is much prized as a songster. It has a loud, clear and melodious thrush-like song, rich in notes and tones.
10.	Blackbird (*Turdus merula*)	:	The bird has melodious songs like Shama, with the habit of mimicking other birds' calls.

Contd.

* Bhatt and Sethi (2006) report that this bird has year-to-year variation in the song, adding a number of new phrases every year with proportionate deletion of others.

	BIRD		CHARACTERISTICS
11.	Hill Myna (*Gracula religiosa*)	:	This bird can sing a melodious whistling song. Salim Ali (1987) reports that this bird is an accomplished mimic and a talker.
12.	Red-whiskered Bulbul (*Pycnonotus jacosus*)	:	Unlike Redvented Bulbul, this bird has joyous musical notes that are readily distinguishable.
13.	Blcak-crested Yellow Bulbul (*Pycnonotus melanicterus*)	:	Three birds keep in touch with one another by low churring notes and songs similar to Redwhiskered Bulbuls, as well as other Bulbuls in general.
14.	Bluethroat (*Luscinia svecica*)	:	In the breeding season during summer, the male bird delivers a sweet loud and rich song.
15.	Malabar Whistling Thrush (*Myiophonus horsfieldii*)	:	Salim Ali (1987) reports that, in the breeding season during summer, the male bird has a remarkably human whistling song, rambling aimlessly up and down the scale.
16.	Grey-winged Blackbird (*Turdus boulboul*)	:	This bird has a great variety of rich and melodious songs which are far-carrying. According to Salim Ali (1987), the song consists of one soft introductory note, followed by four high-pitched ones on a descending scale. It is one of our finest songsters and a favourite cage bird.
17.	Common Nightingale (*Luscinia megarhynchos*)	:	Kiefer and Hultsch (2006) report that the Nightingales are very good learners of new song types with increasing age. Their whistle-songs meant for attracting mates are much more varied in older males than in the younger birds.

12

Sense Perception

The great naturalist-philosopher Aristotle (384-322 B.C.) stated "Nothing is in the mind that does not pass through the senses". How an external stimulus of the environment awakens an internal stimulus of the nervous system for an appropriate reaction is a matter of great scientific interest. Some perception is a phenomenon that is processed through sensory cells, or, more scientifically, through the receptor cells of different kinds which specifically arouse different sensations like hot, cold, sweet, sour, touch, pleasure and pain. The receptor cells are highly selective for specific forms of stimulus energy.

Avian sense perception, like that of the higher vertebrates, is very well-developed, particularly with respect to the senses of sight and hearing.

The skin of birds being completely invested with feathers, the tactile sense organs are rather poorly developed. Some sensory nerve-endings are, however, found scattered in the bill of birds, particularly in ducks and geese. In the parrots, owls and pigeons, at the base of nostrils of the upper beak, is located a soft, fleshy or glandular waxlike structure called 'Cere' or 'Operculum'. It is a sensory organ believed to be tactile in function. In some birds, the rictal bristles that develop at the rictus of the bill are sensory organs, although without nerve-endings but with sensory cells at the base of the feather follicles.

Although the entire bill of birds is usually thick and hard, most waders (Charadriiformes) have a relatively soft and leathery bill. In ducks and geese (Anseriformes), the bill is hard only at the tip, forming the maxillary and mandibular nails, respectively. The nail and its associated dermis form a complex sensory organ known as the **'bill tip organ'** of which one lies in the upper and one in the lower beak. The bill tip organ plays an important role in the sensory discrimination between food and other objects. The type of receptor cells present in the 'bill tip organ' are the mechanoreceptors, **Herbst's corpuscles'**. These are highly sensitive to vibration and are capable of following very fast frequencies (up to 1,000 Hz). Zweers (1982) reports that the bill tip organ of pigeon is a well-developed touch organ of complexly arranged Herbs't corpuscles under the keratin nails of the tips of the beaks. These are the most widely distributed mechano- receptor cells in the avian body, occurring wherever there is a mechanical disturbance—at the bases of the feather follicles, along bones of the wing and pelvic limbs and in the muscles, tendons and joint capsules.

The other sensory corpuscles, also mechanorecepters, are distributed in the dermis, associated with the soft horn near the sides and tip of anseriform bill. These are called '**Grandry's corpuscles**' which consist of a few granular cells associated with numerous axonal endings. These are also present in the dermis of the bill of grain-eating passerine birds.

The perception of smell (except in marine species) is rather poorly developed in birds. The external nares lead into the two nasal cavities separated by a median septum from which three raised folds called turbinals or conchae project into the nasal cavity. Of these, the smaller and posterior-most third concha is only provided with an olfactory epithelium. It is assumed that the deficiency in the sense of olfaction in birds may be largely compensated by their increased development of other senses. **Jacobson's organ** is present only as a transitory embryonic rudiment.

The perception of taste, although closely related with that of smell, is not as poorly developed as the latter. There are plenty of mucus-secreting epidermal glands scattered all over the dorsal and ventral surfaces of the tongue (more on the dorsal surface) and the taste-buds are located just behind the bill-tip, in the roof and the floor of the mouth at the level of the rictus, at the lingual base just beneath the epidermis and in the caudal palate. The study in the mouth, tongue, palate and pharynx of birds like mallard (Berkhoudt, 1977) and pigeon (Zweers, 1982) confirms that the taste-buds in the mouth of birds have a distinct role in the selection/rejection of food.

The perception by photosensitive cells in birds has reached a climax. Avian eyes–which are comparatively larger in proportion to the body size–are very similar in structure in all the birds. The photo-receptive retina is provided with rods and two kinds of cones–single and double. The unique presence of a ring-like pad—the '**pulvinus anularis lentis**'—on either side of the biconvex lens and the presence of a very distinct '**fovea centralis**' increases the acuity of vision in birds. According to Evans and Martin (1993), this 'pulvinus' is largely responsible for the characteristic accommodation of lens in avian eye.

The receptor cells effecting the sense of hearing are called the hair cells or the "**Organ of Corti**" found on the sensory basilar membrane within the 'scala media' of the cochlea. The scala media is filled up with a fluid called the endolymph, whereas the scala vestibuli and scala tympani are filled with perilymph.

The sensory receptors of the Organ of Corti are stimulated either by the movement of endolymph or by that of the basilar membrane and the impulses, transmitted by the auditory nerve to the brain, are finally interpreted as sound. The birds possess a better-developed middle ear in comparison to that in reptiles. The tympanic membrane lies deeper from the surface of the head at the base of a narrow passage.

The internal ear in higher vertebrates, besides hearing, also functions as an organ of equilibrium. Pumphrey (1961) [In: Avian Physiology, 2nd ed., 1965] discusses at length the inner ear structures responsible for equilibrium in the bird. The semicircular canals provide continuous stimulation to the somatic musculature and are probably involved in the production of muscular tonus. Impulses arising in the semicircular canals, and their coordination with other stimuli, lead to compensatory body movement.

The cutaneous receptors for the perception of senses—light, touch, pressure, temperature, pain etc.—consist of either free nerve endings or encapsulated nerve endings of various types. The internal receptors that occur in the muscles, tendons, joints and periostium may be supplied with nerve endings which are responsible for perception of deep pain sensation in these organs.

13

Endocrine Glands

All the glands in the body are classified into two groups—the exocrine and the endocrine glands, based on the site into which their secretion is released with or without the help of a duct. The exocrine glands possess ducts through which their secretion is conveyed to a specific compartment such as the lumen of the gastro-intestinal tract or to the epithelial surfaces of the body. The endocrine glands, on the other hand, are ductless and their secretion is discharged directly into the bloodstream via the extracellular spaces and capillaries or lymphatics. However, certain glands like the pancreas have both exocrine and endocrine portions.

HORMONES

The secretions of the endocrine glands are known as hormones (Gr. *hormone, hormaein*– to excite) which act as 'chemical regulators' of the body. A hormone may be defined as a chemical substance synthesised by a specific organ or tissue generally the specialised ductless glands, and secreted into the blood which carries it to other sites of the body where its actions take place.

The hormones generally fall into two categories: **steroids** and **amino acid derivatives.** They control and integrate many essential bodily functions such as reproduction, organic metabolism and energy balance as well as mineral metabolism. In fact, the nervous and endocrine systems of the body function as a single interrelated system. The hormones are highly specific in their actions at the remote control centres. In the structural classification of hormones, however, four molecular groups have been recognised. These are: (i)**amines** (e.g. Epinephrine), (ii) **prostaglandins** (e.g. Prostaglandin PGE_2 that are cyclic unsaturated fatty acids), (iii) **steroids** (e.g. Testosterone), (iv) **polypeptides** (e.g. vasopressin).

From functional standpoint, the hormones have also been classified into four groups depending on the effects they induce on the target tissues: kinetic effect, metabolic effect, morphogenetic effect and behavioural effect.

Only a trace amount of hormone secreted by the endocrine gland reaches the **target tissue** and reacts with the specific **receptor molecules** generally present in the target tissue to activate the effect very often through specific enzymes.

Besides the above characteristics of the hormones, it is important to note that a single hormone may have multiple effects on a single target tissue or on several different target tissues.

The endocrine glands have some uniformity in their occurrence among the vertebrates, and all of them—except the interstitial cells of Leydig of the testes—develop from epithelial surfaces of the body as groups of cells which sink deep into the connective tissues and gradually differentiate into glandular structures.

As in mammals, the avian endocrine glands comprise the pituitary, thyroid, parathyroid, endocrine pancreas and adrenal glands. The thymus and the bursa of Fabricius are lymphoid structure and were formerly held as non-endocrine in function. But recent experimental evidences assign their indirect endocrine role.

A brief description of the histophysiology of endocrine glands in birds is given:

13.1. PITUITARY GLAND (HYPOPHYSIS CEREBRI)

Although all the endocrine glands have their functions interrelated and coordinated, the pituitary gland occupies an exalted position amongst them due to its multiple controlling functions. The other endocrine glands—which do not directly arise from either the hypothalamus or hypophysis of the brain—have been artificially designated by Assenmacher (In: Avian Biology, Vol. III, ed. Farner and King, 1973) as 'the peripheral endocrine glands'. The pituitary gland not only exerts influence on non-endocrine tissues but it also influences many other endocrine glands and, in turn, is influenced by them.

Location: The pituitary gland is a small peaseed-like structure which lies ventral to the diencephalon of the brain within a round or oval depression of the basisphenoid bone called the *sella turcica*. The gland is connected to the brain by an infundibular stalk. The gland is ovoid, slightly flattened and greyish in colour.

Subdivision: The pituitary gland is subdivided into three main parts: anterior, intermediate and posterior. Microscopically, however, the pituitary gland consists of six parts: **Pars anterior (Pars distalis), Pars tuberalis, Pars intermedia, Pars nervosa (Pars posterior), Median eminence, and Infundibulium.** While the first three parts form the adenohypophysis (glandular division) the last three typically form the neurohypophysis (neural division).

The intermediate lobe of the pituitary gland is absent in the birds and in a few mammals (Whales, Indian elephant, etc.) although the hormone—'**intermedin**'—secreted by this lobe is also present in birds. The hormone appears to be secreted by the hindermost portion of the anterior pituitary where it joins the posterior lobe.

13.1.1. Origin of the Pituitary Gland

The adenohypophysis or the Rathke's pouch arises as a dorsal evagination from the ectodermal epithelium of the primitive mouth cavity and later constricts off as a closed vesicle. The neurohypophysis arises from the infundibulum which originates as a ventral evagination of the embryonic dienephalon, and includes the median eminence of the tubercinerium. The anterior portion of the primitive Rathke's pouch elongates and becomes a layer to form the anterior lobe or the Pars distalis from which the Pars tuberalis is derived. The posterior portion which connects to the neurohypophysis forms the narrow intermediate lobe. In birds, the intermediate zone between the adenohypophysis and neurohypophysis is lacking.

13.1.2. Histology of Different Parts of the Gland

1. Anterior lobe (Pars distalis)

The anterior lobe contains three distinct kinds of cells—the basophil cells, the acidophil cells, which are mainly secretory, and the chromophobe cells. The first two kinds are also designated as the chromophil cells which secrete at least seven different hormones, controlling essential functions of the body.

A. **Chromophobe cells:** These cells are believed to be the precursors of acidophil cells. These cells contain less cytoplasm and no secreting granules.

B. **Chromophil cells:** Both the acidophil and basophil cells, stained by acid dyes and basic dyes, respectively, possess very well-developed cytoplasm and secretory granules that secrete different hormones.

1. **Acidophil cells (Eosinophil or α cells):** These form about 43% of cells of the anterior pituitary and are of two different types, based on the type of hormone secreted by them:

(a) Somatotrophic or growth hormone cells (STH cells): These cells are ovoid to rounded with centrally placed nucleus and the cytoplasm with large, dense granules. These cells are stained in orange G.

The cells secrete somatotrophic or growth hormones.

(b) Luteotropic hormone cells (LTH cells): These cells are ovoid to ellipsoidal in shape. The nucleus is centrally placed and the cytoplasm is packed with large granules. The cells are stained in acid fuchsin. The LTH cells secrete the lactogenic hormone prolactin.

2. **Basophil cells (β cells):** These form about 7% of cells of the pituitary gland and include, mainly, the following 4 types of hormone-secreting cells:

(a) Thyroid stimulating hormone cells (TSH cells): The cells are somewhat irregular in shape, each with a single nucleus. The cytoplasm having plenty of fine granules (100–140μm) is stained in PAS reagent and aldehyde thionine.

These cells secrete thyroid stimulating hormone.

(b) Follicle stimulating hormone cells (FSH cells): These are large, round cells with the nucleus, either irregular in shape or wavy in outline. The cytoplasm possesses dense, spherical granules (150–300μm). These cells are also stained in PAS reagent.

The cells secrete follicle stimulating hormone.

(c) Luteinising hormone secreting (LH) cells and interstitial cells stimulating hormone secreting (ICSH) cells:

These are round or polygonal cells with wrinkled nucleus and dense cytoplasmic granules (100–300μm), The cells are PAS positive.

The cells secrete luteinising hormone in females and interstitial cells stimulating hormone in males.

(d) Adrenocorticotrophic hormone (ACTH) cells:

These cells are somewhat ovoid in shape and with a centrally-placed spherical nucleus. The cytoplasm is dense with large granules.

The cells secrete adrenocorticoptrophic hormone which stimulates secretion of the adrenal glands.

Besides the above principal types of cells of the anterior pituitary gland, it may also contain some chromatophototrophic or melanin stimulating hormone cells (MSH cells). These cells are also basophilic in nature. Certain 'Release Factors' (RF) and 'Release Inhibitory Factors,(RIF) of the hypothalamus are also effective in releasing or inhibiting release of the anterior gland hormones.

The somatotrophic, luteotrophic and adrenocorticotrophic hormones are protein in nature while the thyroid stimulating, follicle stimulating and interstitial cell stimulating and luteinising hormones are glycoproteins.

Additional information about anterior pituitary hormones : Goldsmith and Follett (In: Avian Endocrinology, eds. Epple and Stetson, 1980) report the following experimental data on anterior pituitary hormones.

Studies on the purification, properties and radio-immuno assay of both protein and glycoprotein hormones have been carried out by several authors, using birds like chicken, turkey, Japanese Quail etc. (Cunningham, 1969; Farner et al, 1974; Harvey and Scanes, 1977). It has been observed from the tests with plasma LH and FSH using blood samples taken from the Japanese Quail killed after exposure to long and short days that in all cases, the male quail exposed to long days showed higher plasma LH and FSH concentration than short-day males.

As regards thyrotrophins, recent investigations have made it possible to measure **thyroxine** release *in vivo* in adult, sexually immature female quail in response to a single subcutaneous injection of TSH.

Prolactin too has been purified from pituitary glands of chicken and turkeys. It has been reported that prolactin acts synergistically with steroid hormones (Goldsmith and Follett, 1980). The authors have quoted that defeathering, vascularisation and epidermal hyperplasia of the blood patch are regulated by prolactin in conjuction with oestrogen and, perhaps, other steroids. Prolactin may play a key role in the regulation of incubation behaviour. It has been demonstrated that in various species of birds (e.g. mallards, black swans)—in which both male and female birds sit on eggs during incubation in the natural breeding season—the plasma prolactin concentrations are high in both the sexes. The crop glands of pigeons and doves favourably respond to exogenous prolactin secretion. Their plasma prolactin concentrations are low during courtship, egg laying and early incubation, but increase markedly between the seventh and fifteenth days of incubation when the crop weight increases considerably.

2. Pars tuberalis

This part is a modification of the Pars distalis and lies adjacent to it and to the Pars intermedia. Actually it lies very close to the infundibular stalk, being separated from it only by fibrous connective tissue.

This part, consisting mainly of cuboidal and columnar cells, contains lipid and glycogen droplets.

3. Pars intermedia

It is a narrow region placed between the Pars nervosa and the Pars distalis. It is closely associated with the Pars nervosa and separated from the Pars distalis by interglandular clefts. This part consists of basophilic polygonal cells called melanocytes containing plenty of secretory granules within the cytoplasm. The secretory granules of the intermediate lobe secrete **melanophore stimulating hormone** (MSH). However, this part , as such, is not discernible in the avian anterior pituitary. But the melanophoric hormone is secreted as usual—possibly from the hindermost region of the Pars distalis.

4. Posterior lobe (Pars nervosa)

Histologically, this part is not an endocrine gland in the strict sense, as the hormones referred in connection with this gland are produced by the neurosecretory cells located in the supraotic and paraventricular nuclei of the hypothalamus.

The various kinds of cells forming the posterior lobe are pituicytes, mast cells, Herring bodies, non-medulated nerve fibres and blood vessels. The hormones secreted by the posterior lobe are **oxytocin** and **vasopressin.**

5. Median eminence (Eminentia mediana; NAA, 1993)

A small eminence in the form of a protruberance is observed in the infundibular region, just ventral to the anterior hypothalamus. This is the median eminence which consists of hypothalamic cells, neuroglial cells and blood vessels. Dubois et al (1974) describe the distribution of somatostatin in the median eminence of mammals, birds and amphibians.

6. Infundibulum

The pituitary gland is attached to the ventral diencephalon of the brain through a narrow infundibular or hypophysial stalk called the infundibulum. This region is formed of neuroglial cells, sinusoids and neurosecretory cells with long axons.

Benowitz and Karten (1976a) have described 'Tractum infundibularis' and other tracts in the region of the infundibulum.

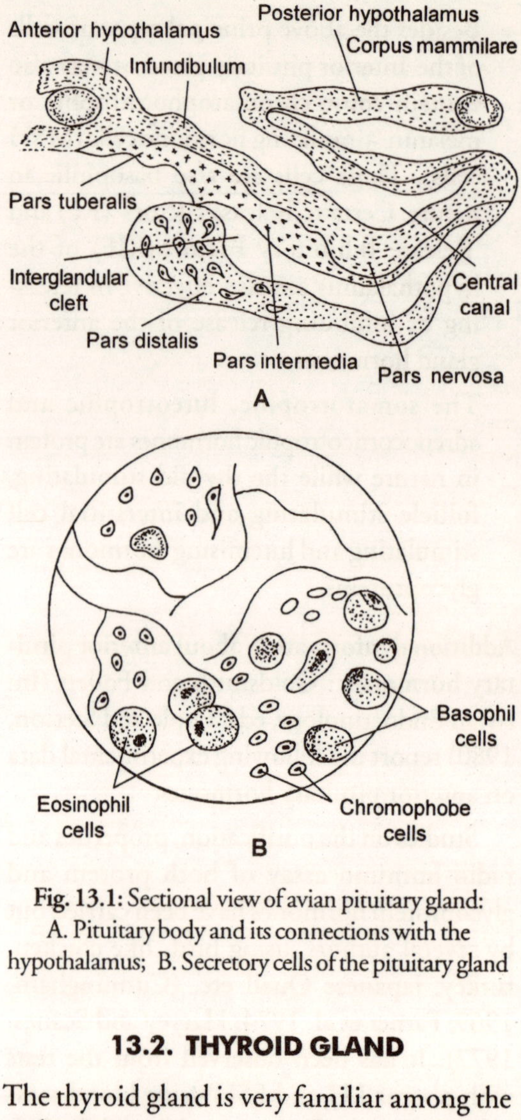

Fig. 13.1: Sectional view of avian pituitary gland: A. Pituitary body and its connections with the hypothalamus; B. Secretory cells of the pituitary gland

13.2. THYROID GLAND

The thyroid gland is very familiar among the endocrine system in vertebrates. Its evolutionary history can be traced back to the endostyle of the lower chordates.

Location: The two lobes of the thyroid gland are situated at the root of the neck lying ventral and lateral to the trachea, just posterior to the larynx and very close to the common carotid artery and the jugular vein on either side.

Origin: The thyroid gland arises in the embryo as a single midventral endodermal evagination of the primitive pharyngeal floor. The gland later divides into two separate lobes by the fifth day of incubation.

MORPHOLOGY

Each lobe of the thyroid gland is more or less rounded or oval in shape, or it may be even lenticular–as in the pigeons. The two lateral lobes are connected with each other at their caudal ends by a narrow region of thyroid or connective tissue called the isthmus which passes over the ventral surface of the trachea. Dense vascular network of the parenchyma provides the red colour of the gland–which may be lighter or darker according to species. Blood supply to the gland is provided by one to three branches from the common carotid artery (Glenny, 1951), or from branches of the common carotid artery (Bhaduri and Biswas, 1945). The number of thyroid veins, though variable, is generally 2-5. According to Watanabe (1960), Malinovsky (1962) and Baumel (1975), the avian thyroid gland seems to have a parasympathetic vagal component. The gland is invested by a thin, dense connective tissue capsule.

HISTOPHYSIOLOGY

Histologically, the gland consists of a large number of rounded closed vesicles called thyroid follicles of various sizes, separated from one another by delicate strands of connective tissue. Each follicle is lined with a single layer of cuboidal epithelial cells enclosing a central lumen filled with a mass of homogeneous viscid iodine-containing substance called colloid. Assenmacher (1973) reported that midway in the incubation period, the cylindrical cords of epithelial cells begin to break-up in small cell groups that acquire central lumina as a result of secretion and coalescence of colloid droplets originating in the follicular cells. During the latter half of incubation the follicles grow in size and number.

A very distinct anatomical characteristic of the avian thyroid is the absence of calcitonin cells or parafollicular cells in the interstitial tissue. This is because the site of origin of these cells remains separate from the thyroid gland throughout life in birds. In pigeons and doves, however, the calcitonin cells have been described only within the follicular epithelium, but not in the parafollicluar cells.

The electron microscopic structure of the follicular cells reveals that all the different organelles and structure involved in thyroglobulin formation and storage as well as in thyroid hormone production are very similar to that described in mammals (Sakai, 1960; Fujita, 1963; French and Hodges, 1977).

The plasma membrane of the cell is about 7 nm thick and is typically trilaminar, formed of two dense and a less dense layer similar to that of other cytoplasmic membrane systems. The dense granules are formed in the Golgi field, while the less dense granules are derived from the endoplasmic reticulum. From the apical plasma membrane, a number of microvilli project into the colloid. The nucleus is large and spherical containing 2-3 nucleoli. Mictochondria and rough endoplasmic reticulum are well-developed. Golgi bodies are commonly supranuclear and paranuclear. Smaller electron-dense granules, 100—500 nm in diameter, seen in the apical part of the cell are considered the primary lysosomes. Assenmacher (In: Avian Biology, Vol. 3, Ed. Farner and King, 1973) states that after TSH treatment, a distinct endoplasmic reticulum and large droplets of low density may be seen close to the basal plasma membrane, suggesting a direct secretion of droplets and endoplasmic sacs into the interfollicular space from the basal part of the follicular cells.

Some authors distinguish the epithelial cells of the gland into two types: (i) the **chief cells** (actively secreting or pre-secretory cells) which predominate in the normal gland; and (ii) the **colloid cells** which represent the collapsed, degenerating post-secretory cells. French and Hodges (1977) have given a more complete description of the various vesicles and dense bodies producing thyroglobulin and thyroid hormones.

Hormones: The colloid is largely composed of an iodised glycoprotein called thyroglobulin. The other iodine-containing compounds are di-iodo-tyrosine and thyroxine. Thyroxine, a derivative of thyroglobulin, is an alpha-amino acid related to tyrosin and contains 65% by weight of iodine. By using radio isotopes, two classical thyroid hormones–triiodothyronine and thyroxine–have been identified. Of the two, thyroxine is more predominant in a number of species (domestic fowl, Japanese quail, duck, pigeon, etc.) In domestic fowl, however, it is reported that only thyroxine has been identified in the thyroid gland. The biosynthesis of thyroid hormones takes place within the polypeptide chain of thyroglobulin which is the major iodoprotein of thyroid gland.

Functions: The physiology and biochemistry of the thyroid gland have been reviewed by Falconer (1971), Assenmacher (1973), and Ringer (1976). The thyroid functions in birds are very similar to those in mammals, except some quantitative differences. At the thyroid gland level, the birds have a higher total thyroid 127I content and a higher thyroid T_4 127I concentration. The gland is characterised by a prominent iodine storage. At the plasma level, avian thyroid metabolism shows some peculiarities. In the absence of a thyroxine binding globulin (TBG) in birds, the only available binding proteins are contained within the prealbumin and albumin. The pigeon is the only species that binds T_3 to prealbumin.

In birds, triiodothyronine has more potency than thyroxine in blocking TSH release.

The functional roles played by thyroid hormones are summarised:

i. **Growth:** In birds, as in other vertebrates, the thyroid hormones are essential in regulating balanced growth of the body.

ii. **Heat production:** One of the chief functions of the thyroid gland in adult birds is to maintain normal metabolic rate of the body. The normal heat production in the body is highly reduced by removal of the thyroid gland. The acceleration of metabolism is stimulated by thyroid hormones and leads to a rise in heat production.

iii. **Moulting of skin and feathers:** Recent investigation shows that, at 14 days of incubation, the skin epithelium of an untreated embryo has only three undifferentiated cell layers, whereas embryos treated with thyroxine show seven cell layers with distinct germinative and corneal layers. Thyroid hormone has a significant role on the plumage of adult birds—in the onset of moult and on the development of young feathers.

iv. **Action on carbohydrate:** It has been observed that thyroxine reduces liver glycogen level and produces mild hyperglycemia in the pigeons and in the same birds, the removal of the thyroid causes hypoglycemia.

v. **Reproduction and development:** The thyroid hormones play significant roles in the development and maturation of various vertebrate groups. In male domestic birds, the thyroid hormones, injected in moderate doses, promote growth of the testes. In the female domestic fowl, removal of the thyroid gland inhibits maturation of the gonads and decreases egg production. However, the developmental effects of thyroid hormones occur only in the presence of growth hormones (GH), and vice versa.

Evidences suggest that there is a distinct gonad-thyroid relationship. The annual cycle of thyroid function in birds has a relation with the reproductive cycle. Although there is no evidence of diurnal or circadian rhythms in thyroid functions, the avian thyroid gland has seasonal functional fluctuations.

vi. **Thyroid activity in migratory birds:** Migrating birds possess very active thyroid glands compared to non-migratory resting birds.

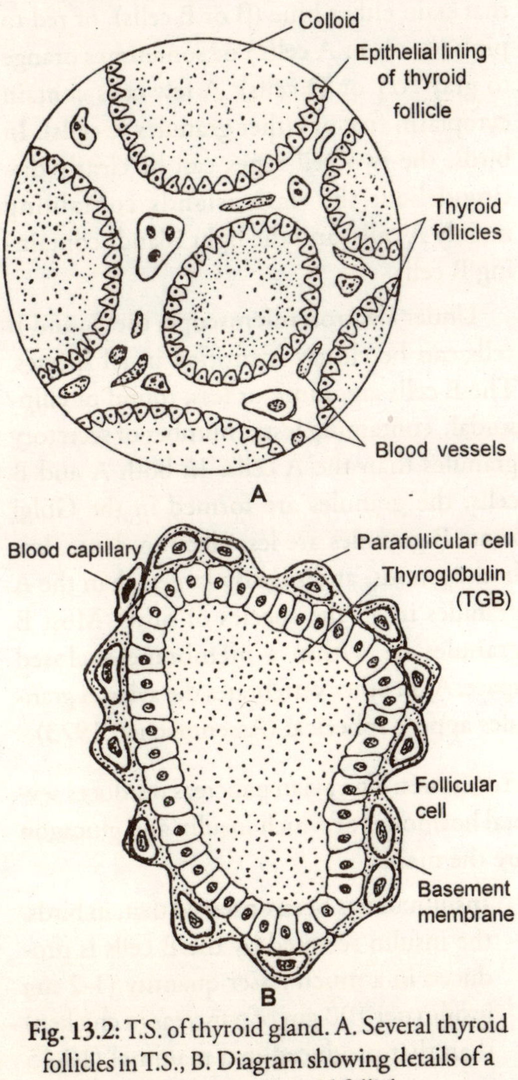

Fig. 13.2: T.S. of thyroid gland. A. Several thyroid follicles in T.S., B. Diagram showing details of a single human thyroid follicle

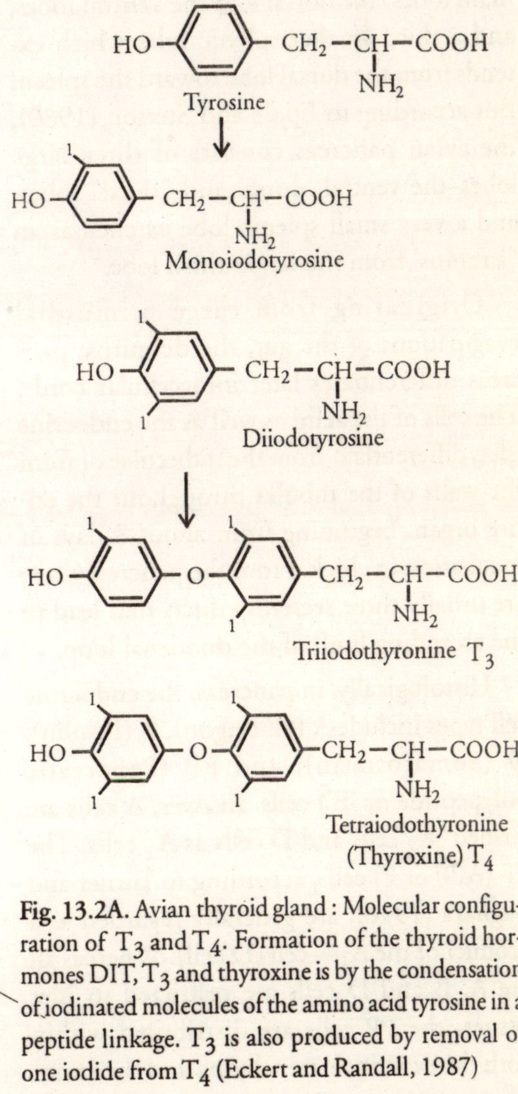

Fig. 13.2A. Avian thyroid gland : Molecular configuration of T_3 and T_4. Formation of the thyroid hormones DIT, T_3 and thyroxine is by the condensation of iodinated molecules of the amino acid tyrosine in a peptide linkage. T_3 is also produced by removal of one iodide from T_4 (Eckert and Randall, 1987)

13.3. ENDOCRINE PANCREAS

The avian pancreas, as in other vertebrates, is a compound tubuloacinar gland which lies transversely as a flat elongated body between the two limbs of the duodenal loop. Different authors have differently described the morphology of the pancreas and its subdivision into lobes. Copenhaver, Kelly and Wood (1978) and Tortora and Anagnostakos (1986) agree, in general, that the pancreas consists of a head, body and tail. Assenmacher (1973) describes that the pancreas consists of two main lobes, the dorsal and the ventral lobes, and a third smaller splenic lobe which extends from the dorsal lobe toward the spleen. But according to Epple and Stetson (1980), the avian pancreas consists of three large lobes–the ventral, dorsal and "third" lobes, and a very small splenic lobe extends as an "isthmus' from the larger third lobe.

Originating from three primordial evaginations of the gut, the definitive pancreas differentiates later into cellular cords. The cells of the acini as well as the endocrine islets differentiate from the trabeculae or from the walls of the tubules throughout the entire organ, beginning from about 8 days of incubation in chick. From the pancreas there are usually three secretory ducts that lead to the ascending limb of the duodenal loop.

Histologically, in pancreas, the endocrine cell types include A (Glucagon), B (Insulin), D (Somatostatin), and PP (Pancreatic polypeptide or 'F') cells. In Aves, A cells are termed A_2 cells and D cells as A_1 cells. The PP (APP or F) cells, according to Turner and Bagnara (1976), are generally regarded as a product of the A_1—cell (D cell). Whereas all the A, B and D cells are restricted to islet tissues, the PP cells are distributed within both the exocrine parenchyma and the endocrine islets, usually at their periphery. Although the splenic lobe is very small in size, it contains large endocrine cells, particularly A and D cells.

Under light microscope, the avian **islets of Langerhans** appear as, more or less, circular aggregate cells with a diameter ranging from 50 to 200μ. A thin layer of connective tissue extends into the islets and divides them into lobules. According to Epple (1968), Falkmer and Patent (1972), the majority of islet cells contain number of small granules that stain either blue (β or B cells), or red to pink (α, α_2 or A cells) or sometimes orange to gray (α_1 or D cells). A few cells contain cytoplasm free of either granules (c cells). In birds, the two cell types can be clearly distinguished: the dark islands containing mainly A cells, and the light islands containing B cells.

Under electron microscope, the A and B cells can be clearly identified in all species. The B cells are, more or less, round or ellipsoidal, containing lesser number of secretory granules than the A cells. In both A and B cells, the granules are formed in the Golgi zone. B granules are less electron dense, less homogenous, and slightly larger than the A granules in all the species of birds. Most B granules lie within a membrane-enclosed space. As in other vertebrates the avian A granules appear spherical (Assenmacher, 1973).

Hormones: The avian pancreas produces several hormones of which insulin and glucagon are the main:

1. **Insulin:** It has been observed that, in birds, the insulin secreted by the B cells is produced in a much lesser quantity (1-2 mg insulin per 100 gm of pancreas in chicken) than that produced in a mammal (10-15 mg per 100 gm). Assenmacher (1973)

reports that avian insulin is more effective in promoting hypoglycemia, when compared with equivalent amounts of mammalian hormone.

2. **Glucagon:** As in mammals, the pancreas and the gut are the two main sources giving rise to glucagon. From experiments with a number of birds (e.g. duck) it appears that two forms of glucagon are present—one with mol. wt. 3,000, similar to that of mammalian glucagon and the other with 6,000—double the mol. wt. Intestinal extracts contain large amounts of the latter type (large mols), but none of the small type which is mainly responsible for maintaining the blood glucagon level.

3. **Somatostatin:** It is secreted by delta cells (D cells) as well as from the gut and is termed 'Somatostatin' or the 'Growth Hormone Inhibiting Hormone' (GHIH) as it prevents the release of somatotrophin or the growth hormone. It is also termed as 'Somatotrophin Releasing Inhibitory Factor (SRIF). It is a tetradecapeptide.

   Epple and Stetson (1980) comment that this polypeptide may be defined more accurately as a 'useful diffusion substance' rather than a true hormone. The birds are unique among vertebrates in the possession of much greater quantity and higher concentration of pancreatic somatostatin.

4. **Pancreatic polypeptide (PP):** The PP or F cells are located within the endocrine islets along with insulin, glucagon and somatostatin secreting cells, as well as sparsely distributed throughout the exocrine parenchyma. Unlike in most mammals, PP cells are not found within the avian gut proper. Avian pancreatic polypeptide (APP) is glycogenolytic (without an effect on blood glucose). According to Hazelwood (1973), it is also a powerful secretogogue to the proventriculus, releasing large amounts of gastric fluid, hydrogen ion, pepsinogen and total protein. In non-fasted birds, APP appears to be an anabolic hormone, as it promotes net synthesis of lipid in the liver.

   The following schematic representation (Epple and Stetson, 1980; In: Avian Endocrinology) explains the pancreatic hormone responses in birds:

   ARG = Argenine; ACH = Acetylcholine;

   + = stimulate; – = inhibit

Two principal roles of pancreatic hormones are briefly mentioned :

1. **Carbohydrate metabolism:** Recent findings demonstrate the role of two pancreatic hormones (Insulin and Glucagon) in avian carbohydrate metabolism. The hypoglycemic role of insulin and the hyperglycemic role of glucagon have been well-established in domestic birds. The glycogenic effect of insulin and the glycogenolytic effect of glucagon have been demonstrated in pancreatectomised ducks and fowls.

2. **Lipid metabolism:** Glucagon shows distinct lipolytic action in birds. The stimulatory effect of glucagon on the concentration of free fatty acids in the plasma *in vivo* in the domestic fowl has been well-established.

As in mammals, the pancreatic hormone secretion is mainly regulated by the blood sugar level in birds.

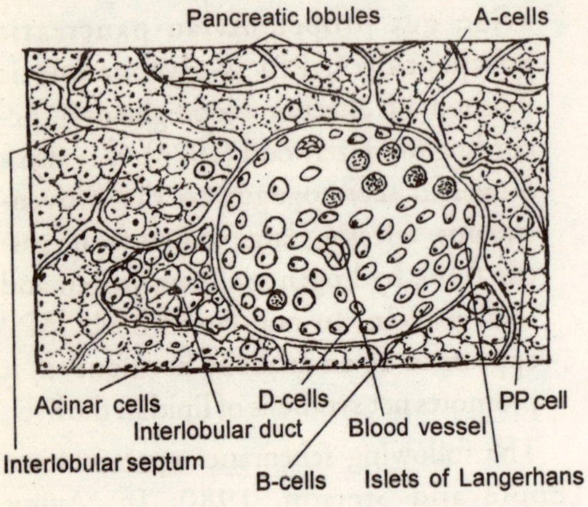

Fig. 13.3: Section through the pancreas of a rat showing Islets of Langerhans within endocrine pancreas surrounded by pancreatic lobules and acinar tissue (based on and modified from Turner and Bagnara, 1976)

13.4. ADRENAL GLANDS

MORPHOLOGY

The adrenal glands in birds are paired, oval, pear-shaped or triangular organs, yellow, orange or dark reddish-brown in colour. They lie on each side of the dorsal midline, just anterior to the postcaval vein and to the cephalic lobe of the kidney.

Holmes and Cronshaw (In: Avian Endocrinology, ed. Epple and Stetson, 1980) report that the glands receive arterial blood, either from a branch of the dorsal aorta or via branches of the renal and sixth intercostal artery and the venous blood is returned to the vena cava, directly or indirectly.

As in reptiles and mammals each adrenal gland is formed of two components– the cortex (adrenocortical or interrenal tissue) and the medulla (chromaffin tissue). These two components have different origins and functions. Unlike in mammals, no centralised medullary zone is observed in birds. Rather, the chromaffin tissues are distributed throughout the gland as numerous islands of cells which may be arranged differently in different parts of the gland, such as towards outer portion of the gland as in Yellow Warbler (*Dendroica aestiva*), or scattered throughout the adrenocortical tissue as in Brown Pelican (*Pelican occidentalis*), etc.

Adrenal cortex: As to the origin, the cortical portion differentiates first, as its primordium buds off from the coclomic epithelium, ventro-medially to the mesonephros. The prospective cortical cells then leave the epithelium and form paired masses of cells from either side of the aorta. Later, the cortical cells become arranged in cords which lie in a double row of columnar epithelial cells with the long axis of cells arranged in the transverse plane of the strand. The medullary cells appear a few days later from the primordium of the sympathetic nervous system. According to Romanoff (1960), these cells migrate singly and ventrally between the aorta and cortical tissue and then separate into groups of variable size around blood vessels.

In the adult animal, the ratio of cortical to chromaffin tissue is about 2 to 1. Unlike the mammalian cortical cells, the avian cortical cells have a distinct polarity in the distribution of cytoplasmic organelles. The seasonal and environmental changes induce increase or decrease in adrenal gland weight in mature birds. Such variations reflect growth and atrophy of the adrenocortical tissue, as observed in the pigeon, chick, blackbird, house sparrow etc. The variations in the size of the adrenal gland may also be due to their changing habitat. Thus marine birds have larger adrenal glands than terrestrial and freshwater birds, whereas, brackish water birds have the glands of intermediate size.

Fine structure of cortical tissue: In the arrangement of the avian cortical tissue, most authors distinguish two different zones—a thin subscapular zone (SCZ) and a thicker inner zone (IZ) e.g., in the Pigeon (Sinha and Ghosh, 1961).

The structural zonation of the adrenocortical cells on the basis of lipids and alkaline phosphatase has been confrimed by Ghosh (1962) who observed such zonation in seven out of eleven species. Wright et al (1966) also observed the same zonation in ducks, but experiments with many other birds were not confirmatory.

Lehninger(1970) observed mitochondria in the cells of both the subscapular zone and the inner zone containing electron-opaque granules, similar to calcium deposits. An extensive system of smooth ER and a comparatively smaller amount of rough ER are present in both the types of cortical cells. The enzymes responsible for the production of steroid hormones reside in the ER and the mitochondria. The cells of both the subscapular zone and the inner zone possess microtubules which concentrate more in the regions of the centrioles, nuclei and the peripheral cytoplasm. The plasma membrane of adjacent cells run parallel to one another. Pearce et al (1977) observed several types of junctions between adjacent plasma membranes—intermediate and gap junctions.

ACTH injections in cortical tissue induces depletion of the lipid droplets. Such depletion is associated with increased density of mitochondria, increase in smooth surface reticulum, as also in the size of the Golgi apparatus etc.

Functions of adrenal cortex: Adrenal function in birds, as in other vertebrates, is of vital importance. In mammalian adrenal cortex, zona glomerulosa synthesises aldosterone, whereas zona fasciculata produces mainly glucocorticoids. Holmes and Cronshaw (1980) comment that although various authors have claimed similar functions of the avian adrenal cortex, such similarities may not be conclusive.

Corticosterone is the main corticosteroid secreted by the avian adrenal cortex. Aldosterone is the other important corticoid hormone. The avian plasma corticosteroid level is generally lower than that of mammals. The following functions of adrenocortical hormones in birds are significant:

1. **Role in carbohydrate metabolism:** Since the time of Riddle (1937), many works have shown that corticosterone and hydrocortisone induce hyperglycemia together with increased accumulation of glycogen in the liver in fowl.
2. **Role in lipid metabolism:** Corticosterone and hydrocortisone also induce lipogenesis in the fowl as evidenced by the increase in plasma level and fat deposition in the liver.
3. **Role in protein metabolism:** Corticosteroid hormones inhibit growth in chickens.
4. **Action on kidney:** According to Assenmacher (1973), the corticosteroid hormone—aldosterone—present in birds is very effective in causing sodium retention and it considerably lowers the volume and sodium concentration of urine. Holmes and Cronshaw (1980) state that although birds do possess a renin-angiotensin system and secrete aldosterone, the role of angiotensin is little known.
5. **Regulation of extra renal electrolyte excretion:** Adrenocortical hormones also regulate extrarenal electrolyte excretion in birds with active nasal glands present in a

number of salt water species. The nasal glands in the birds secrete sodium chloride solution which is hypertonic to the plasma. The nasal salt gland allows the animal to remain in water balance, even for prolonged period of high salt diet.

6. **Role in reproduction:** The removal of adrenal gland causes atrophy of the testes and inhibits the development of the right gonad in ovariectomised chickens.

Regulation of Adrenal cortex: The adrenal cortical function is regulated by the pituitary gland and the hypothalamus. Diurnal and annual seasonal rhythms, as observed in mammals, are also exhibited by many species of birds. It has been stated that the diurnal periodicity in adrenal cortical function provides a better clue than seasonal rhythms for the study of photoperiod. Like photoperiod, temperature and feeding also have been correlated with adrenal cortical function.

Adrenal medulla: Ghosh (1980) in his article on adrenal medulla (in: Avian Endocrionology, ed. Epple and Stetson, 1980) reports that the birds possess an adrenal gland in which interrenal (cortical) and medullary tissues are completely intermingled.

Electron microscopic study of adrenal medulla in embryonic chick reveals two types of cells representing epinephrine (adrenaline) and norepinephrine (noradrenaline) containing cells (Hall and Hughes, 1970). Histochemical investigations have shown that there are marked species differences in the norepinephrine–epinephrine ratio (Ghosh, 1962, 1973). Medullas of Pelicaniformes, Ciconiiformes and Galliformes mainly produce norepinephrine, whereas Passeriformes mainly secrete epinephrine. Columbiformes and Cuculiformes secrete both the types of medullary hormones. Such variations reflect a phylogenetic diversity in hormonal secretion.

Chemically, of the two catecholamines, epinephrine and norepinephrine, the former is probably a derivative of the amino acid tyrosine and is dehydroxy-methyl-amino-ethyol benzene. The latter has the same chemical nature except the methyl group:

Tyrosine
↓ Oxidation
Dihydroxyphenyl-alanine
(DOPA)
↓ Carboxylation
Dopamine
↓ Oxidation
Norepinephrine
↓ methylation
Epinephrine
($2HO.CH.OHCH_2.NH.CH_3$)

Acetal phosphatides,vicinal glycol containing neutral polysaccharides, weakly acidic non-sulphated muco-substances, phosphatases, and choninesterases appear to be the basic cytochemical unit of the avian adrenal medulla. In birds, both epinephrine and norepinephrine are equally effective as blood sugar enhancing hormone.

Functions of adrenal medulla :

1. **Carbohydrate metabolism:** In the duck, goose, chicken and pigeon the hyperglycemic effect of epinephrine was known for many years. Ghosh (1973) observed that norepinephrine was an equally effective hyperglycemic agent as epinephrine in pigeons.

2. **Lipid metabolism:** It had been observed that norepinephrine induces a rise in plasma-free fatty acids in neonatal fowl from hatching to 2 weeks of age and declines considerably by 8 weeks. In adult fowls, ducks, turkeys, etc. none of the medullary hormones releases free fatty acids from adipose tissue.
3. **Thermogenic effect:** According to Freeman (1970), neither epinephrine nor norepinephrine arrests the fall of body temperature in young chicks treated with cold and anesthesia.
4. **Blood pressure:** In chickens, both epinephrine and norepinephrine cause appreciable rise in the systolic and diastolic blood pressure.
5. **Role in reproduction and migration:** Ghosh (1980) emphasises the important physiological role played by the adrenal medulla in avian reproduction and migration. According to Ghosh and Bhakta (1980), the high norepinephrine storage just before breeding migration helps in meeting the high energy requirement through lipolysis of stored fat during migratory flight.

13.5. PARATHYROID GLANDS

Origin and location: As in the mammals, the parathyroid glands arise from the dorsal aspect of the 3rd and 4th branchial pouches. These cell's then thicken into solid masses, the primordia of the parathyroid glands of which two are formed on either side close to the posterior pole of the thyroid. The number and location of the four parathyroid glands remain more or less the same in most species of birds. However, partial fusion between the glands on either side has been observed in some (e.g. domestic fowl).

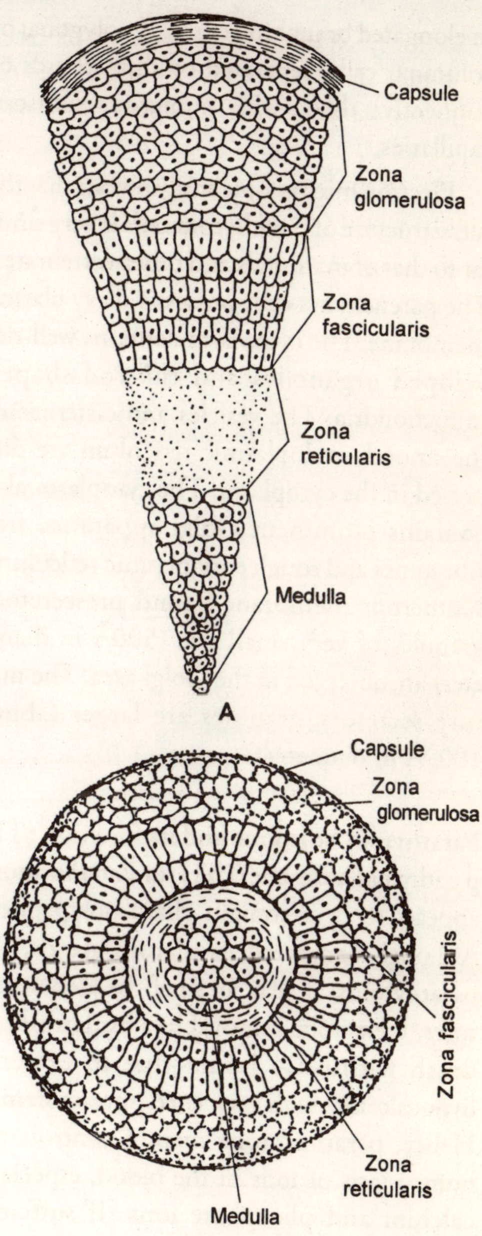

Fig. 13.4: A. Portion of an enlarged sectional view of adrenal gland; B. Cross-sectional view of the adrenal gland

Morphology: In light microscopic view, it is observed that the avian parathyroid gland can be clearly distinguished from the mammalian gland by the absence of oxyphilic cells between the chief cells. The latter are arranged in elongated branching cords of polygonal or columnar cells, separated by thin strands of connective tissue with numerous sinusoid capillaries.

Electron microscopic study reveals that the ultrastructure of parathyroid cells is very similar to that of mammals and other vertebrates. The parenchyma cells possess a wavy plasma membrane. The cytoplasm contains well-developed organelles and the rod-shaped mitochondria. The vesicles and cisternae of the smooth endoplasmic reticulum are dispersed in the cytoplasm. The cytoplasm also contains prominent Golgi apparatus, free ribosomes and rough endoplasmic reticulum. Numerous membrane-bound presecretory granules of very small size (500Å in diameter) are observed in the Golgi area. The mature secretory granules are larger (about 1000Å in diameter).

Parathyroid hormone and its function : The parathyroid hormone (PTH) or **Parathormone** appears to be a straight chain polypeptide. Although compared to thyroid glands, the parathyroids appear as small beaded structures, their total removal in a bird causes its death within 24 hours–due to extreme hypocalcemia with associated phosphoremia. Hence, parathormone mainly controls the homeostasis of ions in the blood, especially calcium and phosphate ions. If suffcient amount of Vit. D is present, parathormone increases the rate of calcium, phosphate and some magnesium absorption from the gastrointestinal tract into the blood. Parathormone is also responsible for the activation of Vit D.

Another vital function of parathormone is to increase the number and activity of osteoclasts or bone-destroying cells which breakdown the bone tissue, thus releasing the calcium and phosphate into the blood.

PTH secretion is not controlled by the pituitary gland. The pinpointed chief functions of the parathormone are to increase the levels of blood calcium and magnesium and to decrease the level of blood phosphate.

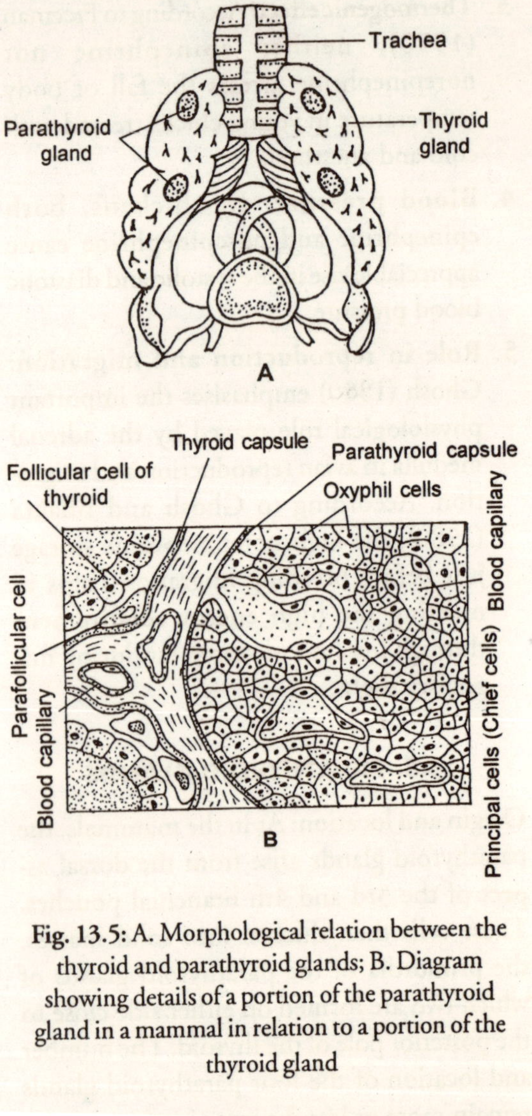

Fig. 13.5: A. Morphological relation between the thyroid and parathyroid glands; B. Diagram showing details of a portion of the parathyroid gland in a mammal in relation to a portion of the thyroid gland

13.6. ULTIMOBRANCHIAL BODIES

Origin and location: According to Assenmacher (1973), the paired ultimobranchial bodies arise from the fifth and sixth branchial pouches. In adult mammals, the ultimobranchial bodies are embedded within the thyroid gland, forming there in the parafollicular C cells. But in most avian species (except in the pigeons where the two glands are, perhaps, partly incorporated with the thyroid gland), both the glands remain distinct. In birds, the ultimobranchial bodies lie typically, one on either side, near the origin of the subclavian and common carotid arteries and in line with the two parathyroid and the thyroid glands.

Morphology: Under light microscope, the ultimobranchial bodies appear cystlike with varying shapes and sizes. Within the cavity of the gland, cords of parenchymal cells are observed. Among the aggregation of cells, numerous blood vessels are present.

In electron microscopic view, the cells bordering cavities of the gland are columnar or cuboidal with short microvilli. Actually, two distinct cell types are present here—the dominant light cells containing pale cytoplasm and some scattered dark cells having dense cytoplasm.

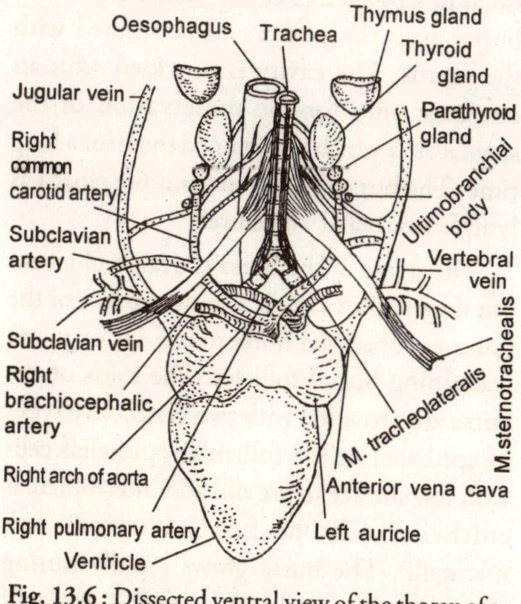

Fig. 13.6 : Dissected ventral view of the thorax of a domestic fowl showing location of the thymus, thyroid, parathyroid and ultimobranchial body in relation to the oesophagus, trachea, heart and major blood vessels (based on and modified from Turner and Bagnara, 1976)

Hormone—nature and functions: Calcitonin is the specific hypocalcemic hormone secreted by the ultimobranchial bodies. Chemically, calcitonin appears to be a polypeptide. Its mol. wt. ranges between 4,300 and 4,600 as against 4,000 in mammals. Although detailed information about the functional significance of calcitonin is not available, its hypocalcemic effect has been well-established. The ultimobranchial bodies in birds containing high concentration of calcitonin, are, in all probability, responsible for the regulation of calcium metabolism. According to Sunkiss (1961) birds are able to remove calcium from blood five times more rapidly than mammals. However, calcitonin has no major role in the growth of the skeleton.

13.7. THYMUS

To consider thymus as an endocrine gland was questionable since long back. In the last 3 decades, however, enzymatic and other studies on the avian and mammalian thymus have revealed possible endocrine role of this gland.

Hammond (1954) reports that thymus has its origin from the 3rd and 4th pharyngeal pouches, similar to the origin of the parathyroid gland. The primordial epithelial cells of the thymus form a cord that gives the appearance of the broken chains of lobes. The **thymocytes** or the **thymic lymphocytes** appear secondarily within the syncytial mass of epithelial tissue (Romanoff, 1960).

Morphology: The avian thymus can be distinguished from that of the mammalian counterpart in the location and number of thymic lobes. The avian thymus is an elongated gland consisting of several lobes (seven lobes in general) which lie along the jugular vein and the vagus nerve on either side, extending from the third cervical segment downward to the thoracic cavity.

The thymus in birds, as also in other vertebrates, increases in size until sexual maturity and then gradually regresses. Each of the thymus lobules in adult birds has distinct dense cortical and loose medullary zones. Payne (1971) reports that the posterior-most lobe of the thymus is intermingled with the thyroid. As in mammals, the size of the thymus in birds is influenced by a number of hormones; thus thyroxine accelerates enlargement of the thymus while cortisone, corticosterone and deoxycorticosterone cause atrophy of the gland.

Functional role of thymus and thymic hormone: The role of avian thymus in the formation of lymphocytes is well-known. Enzymatic studies suggest an endocrine role for the thymus and the bursa of Fabricius. Recent observations indicate two immunological systems—a thymus-dependent system, and a bursa-dependent system. The thymus-dependent system is formed of thymus and the circulating small lymphocytes. Lysophospholipase, ligase, acyl-coA-hydrolase, and phospholipase B are equally present in thymus and bursa in 8-10 week old birds (Epple and Stetson, 1980). Between thymus cortex and medulla, high concentration of Glucose-6-phosphate dehydrogenase and lactate dehydrogenase were actively present. These evidences suggest the possibility of endocrine-like cells in the thymus. The enodcrine-like cells have been actually observed in the chicken thymus, and these cells are very similar to those which synthesise polypeptide hormones.

So, it is quite possible that the avian thymus may be associated with the production of one or two thymic hormones, like the mammalian thymosin (Assenmacher, 1973).

13.8. BURSA OF FABRICIUS

Bruce Glick (In: Avian Endocrinology, ed. Epple and Stetson, 1980) comments that the identification of secretory substances and novel cells leads one to consider the bursa and thymus as lymphoid extensions of the endocrine system. Hieronymus Fabricius first described this obscure gland. It is a peculiar paracloacal derivative in birds.

Location, Origin and Morphology: It is located in a purse-like cavity (hence the name, bursa) near the podex and connected with the uterus. The cavity is provided with an entrance only for the preservation of the spematozoa which are retained there for a long time. The bursa is derived from the cloaca as lymphoepithelial structure.

The shape of the bursa is oval in chicken, but finger-like in duck. On the surface of the bursa are observed numerous folds or plicae containing bursal follicles. The folds of the bursa are provided with two distinctive types of epithelial cells—follicular epithelial cells with few surface microvilli and interfollicular epithelial cells profusely covered with microvilli. The bursa grows rapidly during the period betwen hatching and 3 weeks of age.

Functional roles: It has long been clearly established that the bursa confers immunological competence to birds (Chan et al, 1955). It has been stated that the bursa is responsible for the maturation of the lymphoid cells which produce humoral antibodies. The bursa has lower amounts of acyl transferase than the thymus and contains **sphingomyelin** and **Lecithins.**

There are different relationships between the bursa and other endocrine glands. An inverse correlation exists between the growth of the bursa and that of the gonads and of the adrenals. The steroids of the testes and the enlarged adrenals cause regression of the bursa. The administration of adrenocorticotrophic hormone (ACTH) by injection in domestic fowl (Glick, 1959) and in the pigeon (Bhattacharyya and Ghosh, 1970) results in considerable reduction in size of the bursa. Thus a peculiar relationship between the bursa and the adrenal gland has been suggested by many investigators.

13.9. TESTIS

The testis is formed of a large number of seminiferous tubules and the interstitial **cells of Leydig** present in the spaces between the tubules. The tubules lack blood vessels which, however, are abundantly present in the interstitial spaces. The **Sertoli cells** and the peritubular myoid cells form a barrier between the blood-vessels and the tubules of the testis. The specialised junctions between adjacent sertoli cells are usually tightly closed and these form an efficient blood-testis barrier, dividing the tubular epithelium into two compartments–a **basal compartment** containing spermatogonia and an **adluminal compartment** containing spermatocytes, spermatids and the spermatozoa.

The cells of Leydig are an important source of steroid hormones. This has been established by histochemical tests and ultrastructural studies. The lipid droplets decrease in number when the steroid secretion is maximum. The sertoli cells are also steroid-secreting. During transformation of the spermatids into spermatozoa, the spermatids lose cytoplasm and, for the time being, aggegate around the sertoli cells.

The Leydig cells contain smooth endoplasmic reticulum, a distinct Golgi complex, some lipid droplets and plenty of lysosomes. The smooth ER contains several important enzymes which actively participate in the secretion of some steroid hormones.

The Sertoli cells, too, have almost similar contents as the Leydig cells, besides possessing sufficient mitochondria.

In both birds and mammals, the androgens or the male hormones (**Testosterone,** dihydrotestosterone) are synthesised by the interstitial cells of Leydig as well as by the Sertoli cells. They also arise from the adrenal cortex and the ovaries and most possibly from the placenta (Turner and Bugnara, 1976). The androgens play a vital role in controlling the secondary sex characteristics of the male and in smooth functioning of the accessory reproductive ducts and glands.

An androgen-binding protein (ABP), secreted by the Sertoli cells, joins the testosterone or the dihydrotestosterone which enhance the proliferation and differentiation of the germ cells. The FSH of the anterior pituitary stimulates the Sertoli cells to secrete the ABP. The ABP leaves the testis through the efferent duct fluid which, finally, transports the bound androgen to the epididymis where the hormone may assist in the metabolic process for the maturation of the sperm.

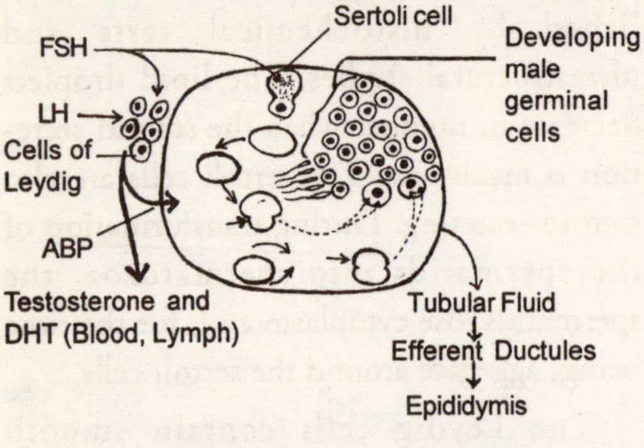

Fig. 13.7: Role of Sertoli cells and cells of Leydig in the synthesis of male hormones (Androgen, Testosterone and Dihydrotestosterne) which enhance male gametogenesis. ABP, Androgen binding protein; DHT, Dihydrotestosterone; FSH, Follicle stimulating hormone; LH, Luteinising hormone; T, Testosteron (Based on and modified from Turnar and Bugnara, 1976)

13.10. OVARY

The large avian follicles do not lodge within the **ovarian stroma** as found in the mammals. They are borne on follicular stalks. Also–unlike in the mammals—there is no antrum or the follicular fluid. However, as in the mammals, the egg is surrounded by the vitelline memebrane, the membrana granulosa and theca folliculi. A semicircular scar–the stigma or the cicatrix—develops opposite the follicular stalk, marking the site of the rupture of the egg from the follicle. Each follicular stalk bears a large number of ovarian follicles of variable sizes—from microscopic to that of a large egg ready to ovulate.

Among the vertebrates, the avian egg becomes mature with adequate yolk materials in the quickest possible time. The walls of the growing ovum and the follicles are supplied with an extensive network of blood vessels. In contrast to the expulsion of eggs in the mammals, the eggs in birds are released from the ovary—everyday for quite a long period of time.

Avian ovary is distinguished by the absence of **Corpora lutea**, but both estrogen and progesterone required for the normal reproductive cycles in birds have been detected (Turner and Bugnara, 1976). Evidence has also been gathered fom the ovary of the domestic fowl that androgen is also secreted by the follicles. In a number of species the post-ovulatory follicle persists for quite a long period for which it is presumed that such follicle may perform some endocrine function.

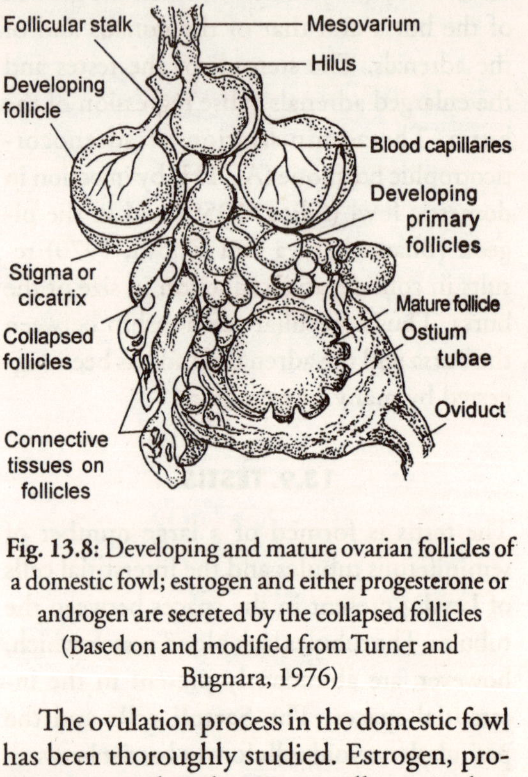

Fig. 13.8: Developing and mature ovarian follicles of a domestic fowl; estrogen and either progesterone or androgen are secreted by the collapsed follicles (Based on and modified from Turner and Bugnara, 1976)

The ovulation process in the domestic fowl has been thoroughly studied. Estrogen, progesterone and androgen are all required to stimulate copious secretion of albumin around the ovum in the region of magnum of the oviduct. Estrogen stimulates certain cell types which produce the egg-white proteins called **ovalbumin** and **lysozyme** and progesterone stimulates secretion of another protein—**avidin**. All these must happen before the ovum enters into the magnum and then only the secretion of albumin occurs around the ovum. Ovulation ceases to occur if there is any egg in any part of the oviduct.

14

Reproduction

Like the other vertebrate groups, the avian gonads are responsible both for (i) production and maturation of gametes and for (ii) secretion of steroid sex hormones which control the devlopment, growth and functions of accessory sexual organs and the secondary sexual characters.

Both these functions are again regulated by two very distinct gonadotropic hormones, namely, the Follicle Stimulating Hormone (FSH) and the Luteinizing Hormone (LH), both secreted by the pars distalis of the anterior pituitary gland. While the former mainly control the gametogenetic activity of the germinal epithelium (as observed in male Japanese Quail—FSH stimulates the spermatogenetic activity of the semniferous tubules), the latter regulates the steroidogenic activity of the interstitial Leydig cells (Brown and Follete, 1977) in male, and, in the female, induces ovulation and the formation of the ovarian corpora lutea (Lofts and Murton, In: Avian Biol., Vol. III; ed. Farner and King, 1973).

The primordial germ cells forming the male and female gonads are derived from the embryonic splanchnopleure and migrate in the blood to the location of the gonads and become the presumptive germinal epithelium. The left presumptive gonad is endowed with much greater number of primordial germ cells than the right, and hence, the asymmetry of gonadal development persists throughout life. In males, the embryonic primordia of the gonads develop into paired testes, but in the females of many species, only the left gonad develops into a functional ovary, the right one atrophies or remains inactive. However, Lofts and Murton (1973) report that many species of adult birds possess two functional ovaries (e.g., Fam. Accipitridae, Falconidae and Cathartidae), but even in these species, the right oviduct is not present. Functional right ovary is also present in certain species of pigeons and in the Herring Gull.

The male and female reproductive organs in birds have been described here under the following broad heads :

A. **The male and female gonads.**

B. **The male and female accessory reproductive organs.**

C. **Hormonal control of the testis and the ovary.**

D. **Secondary sexual characters.**

E. **Reproductive behavior.**

14.1. THE MALE AND FEMALE GONADS

1. The male gonad: Testis

1. **General structure:** In male birds, the gonads are the paired testes which, unlike in the mammals, are located in the body cavity, ventral to the anterior end of the kidneys. Each testis is attached to the dorsal body wall by a short band of mesentery called mesorchium. The spermatic arteries and the testicular veins supply arterial blood to and drain out venous blood from the testis. The left testis is larger than the right.

Each ovoid testis is enclosed by a fibrous coat, the tunica albuginea having a serous outer sheath, the tunica vaginalis. The testis size varies greatly in seasonally breeding birds. Internally, each testis is formed of highly convoluted and anastomosing network of seminiferous tubules. These are lined by the germinal epithelium containing developing germ cells and non-germinal Sertoli cells. The development of the germinal epithelium progresses in three distinct phases, such as (i) spermatogonial multiplication and formation of primary spermatocytes. (ii) their division through meiosis to produce secondary spermatocytes which mature into spermatids, and (iii) a phase of elongation and transformation of the spermatids to produce mature spermatozoa. The avian spermatozoon consists of a sperm head, middle piece and a long propulsive tail.

2. **The interstitial cells** : The interstices between the convoluted seminiferous tubules are filled with areolar connective tissue containing blood capillaries, lymph spaces, and the large, glandular interstitial Leydig cells which produce steroid sex hormones (male androgen–testosterone). These cells show well-defined seasonal, morphological, ultrastructural and histochemical changes.

The Leydig cell cycle can be stimulated by long photoperiods and in Japanese Quail, Nicholls and Graham (1972) have recorded the transformation of a fibroblast-like cell to the secretory form. The luteinising hormone (LH) stimulates Leydig cell differentiation in birds. The seasonal secretory cycles of the interstitial cells involve a rhythmical accumulation and depletion of cholesterol-positive lipoidal material.

3. **The sertoli cells:** Like the interstitial cells, the Sertoli cells of the tubule also undergo cyclic seasonal changes involving an accumulation and depletion of cholesterol-rich lipoidal material, similar to that seen in the adjacent interstitial Leydig cell, Ultrastructurally, the Sertoli cells show fine structure associated with a steroid-producing tissue. Various experiments indicate an endocrine role of Sertoli cells. With the termination of spermatogenesis, the seminiferous tubules regress and the Sertoli cells become profusely lipoidal. With the reappearance of spermatogenetic activity, a rapid depletion of lipids in Sertoli cells takes place.

2. The female gonad: Ovary

1. **General structure:** Unlike in other vertebrate groups, the reproduction in birds is exclusively oviparous. The left ovary, which becomes the functional gonad in most species, is attached to the body wall by a short mesovarium in close proximity of the kidneys.

The ovary undergoes the same great variation in size and weight as found in the testes. It has been observed that in the European starling (*Sturnus vulgaris*), the weight of the ovary has a wide fluctuation from 8mg at the time of maximum regression to 1400mg at the peak of the breeding season. In the non-breeding season, the ovary is a compact, flattened organ with small follicles which give the surface a granular apperance. But as it gradually develops into the breeding condition, the follicles become distended and they bulge from the surface of the ovary which now appears as bunch of grapes. The largest follicle remains suspended from the surface by a narrow isthmus of tissue—the pedicle. The surface of the mature follicle has a scar-like area, the stigma, which marks the region of rupture of the follicle during ovulation.

Histologically, as in other vertebrates, the ovary consists of an outer cortex containing the developing ovarian follicles and an inner highly vascular medulla forming the ovarian stroma composed of connective tissue. The cortex is bounded by the germinal epithelium which, by the time of hatching, produces several thousands of primary oocytes. Further development of oocytes ceases in adult birds. In the development of ovarian follicles, it is observed that only a small number of follicles mature fully, the rest undergo follicular atresia which is a characteristic feature of the avian ovary.

Each ovarian follicle consists of a single layer of granulosa cells surrounding the oocyte, but, with maturity, a thecal tissue in several layers is formed around and outside the follicle. The theca becomes richly vascularised. The granulosa cells become separated by intercellular spaces filled with perivitelline substance. The cell membrane of the oocyte gives off villi-like elevations forming the zona radiata. The oocyte completely fills the developing follicle so that no antrum or cavity is left. The follicular cavity, on the other hand, is very prominent in mammals. At the advent of the breeding season, some follicles attain full maturity and bulge from the surface of the ovary. This increased development is due to rapid deposition of yolk and its passage through zona radiata into the egg. In the pigeon the follicular diameter increases from 5.5 mm to 20 mm.

The folicular wall collapses after expulsion of the egg. In the hen, the granulosa cells become inflated with lipids and remain so for about 72 hours after ovulation, but no corpus luteum develops. Even without the corpus luteum, the ovary in birds, like in the mammals, secretes androgen, oestrogen and progesterone.

2. The interstitial cells: Lofts and Murton (1973) report that the ovarian interstitial cells arise from the stromal tissue and are the source of androgenic secretions within the female gonad. The androgenic hormone controls the development of comb of the domestic fowl in both the sexes.

3. The development of the follicle: The thecal cells in the mammalian ovary secrete **oestrogen** and the granulosa cells secrete **progesterone.** In birds, with sexual maturity, the remarkable increase in the size of the oviduct is considered to be the effect of oestrogen. The oviducts become narrower and constricted when follicular development decreases. Oestrone, Oestradiol-17B and oestriol are all found in the ovarian tissue of birds. The deposition of large quantity of yolk within the oocyte also depends on oestrogen secretion. Furr (1969 b) has confirmed that the maximum amount of progesterone is also found in the follicles, and only very little occurs in the ovarian stroma.

4. The postovulatory follicle: The postovulatory follicle and the ruptured follicles contain progesterone, but for a short while as rapid degeneration of the follicles follows.

5. The atretic follicles: The thecal tissue of the atretic follicles produces sufficient cholesterol which is kept as reserve by the time of the ovulation period in seasonally breeding birds.

6. Yolk storage: Before ovulation, the greater increase in the size of the follicles is due to the accumulation of large quantities of yolk in the maturing oocyte. In avian oocyte, the yolk contains proportionately much greater amount of lipid (33%) than protein (16%), as compared to that in other vertebrates.

ENDOCRINE CHANGES IN OVARIAN FOLLICLES

Initially, as the ovarian follicle enlarges, it produces increasing quantities of oestrogen, but as the follicle becomes still larger, progesterone and testosterone synthesis increases. Progesterone exerts its positive feedback action on the hypothalamus of the brain and not on the pituitary gland to induce ovulation in the domestic hen.

14.2. THE MALE AND FEMALE ACCESSORY REPRODUCTIVE ORGANS

General: Excluding the testis and the ovary, those parts of the reproductive system that also function as essential organs in reproduction are known as accessory reproductive organs. The sperm ducts in the male and the oviduct in the female, together with any special cloacal differentiation, fall under this category.

Other morphological and behavioural differences between the two sexes—such as plumage colour or pattern, nature of wattles, bill colour or the voice differences—arise due to secondary sexual characters.

Origin of reproductive ducts: Two pairs of ducts—derived from either the pronephric or mesonephric duct—develop in both sexes. These are the Müllerian and Wolffian ducts which run from the gonad to the urinogenital sinus. In the male adult bird, the Wolffian ducts are transformed into vasa deferentia and epididymes, whereas, in the female, these ducts generally remain vestigial. Here the left Müllerian duct becomes the oviduct, while the right one in the female—and both these ducts in the male—remain vestigial. According to Morgan and Kohlmeyer (1957), the occasional persistence of the right Müllerian duct in the female appears to be genetically controlled.

i. The male reproductive ducts: The spermatic tubules from the testis form a convoluted network through which the mature spermatozoa are conducted into a smaller number of rete tubules and, thence, to the highly convoluted vasa efferentia. With the advent of the reproductive season, the rete tubules become enlarged and more distinct. According to Bailey (1953), the spematozoa are rarely observed in the rete tubules. The vasa efferentia—which also enlarge in the breeding season and may become secretory—leave the **tunica albuginea** and form a long coiled tube—the **epididymis.**

The epididymis is a compact structure that adheres closely to the body of the testis and lies adjacent to the kidney. The epithelial cells of the epididymis are columnar and ciliated, and secrete the seminal fluid. The spermatozoa are not stored in this organ but pass into vasa deferentia, the longer paired muscular tubes that lead downward to the urodaeum of the cloaca. The vasa diferentia, too, become enlarged and more convoluted during the breeding season. Each vas deferens expands to form a seminal vesicle which, during the breeding season, lies heavily impregnated with spermatozoa. A muscular sheath presses the posterior wall of the sperm sac to expand into the cloaca as erectile papillae that help expulsion of the sperms during copulation.

The cloaca in certain birds is modified into a penis-like structure. In waterfowl, the vascularized sac-like penis may be protruded and retracted. Penis-like structure occurs in tinamous and in all ratite birds, as well as in ducks and geese. It has been observed that the penis of the mallard duck increases six times during the breeding season.

Pearson (1955) reports that, in all ratite birds, the male usually incubates the eggs and cares for the young. The Ostrich, Greater Rhea and the other ratites perform cooperative laying in which several females lay their eggs in the same nest in which incubation is attended exclusively by the male.

ii. The female reproductive ducts: It has been mentioned that the right oviduct and the right ovary in birds usually remain vestigial. Hence, in the female, only the left oviduct becomes highly differentiated in producing the complex shelled and pigmented eggs.

The tubular oviduct is attached to the bodywall by dorsal and ventral ligaments and double folds of the peritoneum, covering the oviduct.

The oviduct has five distinct regions: the **infundibulum, magnum, isthmus, shell gland** (uterus), and **vagina**. The ovum expelled from the ovary is possibly guided into the funnel-like infundibulum by the ventral ligament. The fertilisation of the ovum occus in the upper thin-walled neck of the oviduct, before albumin is added to the yolk.

The middle part of the oviduct is the highly glandular magnum where the white albumin is formed and deposited around the ovum in 4 layers. In the hen, this process takes about 3-4 hours. The egg now passes into a less glandular and more muscular isthmus where the two shell membranes are secreted. These membranes are formed of protein fibres cemented with albumin. The actual shell is secreted in the form of about 97% of calcium salts, supported on an organic matrix of protein fibers. The shell gland—also known as the uterus—is a dilated saclike structure which leads to the cloaca though a short narrower vagina.

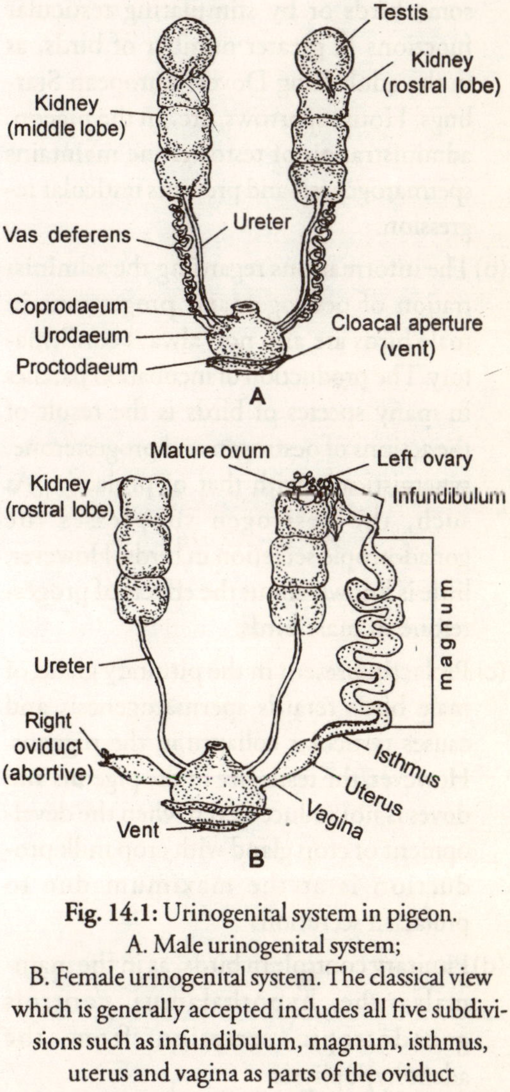

Fig. 14.1: Urinogenital system in pigeon.
A. Male urinogenital system;
B. Female urinogenital system. The classical view which is generally accepted includes all five subdivisions such as infundibulum, magnum, isthmus, uterus and vagina as parts of the oviduct

In domestic fowls, oestrogen alone causes growth of the albumin-secreting glands and the secretion of **ovalbumin**. Progesterone stimulates synthesis of the egg white protein–**avidin**.

14.3. HORMONAL CONTROL OF THE TESTIS AND THE OVARY

1. Hormonal control of the testis

(a) It has been reported that the administration of androgen may act both ways, either regressing the function of testis in some birds or by stimulating testicular functions in greater number of birds, as in the adult Ring Doves, European Starlings, House sparrows, etc. In the pigeon, administration of testosterone maintains spermatogenesis and prevents testicular regression.

(b) The informations regarding the administration of oestrogen and progestrone in male birds are also not always confirmatory. The production of incubation patches in many species of birds is the result of the actions of oestrogen and progesterone, synergistically with that of prolactin. As such, the oestrogen suppresses the gonadotropic secretion in birds. However, little is known about the effects of progesterone in male birds.

(c) Prolactin present in the pituitary gland of male birds retards spermatogenesis and causes testicular collapse in the pigeons. However, the testis size in the pigeons and doves is not reduced even when the development of crop gland with crop milk production is at the maximum due to prolactin secretion.

(d) Pituitary control: In birds, as in the mammals, the hypothalamus controls gonadotropic secretion from the adenohypophysis.

(e) According to Lofts and Murton (1968), the different kinds of photoperiodic r :-sponses in birds have led to a variety of annual breeding patterns. The photoperiodically controlled breeding cycles may be considered as an adaptive mechanisn.

2. Hormonal control of the ovary

(a) The administration of androgen in pigeons and House Sparrow enhances follicular growth and the follicular epithelium becomes stratified.

(b) In Ring Doves and in the domestic hens, exogenous oestrogens delay ovulation. Eoestrogen stimulates the liver to produce the protein and lipid precursors of yolk.

(c) In domestic hens, administration of progesterone inhibits egg-laying and, if given in large doses, results in an increase in follicular atresia.

(d) The effects of prolactin on the ovary are less significant than on the avian male gonad. However, in some birds, this hormone inhibits follicular devlopment with the resultant atresia.

14.4. SECONDARY SEXUAL CHARACTERS

It has been mentioned that the permanent plumage differences in sexual dimorphism may have a genetic basis in which case hormone is unable to change the inherited sex-type. According to Witschi (1961), there are ample examples in birds of the endocrine basis of plumage changes and acquisition of various nuptial decorations. The development of the comb and wattles of newly-hatched chicks depends on androgen and can be stimulated by testosterone injection.

The colour of the bill of birds may be of two main types : (i) That produced by the deposition of brown or black melanins, and (ii) The yellow, red and orange caretenoids that are absorbed from the food of birds. Again, some birds change the colour of their bill at the advent of the breeding season (e.g., European starling, House sparrow) under the influence of androgen.

The ventral apterium in some birds become highly vascular and very thick before or during egg laying. This featherless tract—now known as 'incubation patches' or brood-patches—comes in close contact with the eggs during incubation and provides necessary warmth for the development of the embryos. The incubation patches usually develop in the females, but in a number of species, they may develop in both the sexes (e.g., Starlings and Swallows) and,in rare species, only in the males where they exclusively incubate the eggs (Red-necked Phalarope; Charadriiformes).

14.5. REPRODUCTIVE BEHAVIOUR AND THE ROLE OF HORMONES

So diverse is the reproductive behaviour in different species of birds that it does not conform to any generalised pattern. However, the reproductive behaviour in birds exhibit certain phenomena in common, such as acquisition of a territory by the male, the advertisement of the territory through different song patterns which culminate in courtship behaviour by both the sexes, mating, nest building by the pair, egg laying, incubation, etc.

Tinbergen (In: The Macdonald Encyclopedia of Birds of the World; Little, Brown and Company, UK; 1995, Pp.1-512) attributes several functions to the courtship procedures: the synchronization, coordination and orientation of the behaviour of the two sexes in time and space. The **'bowing display'** is a common practice in the courtship of the male feral pigeons. In Pied Avocets, both birds preen their feathers and then the female lowers herself, thereby inviting the male to mate. In Herring Gulls, both sexes raise their heads several times, each with a cry, and then the male approaches the female and mating takes place. In Grey Heron, a loud cry, stretching the neck backward and bowing movement precede mating. The nuptial parade of the Great Crested Grebe is very interesting. After a series of attitudes in the parade each bird seizes a clump of water weed in its beak and takes an upright breast-to-breast position in the water. The male Bowerbirds of New Guinea and Australia build small "bowers" of twigs, grass, leaves, etc. for their courtship.

As to the hormonal effects in courtship, there is good evidence that androgens influence bowing behaviour in both Ring Doves and Feral Pigeons. In bowing display, the male shows submissive posture by crouching, with the tail fanning out and the head lowering. In the selection of a mate, the female ensures that she has adequate protection from her mate and a well-defended territory. Male's behaviour in this respect is rather aggressive as he deposits his sperms immediately before another male can do so. The male ensures protection for the eggs and the young.

With the increase in progesterone level, the androgenic effect is inhibited. As a result, the sexual aggressiveness on the part of the male is suppressed and both the male and female now practise mutual caressing and start nest building. In the latter function, oestrogen becomes more effective than progesteron. The behavioural acts in nest building by the male stimulate gonadal hormone secretion from the ovary which successively leads to oviduct development, ovulation, oviposition and the onset of incubation.

In short, the role of hormones in the reproductive behaviour of birds is of great importance. The following scheme summarises the progressive course of events leading to increased prolactin secretion which is involved in the initiation of broodiness :

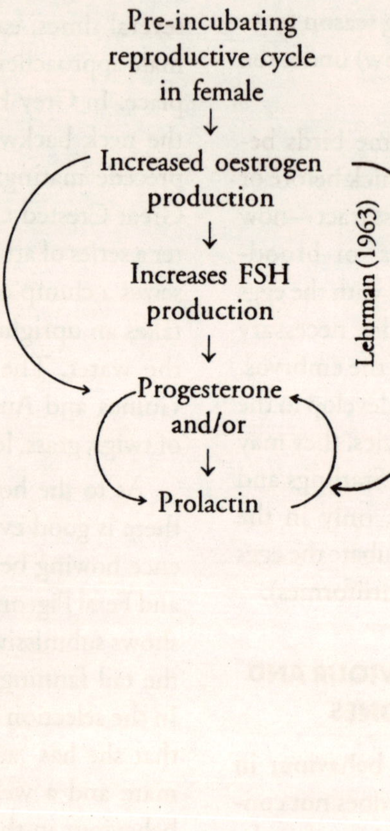

15

Nest Building, Incubation of Egg, Care of Nestling and Brood Parasitism

Gianfranco Bologna (In the Macdonald Encyclopedia of 'The Birds of the World; Little, Browns & CO. London, 1995), mentions the birds' nest as 'A Jewel of Nature'. Indeed, it is not an exaggeration that bird's nest is a remarkable expression of instinctive talent.

For laying and incubation of eggs and for nursing and feeding the newly-hatched youngs, almost all species of birds build varieties of nests. Usually, both male and female participate in building their nest. From a very haphazard and irregular type of nest, many intermediate forms and designs, and some nests with amazingly skilful architecture are observed in nature. The nest building materials, methods and patterns are usually species-specific.

Quite different types of nests are built by different groups of birds. Salim Ali (The Book of Indian Birds, 11th ed; 1979) mentions the following main types of birds' nests as found in India:

1. **Simple scrapes:** Birds like quail, sandgrouse, oystercatcher, common peafowl, jungle fowl, sandpiper, terns and many others just collect grass and leaves on the ground and build some form of irregular nests with them without any lining.
2. **Twig nests:** This is a quite common type of nest built by a number of bird families amidst twigs and thickly-set short branches of a leafy tree. The nest is cup-shaped, the inner lining of which is formed of soft grass, feathers, threads, etc. Usually, birds like crows, pigeons, doves, kites, vultures, cormorants, egrets, storks, etc. build such a nest.

Fig.15.1: Platform nest of dead twigs of a Pelican

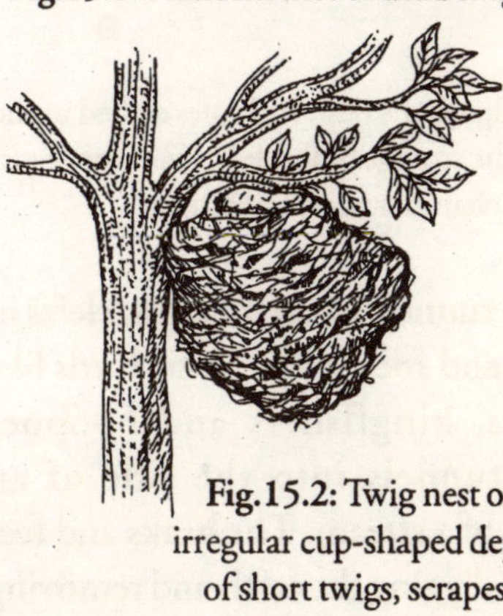

Fig.15.2: Twig nest of a Pigeon: an irregular cup-shaped depression formed of short twigs, scrapes, feathers, etc.

Fig.15.3: Saucer-shaped stick platform nest of white stork made of mud and feathers

3. **Tree-holes:** Birds like woodpeckers, parakeets, barbets, hornbills, owls, some mynas including hill mynas, and most of our resident ducks build nest with or without lining in tree-holes which may be either natural or excavated.

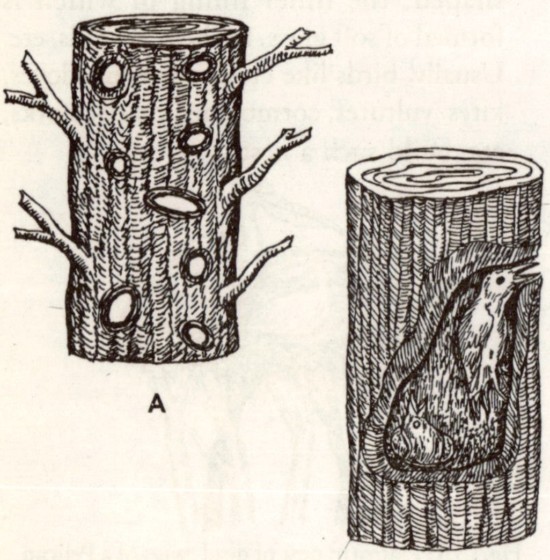

Fig.15.4: Woodpecker's nest. A. Holes drilled in the tree trunk for the storage of food; B. Woodpecker's nest chamber with hatchlings

4. **Excavated tunnels in earthbanks, clefts in buildings and rock cliffs:** Some birds like bee-eaters, kingfishers and hoopoes excavate tunnels into the side of an earthbank of a stream. The beaks and feet are used in digging the earth and removing loose soil, respectively.

Fig.15.5: Kingfisher's nest with excavated tunnnel in earthbank. The Bee-eaters also form the same type of nests, sometimes forming colonies

5. **Mud-nests:** The whistling thrush, some blackbirds, swallows, etc. collect sticky mud and mix it with their saliva. The mud nest thus formed is allowed to dry up gradually.

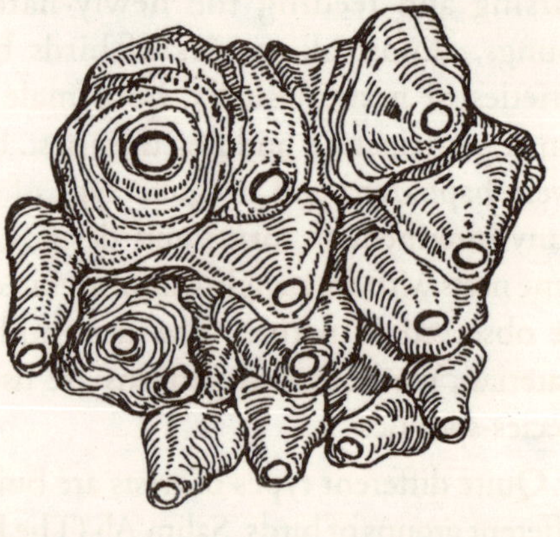

Fig.15.6: Colonial mud nests of cliff Swallow

6. **Cup-shaped nests of grass and fibres:** In the forks of tree-branches, birds like flycatchers, orioles, bulbuls, black drongos, babblers, reed warblers and cuckoo-shrikes, build cup-shaped nest with grass and fibres, plastered with cobwebs (i.e. spider's network of threads). The cobweb is used as a cementing material.

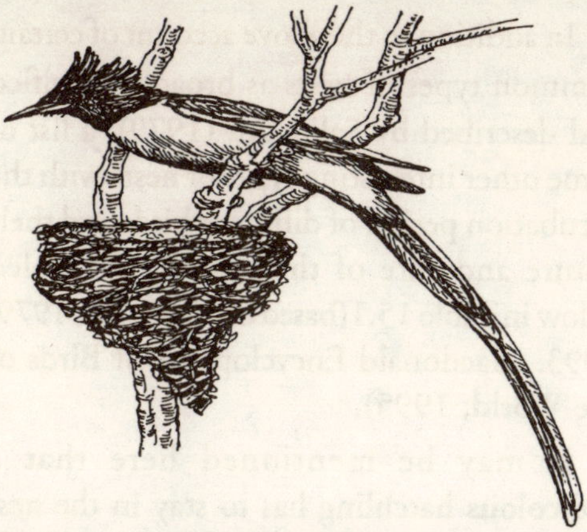

Fig.15.7: Cup-shaped nest of grass and fibres at the junction of a tree-branch built by a Paradise Flycatcher

7. **Dome or ball-shaped nests:** This type of dome-shaped nest, formed of twigs, grass, rootlets, etc. has only one lateral entrance hole which may be prolonged to a short neck e.g., Munias and Rufous-bellied babblers.

8. **Pendant nests:** The vertical and oblong pouch-like nests that look like pendants are found suspended from outhanging branches or twigs at heights between 7 and 12 feet (2-3.6 metres) from the ground. Such nests, skilfully designed and built by the Baya Weaver birds, Sunbirds and Flowerpeckers has only one lateral entrance hole near the top.

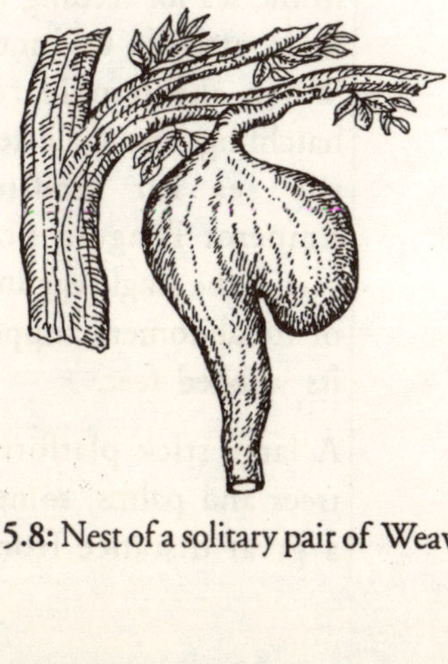

Fig.15.8: Nest of a solitary pair of Weavers

9. **Woven oblong purse:** The oblong purse-like nests built by some of the Wren warblers. The nest is attached to the stem of a tall grass or of a low bushy tree.

10. **Nest in leaves stitched together:** The Tailor bird, Franklin's Wren Warbler, Ashy Wren Warbler, etc. build nest with leaves that are remarkably stitched together in the form of a funnel.

Fig.15.9: Nest of a Tailor Bird: two leaves stitched together like a funnel

Salim Ali and Ripley (1987) mention that there are other nests which are less conventional. The Swifts build half-saucer-shaped nests with their saliva and/or with a mixture of straw and feathers. The nests are attached to the sides of a rock or in dark caves in islands. The Palm Swift has the same type of nest but much stronger with greater number of feathers—the nest is attached with the bird's saliva in the fold on the underside of a 'tal' (palm) leaf. The Pheasant-tailed Jacana builds its floating nest with grass and aquatic vegetation. The Rufous Woodpeckers live in coexistence in the nests of certain tree ants. Salim Ali (1979) comments that there is a clear distinction between the nests built by individual pairs and those built by colonial birds—The Weaver birds, Cliff Swallows,

Storks, Cormorants, Flamingos and Herons are examples of the latter category.

Fig.15.10: A Pheasant-tailed Jacana (Order charadriiformes) on a floating nest

In addition to the above account of certain common types of nests as broadly classified and described by Salim Ali (1979), a list of some other interesting types of nests with the incubation period of different birds and their nature and care of the nestlings is tabled below in Table 15.1 (based on Salim Ali, 1979, 1993: Macdonald Encyclopedia of Birds of the World, 1995).

It may be mentioned here that a **nidicolous** hatchling has to stay in the nest for parental rearing before flying whereas a **nidifugous** one is able to leave the nest shortly after birth.

Table 15.1: Nesting, incubation and parental care of some common birds

Name of Bird	Incubation	Nature and Care of the Nestling	Type of Nest
1. **Royal albatross** *(Diomedea epomophora)*	79 days; incubated by both parents	nidicolous; single nestling, reared by both parents	This large white sea bird, weighing about 7-8 kg and having wing-span ranging between 200-320 cm, builds its nest in colonies on the ground. It is made of mounds of vegetable matter.
2. **Emperor Penguin** *(Aptenodytes forsteri)*	About 90 days; (including warming up period; effective incubation period 62-64 days), incubation done by the male	nidicolous; hatchlings reared by the female	The eggs are laid in autumn when the offshore sea is iced over. The females then return to the sea for feeding while the male attends to incubation. When the female rears the hatchlings the male returns to the sea for feeding. The Emperor Penguin practically carries the single egg in the fold of its abdomen, supported by its webbed feet.
3a. **Grey Pelican** (*Pelecanus philippensis*)	28-30 days	same as above	A large stick platform in tall trees and palms, sometimes at a great distance from water.

Contd.

Name of Bird	Incubation	Nature and Care of the Nestling	Type of Nest
			Usually there are several nests in the same tree.
b. **Brown Pelican** (*Pelecanus occidentalis*)	28-30 days	nidicolous; hatchlings reared by the female	The nests are built in large colonies—on the ground or in low bushes—lined with plant matter, guans or twigs.
c. **Old World White Pelican** (*Pelecanus onocrotalus*)	29-30 days	same as above	It makes large nests of branches and twigs in trees in large colonies.
4a. **Little Cormorant** (*Phalacrocorax niger*)	4 weeks	nidifugous; reared by both the parents	The nest is built in the form of a shallow twig platform in trees, with colonies of egrets, storks, etc.
b. **Galapagos or Flightless Cormorants** (*Nannopterum harrisi*)	23-25 days	same as above	The nest is built on rocks close to the sea, with seaweeds and twigs. This is the only flightless cormorant species the number of which is gradually declining.
5. **Old World Spoonbill** (*Platelea leucorodia*)	21 days	same as above	The bird nests in colonies, usually in canebrakes or on trees in large stick platform made of sticks, seeds and vegetable matter.
6. **Grey Heron** (*Ardea cinerea*)	23-28 days	same as above	Nest in colonies, usually high up in trees, but also occasionally on the ground in bushes. The hatchlings are cared by their parents for about 50-55 days.
7. **White Stork** (*Ciconia ciconia*)	25-30 days	same as above	The nest is a large saucer-shaped stick platform, supported by mud and feathers on chimneys, tops of buildings, tall trees, etc.; uses same nest for many years.

Contd.

Name of Bird	Incubation	Nature and Care of the Nestling	Type of Nest
8. Greater Flamingo (*Phoenicopterus ruber*)	28-32 days	nidicolous	The nest is a truncated conical mound of hard, sun-baked mud about 6-12 inches (15-30 cm) in height with a slight depression on top; the nests are built in hundreds, close to one another.
9. Barheaded Goose (*Anser indicus*)	4 weeks	same as above	The nest is built in a depression in herbage bordering high altitude lakes.
10. Brahminy Duck (Ruddy Shelduck) *Tadorna ferruginea*	4 weeks	same as above	The nests are built in holes in cliffs or in sand dunes, sand banks or even in buildings which may be located near or far from water.
11. a. Black or King Vulture (*Sarcogyps calvus*)	8 weeks (approx.)	nidicolous; nestling fed by both the parents	The nest is a massive platform of twigs on the top of a large tree, about 30-40 feet (9-12 metres) high.
b. Bengal Vulture (*Gyps bengalensis*)	6-7 weeks	same as above	The nest is built somewhat hapazardly in the form of a twig platform on the top of a high tree—banyan, tamarind, etc.
c. Egyptian Vulture (*Neophron percnopterus*)	6 weeks	same as above	The bird nests at various altitudes, from about sea-level up to about 11,800 ft (3,600 metres) in the Caucasus. The crude nest is built with small branches covered with rubbish in holes or in nooks of rocks, but rarely in trees.
d. Bearded Vulture (*Lammergeier*) *Gypaetus barbatus* Himalayan Bearded Vulture (*Gypaetus barbatus aureus*)	53 days (approx.)	same as above; incubation done by both parents	The bird builds its nest in rocky cliffs, hollows and cracks of high altitudinous mountains with the help of sticks, feathers, sheep's wool and rubbish. Each pair has two or three nest sites used in rotation.

Contd.

Name of Bird	Incubation	Nature and Care of the Nestling	Type of Nest
e. Gyr Falcon (*Falco rusticolus*)	28-29 days	nidicolous; nestling fed by both the parents	This Arctic bird of Europe, Asia and America builds nest on sea cliffs and in rocks.
f. Andean Condor (*Vultur gryphus*)	54-58 days	same as above	This is the largest of all vultures, living mainly in the Andes of South America. It lives in the highest parts of the mountain, flying up to an altitude of 21,000 feet (6,400 metres). The nest is built in the crack of a steep, rugged rock.
12. Brahminy Kite (*Haliastur indus*)	26-27 days	nidifugous; incubation done by the female	This bird builds its nest either in a solitary tree or in a building with small dry branches, leaves, pieces of paper and fish bones and the nest may be built up to an altitude of 8,200 feet (2,500 metres).
13. Golden Eagle (*Aquila chrysaetos*)	43-45 days; incubation done by the female	nidifugous; cared by both parents	The nest is a crude one used for several years and is built with huge pile of branches, twigs and vegetable matter—either on the top of trees or in rocks.
14. Red Jungle Fowl (*Gallus gallus*)	19-21 days; incubation done by the female	nidicolous	It builds its nest in a shallow scrape on the ground in thick undergrowth, lined with sticks and dry leaves.
15. Common Peafowl (*Pavo cristatus*)	28 days	same as above	This bird, also known as Blue Peafowl,usually builds its nest in a thick bush, but, rarely, the nest is also built in a hole in a tree branch or in a building.
16. Mallee Fowl (*Leipoa ocellata*)	incubation done by the female	same as above	Mallee Fowl of Australia is unique among birds in its incubation strategies. Some species lay eggs on the sunniest spots on the beach, thus using the sun's heat directly for

Contd.

Name of Bird	Incubation	Nature and Care of the Nestling	Type of Nest
			incubation. Others dig deep tunnels up to 3.9 ft (1.2 m) deep and 6.6 ft (2 m) wide which they cover with dry branches and leaves to maintain the incubation temperature. Some others build huge mounds of leaves and loam up to 16.4 ft (5 m) high and 39 ft (12 m) in diameter within which incubation temperature is provided by the sun and the decomposition of leaves. After 4 months., the female lays up to 35 eggs in this nest at an interval of 4-8 days.
17. Sarus Crane (*Grus antigone*)	28-30 days	nidifugous; reared by both parents	The nest is a huge mass of reed and bush stems and straw in the middle of a flooded paddy field or marsh.
18.a. Indian Whiskered Tern (*Chlidonias hybrida*)	2 weeks	seminidifugous; incubation done by both sexes, but mainly by the female	Nest is built usually in colonies in the form of a circular pad of reeds and rushes in tangles of aquatic vegetation in 'jheels' and swamps.
b. Caspian Tern (*Hydroprogne caspia*)	20-22 days	same as above	The nest is built in colonies in hollows on the sandy ground on islands off Makran coast and sand banks off Mannar.
19. Blue Rock Pigeon (*Columba livia*)	17-19 days; incubated by both parents	nidicolous nestlings fed by both parents with crop milk	Gregarious by nature; the bird nests in colonies in hollows and in caves in cliffs (in the wild).
20. Spotted Dove (*Streptopelia chinensis*)	About 2 weeks; incubation by both parents	nidicolous; fed by both parents	The nest is built usually of two crossed sticks, low down in a tree or bush, or in the ceiling and beams of building.

Contd.

Name of Bird	Incubation	Nature and Care of the Nestling	Type of Nest
21. Kakapo or Owl Parrot (*Strigops habroptilus*)	incubation period undetected; incubation done by the female	nidicolous; nestlings fed by both parents	It is mainly a ground-dwelling bird which excavates its nest in a rotten tree trunk, usually at the foot or amidst roots of the tree. This bird is facing near extinction and is now restricted to the North Island of New Zealand.
22. Lorikeet (*Loriculus vernalis*)	incubation period undetected. Both parents incubate though mainly done by the female	nidicolous; fed by both parents	The nest is a hole excavated by the bird in a rotten branch or tree-stump, although a natural hollow is preferred. The Lorikeet is unique among Indian birds for its habit of roosting at night like a bat, hanging its head downwards.
23. Koel–Old World Cuckoo (*Eudynamys scolopacea*)		nidicolous; hatchlings fed by both parents	Parasitic, using crow's nest. As many as 13 eggs have been found in a crow's nest.
24. Barn Owl (*Tyto alba*)	32-34 days incubation done by both parents	nidicolous; fed by both parents	The nest is built in tree hollows or holes in ruined walls or in space between ceiling and roof of a dwelling house.
25. Great Horned Owl (*Bubo virginianus*)	35 days; incubation done by both parents	nidicolous; nestlings reared by both parents	This bird usually nests in abandoned nests of birds or in the hollows in rocks; sometimes, it may also build nest on the ground.
26. Old World Swift (*Apus apus*)	14-20 days; incubated by both sexes	same as above	This bird spends most of its time in the air, as such, described as the most 'aerial bird'. It nests in holes in roof-tiles in old or new buildings and also in rocks.

Contd.

Name of Bird	Incubation	Nature and Care of the Nestling	Type of Nest
27. White breasted Kingfisher (*Halcyon smyrnensis*) (See Fig. 15.5)	About 3 weeks	nidicolous; nestlings reared by both parents	The nest, typical to that of other kingfishers, is built in the form of a horizontal tunnel about 2-4 feet (0.6-12 metres) in length, dug into the side of an earth bank or nullah. The tunnel ends in a spacious "chamber" which is the actual nest. Digging is done by both sexes.
28. Small Green Bee-eater (*Merops orientalis*)	About 4 weeks; incubation done by both sexes	nidicolous; nestlings reared by both parents	The nest is built in colonies in the form of a long horizontal or oblong tunnel (about $2\frac{1}{2}$-$8\frac{1}{2}$ feet [0.76-2.6 m] long), ending in wide chamber. The chamber may be lined with remains of insects eaten. The tunnel is dug with the help of beaks and legs by both the sexes.
29. Hoopoe (*Upupa epops*)	16-19 days; incubated by the female	nidicolous; fed mainly by the female	It nests in a natural tree-hollow or in a hole in the wall or ceiling of a building lined with straw, linen and rubbish in a filthy manner. Usually a single clutch of 5-8 eggs are laid but, exceptionally, up to 12 eggs may be laid.
30.a. Common Gray Hornbill (*Tockus birostris*)	30 days; incubation done by the female.	nidicolous hatchlings; fed by both the parents	The female confines herself in a large tree-trunk hollow and walls up herself with mud, food remnants and faeces which are first converted into a paste that hardens into a long-lasting wall, leaving only a narrow slit through which the male feeds the female. After hatching, the wall is broken down and both parents feed the youngs.

Contd.

Fig. 15.11: The female common Grey Hornbill walls up herself within the hollow of the tree trunk for egg laying. The large entrance is almost sealed (except a small opening) with excreta of the female, mixed with mud and hardened into pellets

Name of Bird	Incubation	Nature and Care of the Nestling	Type of Nest
b. African Ground-Hornbill (*Bucorvus leadbeateri*) also known as Southern Ground-Hornbill	About 30 days; incubation done by the female	nidicolous; nestlings fed by both parents	The Ground-Hornbills build their nest in trees and are the only Bucerotidae that do not wall up the female into her nest.
31. Golden-backed Woodpecker (*Dinopium benghalense*)	incubation done by both parents	nidicolous; nestlings fed and cared by both parents	The unlined nest is made in the hollow of a tree stem or branch usually high up and excavated by both parents. Some woodpeckers nest in giant cacti by first making round entrance-hole and then excavating inside.
32. Indian small Skylark (*Alauda gulgula*)	11–12 days; incubation done by female	same as above	The nest is built in a cup-like depression on the ground or in a hoof-print lined with grass.
33. Indian Cliff Swallow (*Hirundo fluvicola*)	3–4 weeks; mainly incubated by the female	same as above	Nest in colonies with heaps of mud or 'mud-pots' that are often fused together, appearing like irregular honey combs.

Contd.

Name of Bird	Incubation	Nature and Care of the Nestling	Type of Nest
(See Fig.15.6)			Each nest has a short, tubular entrance and it is attached under a canal bridge.
34. Nightingale (*Luscinia megarhynchos*)	13-14 days; incubated by the female	nidicolous; nestlings fed and cared by both parents	The female builds the nest as a cup-shaped structure of dry leaves and grass in thick vegetation. In courtship, the male attracts the female with his songs, which is often sung at night.
35. Red-billed Quelea (*Quelea quelea*)	13 days, incubation done mainly by the female	same as above	This African bird of Steppe and Savanna nests in colonies, sometimes very large and forming even up to 10 million nests in several hundred acres. The nests are usually round, neatly woven and built in a tree.
36. Tailor Bird (*Orthotomus sutorius*) (See Fig. 15.9)	About 2 weeks; incubation done by the female	nidicolous; fed and cared by both parents	This bird builds its nest with soft fibres and vegetable matters placed inside a funnel of large folded leaf or two leaves whose edges are sewn with a thin thread with unparalleled skill and craftsmanship.
37. Penduline Tit (*Remiz pendulinus*)	13-14 days; incubation done by the female	same as above	The nest is built with grass, leaves, lichens and moss with only one opening. The nest is suspended from tree branches.

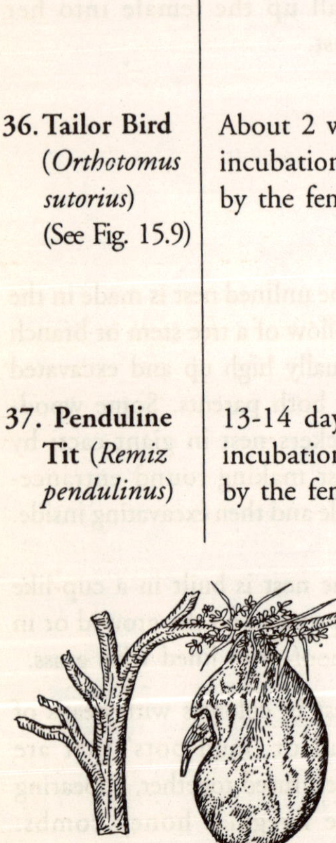

Fig.15.12: Nest of a Penduline Tit having only one opening for entrance/exit

Contd.

Name of Bird	Incubation	Nature and Care of the Nestling	Type of Nest
38. Baya Weaver Bird (*Ploceus philippinus*)	About 2 weeks; incubation done by the female	nidicolous; nestlings cared mainly by the females	The male alone builds the nest or rather several nests (as being polygamous, he has several females at the same time) with paddy leaves and rough-edged grasses into a swinging retort-shaped structure, suspended in clusters from tree-branches hanging over water.

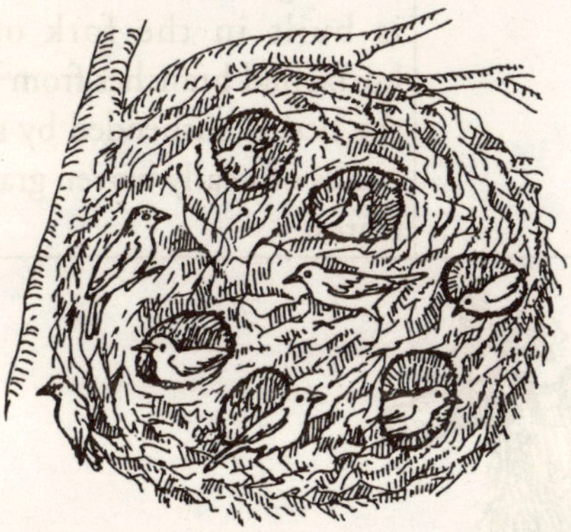

Fig.15.13: Multi-chambered nest of several pairs of Sociable Weavers. Each pair has its own chamber

Name of Bird	Incubation	Nature and Care of the Nestling	Type of Nest
39. Regeant Bowerbird (*Sericulus chrysocephalus*)	incubation period not detected	both incubation and care of nestling possibly done by the female	Hutchinson, an eminent Ornithologist comments (In: Macdonald Encyclopedia, 1995) that in terms of variety and refinement, there is no equal in the animal kingdom. These birds build the so-called 'bower' on a platform with a diameter of about 5 ft (1.5 m) with twigs and plant matter interwoven. In the middle, they build two side-walls with twigs and sticks inserted on the ground and criss-crossed together. The

Contd.

Name of Bird	Incubation	Nature and Care of the Nestling	Type of Nest
			entrance to the bower is in the middle and the side-walls are decorated with pebbles, berries, shells, glass chippings and bits of metal.
40. Golden Oriole (*Oriolus oriolus*)	14-15 days; incubated by both parents	nidicolous; reared and fed by both parents	The female alone builds the nest which is a very distinctive one, made as a cup of dry twigs, grass and vegetable matter. The nest is built in the fork of two horizontal branches from which the nest is suspended by several bands of neatly woven grass and fibres.

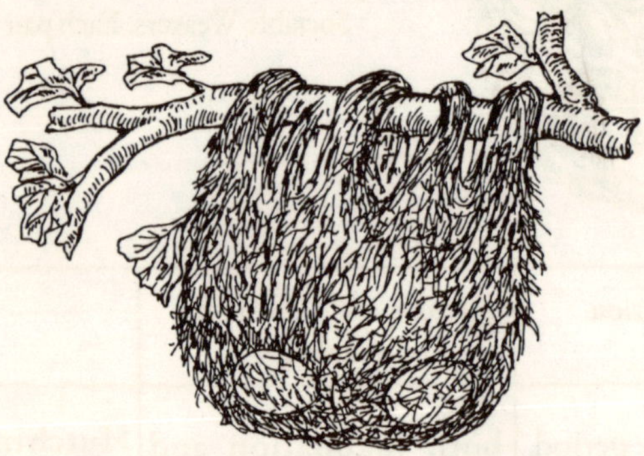

Fig. 15.14: Suspended woven nest of a Golden Oriole (modified from Ward Lock, 1996)

Ratite birds and Tinamou

Name of Bird	Incubation	Nature and Care of the Nestling	Type of Nest
1. Ostrich (*Struthio camelus*)	42 days; incubation done by both parents in turn	nidicolous; hatchlings reared by the male	This largest living bird, which may weigh up to 150 kg, lives in groups, nesting on the ground by digging a hollow in which it lays up to 20 eggs, each weighing about 1.5 kg.

Contd.

Name of Bird	Incubation	Nature and Care of the Nestling	Type of Nest
2. Greater Rhea (*Rhea americana*)	40 days; incubation done by the male	nidicolous; nestlings reared by the male	The male digs a hollow in the ground and covers it with vegetable matter. About 15-20 eggs are laid at intervals of several days.
3. Two-wattled Cassowary (*Casuarius casuarius*)	49-56 days; incubation done by the male	nidicolous; reared by the male	A hollow depression on the ground is filled with vegetable matter to form the nest.
4. Streaked Kiwi (*Apteryx australis*)	75-80 days; incubation done by the male	nidifugous; nestling leaves the nest only after 4 or 5 days	This nocturnal, flightless bird of New Zealand hides in holes or hollows during day time. The nest is built on the ground amongst roots of trees, or in bushes or rocks.
5. Elegant Crested Tinamou (*Eudromia elegans*)	16-22 days; incubation done by the male only	nidifugous; nestling leaves the nest after a day and follows the male	This is a ground-dwelling bird that rarely flies. The males and females are polygamous and polyandrous, respectively. The nest is made on the ground amidst grass and leaves.

For protecting the eggs and youngs from environmental hazards and predators, the birds build their nests with a variety of materials collected from their surroundings. The nests are built either with simple scrapes on the bare ground or in marshland, aquatic vegetation, sand banks, low bushes or in lofty tree branches, expressing superb craftsmanship (e.g. Weaver bird) and decoration (e.g. Bower bird).

As regards site of nests on tree branches, some birds are conservative in maintaining more or less specific heights from the ground every year whereas, for some birds, the height is variable. The House Crow, for example, usually build, its nest amidst thick and leafy tree branches nearly 15-30 feet (4.5-9 meters) above the ground. But the author has observed crow's nests in the thinly sparsed leaves of tall Eucalyptus trees.

BROOD PARASITISM

Some species of birds have a strange reproductive behaviour that they lay their eggs, wholly or partly, in the nest of some other species for the purpose of incubation, feeding, protecting and rearing of the hatchlings by the latter. This phenomenon has been referred to as '**brood parasitism**' or '**nest parasitism**'. Such parasites may be either **non-obligate** or **obligate**, and within these two broad categories several transitional grades may be observed.

NON-OBLIGATE PARASITES

Certain species of birds which are not fully dependent for laying their eggs, incubation and rearing of the hatchlings by other species, but only using nests of the latter are known as non-obligate parasites. Such parasites may be either **egg parasites**—as in the case of certain ducks laying eggs in the nests of some other species, or **nest parasites**—which have been reported in a rather wide variety of birds. In nest parasitism (e.g., a few flycatchers, troupials, finches, cowbirds) some species of birds do not build their own nests but use the nests of other species for egg laying, incubation and rearing of the hatchlings. 'Semi-parasites' is the term used for those egg parasite birds which have their own nests but they lay one or more eggs in the nests of other species. The host bird incubates and rears the young. It has been experimentally observed that most host birds cannot distinguish between the eggs of different species and thus accept the parasitic eggs as their own.

OBLIGATE PARASITES

Van Tyne and Berger (1976) report that obligate parasites are found in five bird families: Anatidae, Cuculidae, Indicatoridae, I'cteridae and Ploceidae.

Examples:

(i) The Black-headed Duck of South America is a total parasite. It lays its eggs in the nest of the Red-fronted Coot, white-faced Ibis and other species living in marshes. The nidifugous ducklings leave the host-nest independently in less than two days. As the Black-headed Duck is least damaging to its host, it has been considered as the most successful and perfect brood parasite.

(ii) Among the cuckoos, quite a number of species are non-parasitic. But several genera of cuckoos are, however, obligatory parasites (e.g., *Clamator, Cuculus, Eudynamis*). There is a wide range of variation in colour and pattern of the eggs of parasitic cuckoos which, quite often, lay eggs in the host-nest containing similar size of eggs. Strangely enough, most cuckoos (not all) throw out the eggs and nestlings from the host-nest except one egg of the parasite. A single cuckoo is thus reared in the nest.

(iii) Some species of Cowbirds (Fam: Icteridae) are parasitic, others are not. It is interesting to note that a species of non-parasitic cowbird (*Molothrus* sp.) builds its own nest yet usually use the nest of some other species for laying and incubation of the eggs. The Screaming Cowbird is exclusively parasitic on the Bay-winged Cowbird. The Giant Cowbird is parasitic on other species of icterids.

The parasitic Cowbirds lack in the development of incubation patches whereas the non-parasitic species develop such patches—being stimulated by prolactin secreted by the pituitary gland.

16

A Brief Classification of Birds

Based on shape, size and structural organisation of the body, the birds possess greater uniformity compared to other vertebrates. One of the reasons for that may be attributed to flight adaptation which has restricted their bodily organisation within a generalised limit.

Avain taxonomy, like that of other vertebrate classes, has a checkered history. With the discovery of progressively increasing number of new species and subspecies at an interval of several years, the Systematists throughout the last few decades have not been very precise in evolving a uniform system of classification of birds; rather—on the basis of new data-bases on morphology, physiology, ecology, palaeontology, biochemistry and immunology–the systematic position of a number of birds, not only at the species or generic levels but also at the higher taxa, remains controversial. The evolution and lineages of avifauna are being searched in recent years not only on the functional analysis of scattered sequences of fossils, but also on the basis of cladistic analysis of osteological form characters (including studies based on functional morphology and trophic diversification) and, more recently, through the application of DNA and DNA hybridisation principles of Molecular Systematics.

Avain Systematics, like that of any other vertebrate class, is intimately related to and developed through the past decades from the concepts on the origin and evolution of birds. With the evolution and divergence of maniraptoran Coelurosauria, such as *Comsognathus*, possibly appeared *Archaeopteryx*, from the late Jurassic (150-170 MY; Ostrom, 1978) of Bavaria, Germany. And, almost at the same geologic time scale (170 MY), appeared an important fossil form, the *Confuciusornis sanctus* from the late Jurassic of China. Zweers et al (1997), while discussing tentative phylogeny of birds, assume that *Confuciusornis* may represent an undiscovered pre-Archeopteryx radiation.

16.1. CLASSIFICATION OF BIRDS DURING LAST 100 YEARS

During the last 100 years, the system of classification of birds has undergone many notable changes, precisely based on many new findings which, from time to time,led to a revision of the systematic position of several groups of birds. Over 100 years ago, in 1898, F. E. Beddard presented a simple classification of birds. The great German biologist E. Stresemann, in his several volumes on Sauropsida, first of all outlined (1927-34) a very convincing classification of birds.

With some changes in Stresemann's scheme, J. L. Peters (1931-51) published a detailed classification of birds in his famous **'Checklist of Birds of the World'**. In Peter's checklist, the order Sphenisciformes (Penguins) was included among ratite birds, comprising 6 orders and the carinate birds comprised 22 orders. The family Phoenicopteridae was separated from the order Ciconiiformes and raised into a separate order 'Phoenicopteriformes'. Peters' successors revised (1964–1970) Peters' checklist of birds that remained as a widely accepted guide during the sixties or even later.

In the meantime, Mayr and Amadon (1951) published a classification of recent birds in 'American Museum Novitates' in which all living birds were classified into 5 ratite and 23 carinate orders. The ordinal names did not terminate with 'formes' (e.g., Galli for Galliformes, Cuculi for Cuculiformes, Laro-Limicolae for charadriifomes). And the family Phoenicopteridae was removed from the order Gressores (Ciconiiformes) and placed in a separate order Phoenicopteri.

In 1934 (revised in 1960), Alexander Wetmore published **'A Systematic Classification for the Birds of the World'** in which he mentioned 7 ratite orders (including 2 extinct orders) and 27 carinate orders. T. J. Parker and W. A. Haswell (1962) in their popular Text Book of Zoology, Vol. II, 6th Ed. presents the classification of birds, based on that of Lowe, Wetmore and others. Parker and Haswell did mention under the subclass Neornithes 3 super orders, namely, Odontognathae with 2 orders, Palaeognathae with 7 orders (including 2 extinct orders) and Neognathae with 22 orders. Among neognath orders, the order colymbiformes is now called Gaviiformes; the order Diatrymiformes has been excluded from recent neognath orders; the order Ralliformes has been included in the order Gruiiformes; the order Telmatomorphormes has been partly included in the order Gruiiformes and partly in the order Charadriiformes, and, lastly, the Swifts and Humming Birds included in the order Passeriformes in Parker and Haswell have long been shifted to the order Apodiformes.

In 1964, J. Fisher and R. T. Peterson (In: The World of Birds, MacDonald, London) suggested 3 subclasses for the class Aves : Saururae (extinct birds), Odontolcae (extinct toothed birds) and Ornithurae (including certain extinct birds and all living birds). The classification outlined in A. L. Thomson's (ed.) **'A New Dictionary of Birds'** (London; Nelson; 1964) was principally based on Peters' 'Checklist'.

From 1980 onward, extensive researches in the field of Avain Systematics based on Biochemistry, Immunology and DNA sequencing (Molecular Systematies) revolutionised the concept of phylogenetic systematics. K. H. Voous (1980) published a different scheme of classification of birds, based on biochemistry, immunology, and, above all, Molecular Systematics founded and established by Sibley and Ahlquist (1972-1992). The classification by Voous recognises only 2 ratite orders (in place of 5 living ratite orders conventionally mentioned) and 23 carinate orders. The 2 ratite orders are Struthioniformes which includes the Rheas, the Casowaries and the Kiwis as Families and Tinamiformes which includes Tinamous.

Among the carinate birds, Voous proposed the orders like Phoenicopteriformes (Flamingos, also recognised as a separate order by Mayr and Amadon, 1951), Cathartiformes (New World Veltures), Accipitrifomes

(Hawks, Eagles, etc.), Pteroclidifomes (Sandgronses), etc.

J. Z. Young (1981; Reprinted 1988) (In: The Life of Vertebrates, 3rd Ed.) mentioned 2 superorders (Odontognathae and Neognathae) under the subclass Neornithes. His first 6 neoganth orders comprise the living as well as extinct ratite orders as described by other authors, but he started with the order Tinamiformes as the first neognath order and, in the order Dinornithiformes, he included both the extinct Moas and the living Kiwis without mentioning any separate order Apterygiformes for the Kiwis. He mentioned 22 carinate orders, but the order Gaviformes was shifted to 11th position.

At about this time Morony, Bock and Farrand (1975) and later Bock and Farrand (1980) contributed valuable information in the 'Reference List of Birds of the World' (published by the American Museum of Natural History).

The classification of birds followed in the present text is mainly based on the phylogenetic classification published by A. S. King and J. McLelland (eds.) (1984) in **'Form and Function in Birds'**. To this classification have been added 2 extinct orders of ratite birds and 2 new orders of carinate birds such as Phoenicopteriformes (formed by isolating the Flamingos from the order Ciconiiformes) and Pteroclidiformes (formed by isolating the Sandgrouses from the Order Columbiformes). Such splitting has been considered quite relevant due to many new facts gathered in recent years. Salim Ali and S. Dillon Ripley (1987; Reprinted 1989), in their monumental works in 10 volumes (In: Compact Handbook of the Birds of India and Pakistan. 2nd Ed., Oxford University. Press: Oxford, New York) described 20 orders of carinate birds represented in India and Pakistan. The two other orders, Sphenisciformes and Colliformes, are not represented in these regions. Hence, this scheme conforms to an old, conventional one.

E. Lack et al (1992) in their **'New Dictionary of Birds'** followed the molecular basis of systematics as proposed by Sibley and Ahlquist (1972-1992) and supported by K. H. Voous (1980).

With the compartively recent publication of **'A Checklist of Birds of the World'** by B. L. Monroe and C. G. Sibley (1993), based on Molecular Systematics, the phylogenetic classification of birds, based principally on palaeontological records, seems to be confronted with a number of contradictions. The classification of birds given in this checklist is based on the earlier publication by Sibley and Ahlquist (1990) on **'Phylogeny and Classification of Birds : A Study in Molecular Evolution'** (Yale Univeresity. Press, Yale). Mindell (1997) (In: Avian Molecular Evolution and Systematics, Academic Press, London) and Cooper and Penny (1997) and others working on Immunology, Biochemistry and using differences in DNA sequencing from mitochondrial and nuclear genomes and using DNA hybridisation techniques support Monroe and Sibley's classification. They have introduced Parvorders, Tribes, etc. and have reduced the total number of orders of neognath birds to 23: (Ratitae 2 orders; Carinatae 21 orders).

Among the carinate birds, the families Cracidae and Megapodidae, earlier included under the Order Galliformes, have been placed in a separate order Craciformes. Many other families mentioned in earlier classification have been placed in Monroe and Sibley's classification under separate orders like Turniciformes (Buttonquails), Galbuliformes (Jacanas, Puffbirds), Buceratoformes

(Hornbills), Upupiformes (Hoopoes), Trochiliformes (Humming birds), and Musophagiformes (Touracos).

On the other hand, many orders like Charadriiformes, Falconiformes, Podicipediformes, Phoenicopteriforms, Pelicaniformes, Procellariformes and Sphenisciformes have all been placed under the order Ciconiiformes (divided into 2 suborders—Charadrii and Ciconii).

Cooper and Penny (1997) state that 'molecular data provide hard evidence for a Cretaceous diversification of modern mammals and birds, followed by a mass survival across the KT boundary'. According to Cooper and Penny, the ratitomorph and gallimorph lineages diverged early, followed by psittacimorphs. Zweers and Vanden Berge (1997-98), however, do not consider any mass survival of birds across the KT boundary; rather, major, well-flourishing avimorph lineages disappeared and only some specific avian ecomorphs that were 'accidentally prepared' by possessing remarkable modifications and adaptive characters like endothermy, feathers, bipedality, flight, and a rhamphothecad prokinetic trophic system could survive across the KT boundary.

Kardong's (1998) classification of birds also claims certain distinctions. Under the class Aves, he has mentioned two subclasses–Sauriurae and Ornithurae, the former subdivided into infraclass Archiornithes and Infraclass Enantiornithes; and the latter into infraclass Odontornithes and infraclass Neornithes.The infraclass Neornithes has been subdivided into two superorders—Palaeognathae and Neognathae, the former including only 4 living ratite orders (the order Apterygiformes being dropped) and the latter containing 23 living carinate orders whose taxonomic positions have been considerably reshuffled. Kardong (1998) has assigned positions of the cuculid, falconid, galliform, columbiform and psittaciform birds earlier than most of the aquatic birds such as the grebes, penguins, albatrosses, pelicans and ducks, and has shifted the positions of the flamingos to Order 11, the gulls and plovers to Order 14, and the loons to Order 15.

16.2. CLASSIFICATION OF BIRDS

As already mentioned, the classification followed in the present text is mainly based on that adopted by King and McLelland (eds.), 1954 and 1980-1989, (In: Form and Function in Birds). As, at present, no uniform system has yet been evolved from a compromise between the three approaches like Cladistic analysis of form characters (Cracraft, 1988), Molecular analysis of DNA characters (Sibley and Ahlquist, 1990; Mindell, 1997; Cooper and Penny, 1997) and evolutionary and functional analysis of fossils (Olson, 1985; Ostrom, 1991; Chiappe, 1995), the system of classification proposed by King and McLelland (1984) has been found to be more acceptable.

The classification mentioned below is up to ordinal level only with distinctive characters and examples:

CLASS AVES

General Characters:

1. The birds are warm-blooded feathered, beaked, biped vertebrates; their forelimbs have been modified as wings which function as the principal organs of flight; and their feet are adapted for walking, running, swimming, climbing, perching on tree-branches and for grasping food-preys.

2. The spindle-shaped body is suitably adapted for flight adaptation.
3. The epidermal feathers are a unique feature in birds; these are light, yet strong and they regulate body temperature and act as a protective covering; the large flight feathers of the wings and tail are specially adapted for flight.
4. The body is divided into four parts—head, neck, trunk and tail; the head more or less round, has its snout prolonged into an upper beak which—together with the mandible or lower beak—form the toothless bill of the bird; neck long; distensible; trunk swollen like middle of the spindle and the tail with long feathers.
5. The skin is dry, loosely held and usually glandless, except the **uropygial gland** (preen gland, oil gland), located dorsally in the uropygium.
6. An elaborate exoskeletal system consists of feathers, beaks, claws and scales. The replacement of reptilian scales by feathers in birds is a remarkable evolutionary acquisition.
7. The endoskeleton is fully ossified; the vertebral and smaller bones are frequently fused and the long bones are hollow and light due to extensions of air sacs within them; lightness, rigidity and flexibility are the key specialisations of avian endoskeleton.
8. The skull is thin, light, smooth and rounded, and the sutures at the joints of skull bones are either absent or very indistinct; presence of a single occipital condyle.
9. The quadrate, pterygoid, palatine and the maxillo-palatine bones form a movable **kinetic complex**, resulting in up and down movements of the upper jaw at the nasal-frontal hinge or at the cranio-facial flexon zone; the lower jaw is formed of 5-6 bones.
10. The cervical vertebrae are characteristically **heterocoelous** (saddle-shaped) and the free caudal vertebrae amphicoelous. The fused Notarium formed of 4-5 thoracic vertebrae and the fused Synsacrum formed of the last thoracic, all lumbar, all sacral and the proximal 5-6 caudal vertebrae and the fused **pygostyle** formed of the last 4-5 caudal vertebrae are also characteristic in carinate birds.
11. In carinate birds, a vertical, flat keel or carina arising midventrally from the broad sternal plate provides spacious origin-sites for the great muscles of the wing. In ratite birds, the keel is either absent or poorly developed.
12. The sabre-shaped scapula and the pillar-shaped coracoid meet in less than 90° in carinate birds, more than 90° in ratite birds. The joining of these two bones with the V-shaped furcula (Wish bone, Merrythought bone) leaves an aperture in-between for the passage of the tendon of the M. supracoracoideus.
13. There are at least a dozen well-developed flight muscles which control movements of the wings, tail and feet in flight.
14. The tongue (the **entoglossum** or the **paraglossalia**) morphology offers great diversity and the tongue shape and size may be correlated with the specific feeding adaptations.
15. The stomach is divided into a glandular **proventriculus** and a highly muscular and somewhat broadly triangular **ventriculus** or **gizzard**.

16. The large heart is completely 4-chambered with 2 auricles and 2 ventricles. The sinus venosus and the conus arteriosus are absent; only the right aortic arch is present.
17. Respiration takes place by two compact lungs rigidly held between the ribs and the dorsal bodywall and connected to thin-walled air sacs extending between the internal organs and into the bones; the voice box is the **syrinx** at the base of the trachea.
18. The urinary bladder is absent and the large cloaca is divided into 3 chambers; excretions semisolid; females usually with only left ovary and oviduct.
19. Fertilization internal; oviparous, eggs with large yolk and covered with hard limy shell laid for incubation; cleavage meroblastic; embroyonic membranes—amnion, chorion, yolk sac and allantois present; hatchlings—either nidicolous (altricial) or nidifugous (precocial), are cared by one or both the parents.

Subclass 1. Archaeornithes

This subclass is also mentioned as **Palaeornithes** by some Ornithologists, and named **"Lizard birds"**.

The extinct birds under this subclass are all grouped into a single order Archaeopterygiformes, named (from Greek *archaios*—ancient and Greek *pteryx*—wings) so after the discovery (in 1861, at Solnhofen, Bavaria, Germamy) of the ancestral bird, *Archaopteryx lithographica*.

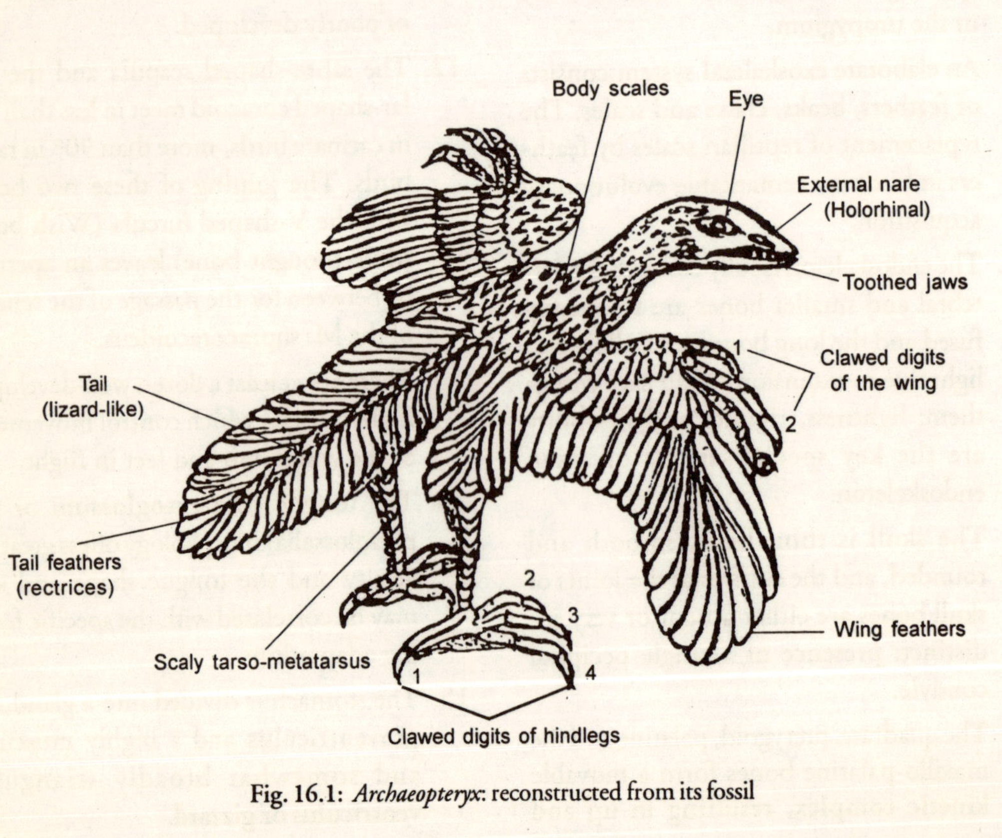

Fig. 16.1: *Archaeopteryx*: reconstructed from its fossil

Order Archaeopterygiformes

1. *Archaeopteryx* and the allied Jurassic birds, such as *Confuciusornis*, as well as similar aviforms of later periods, such as *Sinornis*, *Gobipteryx* and *Patagopteryx*, are all extinct.
2. *Archaeopteryx* was crow-like or slightly larger than crow in size; body covered with feathers and forelimbs modified into wings; carpals and metacarpals separate, each with 3 clawed digits.
3. The upper jaw bears 13 pairs and the lower jaw only 3 pairs of teeth.
4. Long lizard-like body with about 50-56 vertebrae.
5. Long tail formed of 18-23 amphicoelous vertebrae; pygostyle absent.
6. Thoracic ribs without uncinate process; recent studies reveal that Archaeopteryx was possibly with a keeled sternum (Wellnhofer, 1993).

   e.g., *Archaeopteryx lithographica Confuciusornis sanctus, Sinornis santensis.*

Subclass 2. Neornithes

Except a few that are extinct, all birds under this subclass are living.

1. Carpals and metacarpals fused; second finger longest.
2. Caudal vertebrae usually 13 or 14 are compressed; sternum either keeled or flat.

Superorder 1. Odontognathae

These are New World Toothed Birds of the Cretaceous; all extinct; teeth present in furrows on jaws. These birds lived in marine environment.

Order Hesperornithiformes

1. Flightless marine bird with pleurodont dentition.
2. Cervical vertebrae heterococlous, rest amphicoelous.
3. Sternum with a poorly developed keel.

   e.g., *Hesperornis; Baptornis.*

Fig. 16.2: *Hesperornis*, a marine diving fish-eating bird of North America. It lived about 100 m. years ago and was a flightless toothed bird (based on and modified from Van Tyne and Berger, 1976)

Order Ichthyornithiformes

1. Size smaller than Hesperornithiformes; marine bird capable of flight.
2. Cervical and thoracic vertebrae amphicoelous.
3. Sternum with a well-developed keel e.g., *Ichthyornis.*

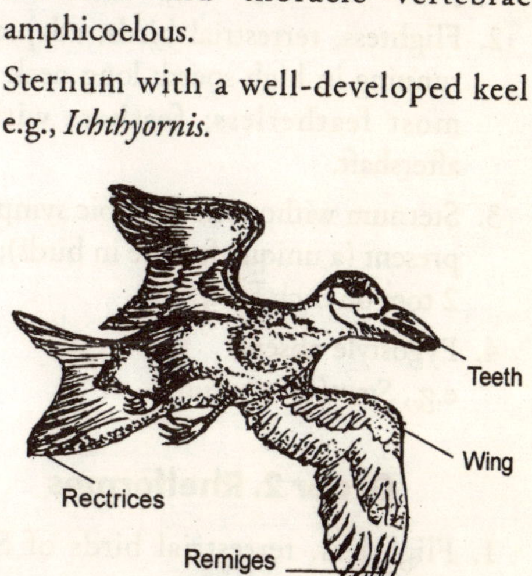

Fig. 16.3: *Ichtyhyornis* (Toothed bird): flying

Superorder 2. Neognathae (For sketches, see chapters 25 and 26)

These are modern birds, devoid of teeth in jaws and distributed throughout Eocene to Recent. These birds have well-developed wings, keeled sternum and pygostyle (except in the birds mentioned under the first 6 orders).

The first 7 orders (mentioned here under Neognathae) are still designated by several authors (Parker Haswell, 1951; Weichert and Presch, 1977 and Bologna 1995) as Palaeognathous birds (as such included under the **superorder Palaeognathae**) that were flightless terrestrial birds without a keeled sternum (except the order Tinamiformes) and the pygostyle; feathers without hooked barbules and the large and broad vomer separates the palatines.

Order 1. Struthioniformes

1. The Ostriches are the largest living birds of Africa and West Asia, above 7 feet tall (up to 3 metres—10ft: MacDonald, 1993) and weighing more than 300 pounds (150 kg or 330 lb).
2. Flightess, terrestrial birds, adapted to running in high speed; long neck is almost featherless; feathers without aftershaft.
3. Sternum without keel; pubic symphysis present (a unique feature in birds); only 2 toes on each foot.
4. Pygostyle absent.
   e.g., *Struthio camelus.*

Order 2. Rheiformes

1. Flightless, terrestrial birds of South America, about $5\frac{1}{2}$ feet (1.7 m) tall and weighing up to 25 kg.
2. Head and neck partly feathered; 3 front toes on each foot; feathers without aftershaft.
3. About 22 transverse horny plates on the front of the legs.
   e.g., *Rhea americana* (Greater Rhea).

Order 3. Casuariiformes

1. The Cassowaries and Emus of Australia are flightless, terrestrial birds; the Cassowary measures about 3.3 feet (1 metre) in height and 85 kg in weight, while the Emu is about 6 feet (1.8 metre) tall and weighs up to 55 kg.
2. Wings small; neck and body densely feathered, feathers with aftershaft nearly equal to shaft; 3 front toes on each foot (Cassowary with very strong legs and the inner toe bears a straight claw which may be up to 4 inches (10 cm) long).
3. Two-wattled Cassowaries bear two smooth caruncles on the front of the neck, one-wattled Cassowary bears one such caruncle.
4. Feathers with aftershaft nearly equal to shaft.
5. Sternum without keel; Emu has no uropygial gland,
   e.g., *Casuarius casuarius* (Two-wattled Cassowary),
   Dromaius novaehollandiae (Emu).

Order 4. Dinornithiformes

1. The Moas of New Zealand were flightless, terrestrial birds of large size and have been extinct about 700 years ago; large-sized bird, about 8 feet (2.4 metres) tall.
2. Wing very small, wing bones reduced or absent; sternum reduced and without keel.

3. Hindlimbs strong, massive and each with 3 or 4 toes; beaks very small; feathers with large aftershaft.

   e.g., *Dinornis* (Moas)

Order 5. Aepyornithiformes

1. These were gaint Elephant birds of Africa and Malagasay, now extinct for several centuries; the bird was about 10 ft (3 metres) tall.
2. Flightless, terrestrial bird with rudimentary wings, short but broad sternum that was without keel.
3. Legs strong, massive and each with 4 toes.
4. Egg size $9\frac{1}{2}$ by 13 inches (24 × 31 cm), largest known animal egg.

   e.g., *Aepyornis* (Elephant Bird)

Order 6. Apterygiformes

1. The Kiwis of New Zealand are comparatively smaller in size (about 2 feet (0.6 m) long); flightless, terrestrial and nocturnal birds that nest in burrows.
2. Bill long and slender (4-8 inches/10-20 cm) with nostrils at end; wing degenerate and sternum without keel.
3. Each foot with 4 toes; body plumage fluffy, hairlike and without aftershaft.

   e.g., *Apteryx australis* (Streaked Kiwi)

Order 7. Tinamiformes

1. The Tinamous of Central and South America are capable of flight and very fast-running terrestrial birds; size similar to that of the Guineafowl (Numididae), up to 21 inches (53 cm) in length and weighing 1-5 lbs (0.5-2.3 kg).
2. Hind toe and pygostyle absent; sternum with a keel. The Elegant Crested Tinamou (*Eudromia elegans*) has a mimetic plumage, with long feathers on the head

   e.g., *Tinamus, Rhynchotus, Eudromia.*

Order 8. Sphenisciformes

1. The Penguins inhabit coasts of the Antarctic as well as subantarctic and cold temperate zones of South Africa and Australia.
2. Size up to 3.8 ft. (1.15 metres) long and weighing about 66 lbs (30 kg) in Emperor Penguins; forelimbs or the so-called wings modified into "flippers", adapted for swimming.
3. Feathers usually small, scalelike with no apteria; feathers longer (1.6 inches/4.2 cm) in Emperor Penguins; bones much compressed, metatarsus incompletely fused; 4 toes, all directed forward; feet completely webbed; thick layer of fat beneath the skin.
4. Can stand erect on metatarsi.

   e.g., *Aptenodytes forsteri* (Emperor Penguin); *Eudyptes crestatus* (Rockhopper Penguin); *Spheniscus humboldti* (Peruvian or Humboldt Penguin).

Order 9. Gaviiformes

1. The Loons inhabit oceans in northern part of Northern Hemisphere (North America, Northern Asia & Europe); adapted for living and feeding on fishes; the Arctic Loon measures about 23-26 inches (58-67 cm) in length, has a wing span of 4 ft (1.2 m) and weighs 4.4-6 lbs (2-2.7 kg).

2. Legs short and at the end of the body; toes fully webbed; tail of 18-20 short stiff feathers; flight swift and direct.
3. Inhabit lakes in the breeding season.
   e.g., *Gavia immer* (Common Loon)
   Gavia arctica (Arctic or Black-throated Loon).

Order 10. Podicipediformes

1. The Grebes inhabit ponds, marshes, bogs and lakes in Asia, Africa, Australia, North America and parts of Europe; the Great Crested grebe (largest of the Old World Grebes) measures about 18 inches (47 cm) long and weighs 590 gms to 1.4 kg.
2. Tail formed by a rudimentary tuft of down feathers; legs far back on body; tarsus compressed, feet lobed; 3 toes of each foot webbed.
   e.g., *Podiceps cristatus* (Great Crested Grebe)
   Podiceps ruficollis (Little Grebe).

Order 11. Procellariiformes

1. The Albatrosses, Fulmars, and Petrels inhabit open sea and islands in the Central and Western Pacific, and in the non-breeding season, in the Southern Hemisphere; the Royal Albatross and the Wandering Albatross have the largest wing-span of all the seabirds [$6\frac{1}{2}$feet /2 metres to more than 10 feet/3 metres and each weighs about 15-17.6 lbs (7-8 kg)].
2. Nostrils tubular; horny sheath of bill formed of several plates; hind toe rudimentary or absent (in diving petrels); plumage compact, oily in texture; wings longs and narrow.
   e.g., *Diomedea exulans* (Wandering Albatross); *Puffinus puffinus* (Manx Shearwater).

Order 12. Pelecaniformes

1. The Pelicans, Cormorants, etc. are included in this group. The Pelicans inhabit seacoasts and estuaries of North, Central and South America, as well as lakes, marshes, lagoons and estuaries of Southern Europe, Asia and Africa, while the Cormorants are practically worldwide on both coastal and inland waters.
2. While the Pelicans measure about 4-$4\frac{1}{2}$ fect (1.2-1.4 m) long, the larger Great Cormorant measures only up to 3 feet (0.9 m).
3. All 4 toes are webbed; nostrils rudimentary or absent; a prominent gular pouch on throat (except in tropic birds); beaks long and stout.
   e.g., *Pelecanus onocrotalus* (Old World White Pelican)
   Pelecanus philippensis (Grey Pelican)
   Phalacrocorax niger (Little Cormorant).

Order 13. Ciconiiformes

The Flamingos (*Phoenicopterus* sp.), previously included under this Order, have been placed in a separate order 'Phoenicopteriformes'.

1. The Herons, storks and Egrets are included in this order; long-necked and long legged watching birds with decorative plumes (Herons) or bare areas on head (storks); little or no web between toes.
2. Inhabit near about water in tropical and subtropical regions; the Adjutant Stork and the Black-necked Stork are about 4 feet (1.2 m) tall, while the white stork is

about $3\frac{1}{2}$ feet (1.1m) and the White necked stork about 3 feet (0.9 m).

e.g., *Ardea cinerea* (Grey Heron)
Bubulcus ibis (Cattle Egret)
Leptoptilos dubius (Adjutant Stork)
Ciconia ciconia (White Stork).

Order 14. Phoenicopteriformes

1. The flamingos are long-legged, long-necked stork-like bird, above 4 feet (1.2 m) high and the pink colour of the legs, wings and of the bill that is bent down at an angle from the middle are diagnostic characters.
2. Inhabit large ponds, lagoons, estuaries and marshlands in general throughout India, Pakistan, Bangladesh, Sri Lanka and also in Africa, Central America and Southern Europe; the food (small animals etc.) contained in the bottom mud is scooped and strained by the curiously bent beaks.

e.g., *Phoencopterus roseus* (Flamingo)

Phoenicopterus ruber (Greater Flamingo).

Order 15. Anseriformes

1. The ducks, geese and swans, though living in aquatic habitat, are very good fliers; the swans are the largest (the Mute Swan about 5 feet (1.5 m) long, followed by the geese (2-3 feet/0.6-0.9 m long).
2. Broad bill covered with soft, cornified epidermis containing numerous sense-pits, with harder 'nail' at tip; margins of bill with many transverse horny ridges (lamellae).
3. Tongue fleshy; legs short; feet webbed; tail short with many feathers.

e.g., *Anas platyrhynchos* (Mallard Duck)
Anas crecca (Common Teal)
Tadorna ferruginea (Ruddy Shelduck, Brahminy Duck)
Anser indicus (Barheaded Goose)
Cygnus olor (Mute Swan).

Order 16. Falconiformes

1. The Vultures, Kites, Hawks, Falcons and Eagles included in this order have a worldwide distribution and are predaceous birds, feeding on various vertebrates, some invertebrates and also on the carrion of varoius animals.
2. Stout bill with hooked upper beak tip that covers tip of the lower beak; mandible sharp-edged; feet adapted for grasping, with sharp, curved claws modified as talons.
3. Most of the birds of this order are called common scavengers of human habitation.

e.g., *Sarcogyps calvas* (Black or King Vulture)

Gyps bengalensis (White backed or Bengal Vulture)

Haliastur indus (Brahminy Kite)
Milvus migrans (Pariah Kite)
Falco rusticolus (Gyrfalcon)
Falco peregrinus peregrinator (Shahin Falcon)

Aquila chrysaetos (Golden Eagle)
Stephanoaetus coronatus (African Crowned Eagle).

Order 17. Galliformes

1. The Galliform birds include the megapodes, guans, grouses, pheasants, guineafowl, turkeys, etc.,—all feed on seeds and grains of cereals and insects; have a worldwide distribtion.

2. Bill short; feathers with aftershaft; feet usually adapted for scratcing and running.
3. Young downy at hatching and nidifugous (Precocial).

e.g., *Gallus gallus* (Red Junglefowl)
Francolinus francolinus (Black Partridge)
Pavo cristatus (Common Peafowl)
Coturnix coturnix (Grey Quail).

Order 18. Gruiformes

1. The Cranes, Rails, Coots, Bustards, Sunbitterns are all included in this order; the Cranes are larger in size (the Sarus Crane and the Crowned Crane have a standing height of about 5 feet (1.5 metres), while the Rails, Coots and others are smaller.
2. They inhabit open marshes, prairies and cultivated plains of Europe, Asia, Africa, and temperate North America; the Indian species are mostly winter visitors.
3. The cranes have some bare skin about head; have strong flight with neck elongate and outstretched; the Rails and Coots have their neck and legs shorter; the rails have a compressed body and the Coots with lobate toes; common in ponds, diving or swimming; *Gallinula* (Moorhen) have toes elongate for walking on lily pads in ponds.
4. Feathers with aftershaft.

e.g., *Grus antigone* (Sarus Crane)
Rallus aquaticus (Water Rail)
Grus leucogeranus (Siberian White Crane)
Fulica atra (Old World Coot).

Order 19. Charadriiformes

1. The shore birds or Waders include the Plovers, Gulls, Terns, the extinct Auks, Avocet, Sandpiper, Snipe, etc. have a worldwide distribution from Arctic islands in the north and throughout palaearctic Eurasia, tropical, sub-tropical and temperate inland as well as coastal regions of Europe, Asia, North America, Central and South America, Africa, Australia and New Zealand.
2. The shore birds probe for food in sand mud or shallow water; some catch fish actively over water, plunging to captive the prey (Gulls and Terns); some have swift flight in air and swifter diving and swimming underwater for catching the prey (Auks, Murres and Puffins).
3. Toes usually webbed, at least at base; plumage dense and firm; the shore birds are long-legged, the gulls are strong-winged and the Auks with 3 toes and legs far back on body; youngs downy when hatched.

e.g., *Pluvialis dominica* (American Golden Plover)
Recurvirostra avosetta (Avocet)
Tringa hypoleucos (Common Sandpiper)
Gallinago gallingo (Fantail Snipe)
Larus brunnicephalus (Brownheaded Gull).

Order 20. Pteroclidiformes

K. H. Voous (1980), based on biochemical evidences and few other authors (Maclean, 1969), removed the Sandgrousses from the order Columbiformes and placed them in a separate Order Pteroclidiformes. Maclean (1969), In: Jour. of Ornithol, 110: 104-107), while discussing Stegmann's (1968) view about systematic position of the Sandgrouses, states that the Pteroclidae were more closely related to the Charadrii than to the Columbidae.

1. Of the size of a pigeon (about 13.4 in 34 cm long), this bird has colour variation among its species. The common Sandgrouse of India is yellowish sandy brown with black band across the breast and brownish black belly, while the black-bellied Sandgrouse of Western Europe, Middle East and North Africa has its upper parts dark with grey, reddish, yellowish and blackish specks.
2. Tail and wings pointed; flight swift and direct; inhabit either barren plains, fallow land or arid regions, semi-desert and grassy steppe.

   e.g., *Pterocles exustus* (Common Sandgrouse)
   Pterocles indicus (Pointed Sandgrouse)
   Pterocles orientalis (Black-bellied Sandgrouse).

Order 21. Columbiformes

1. The Common Pigeons, Doves, Imperial Pigeons, Green Pigeons and the extinct Dodos are included in this order. They have a worldwide distribution, varying in size from 13 inches/33 cm (Common Blue Rock Pigeon) to the large (31-34 in./79-85 cm long) Blue Crowned Pigeon (*Goura cristata*) of New Guinea and nearby islands.
2. They are very good fliers, covering quite a long distance; the extinct Passenger Pigeon of Eastern North America (*Ectopistes migratorius*) was locally migratory.
3. Bill usually short and slender, stout in some (*Ducula, Treron*) and with a thick, soft 'cere' at base; tarsus usually shorter than toes; crop large, producing 'pigeon's milk' to feed the nidicolous young.

   e.g., *Columba livia* (Blue Rock Pigeon)
   Columba palumbus (Wood Pigeon)
   Ducula aenea nicobarica (Nicobar Green Imperial Pigeon).
   Treron phoenicoptera (Green Pigeon)
   Streptopelia chinensis (Spotted Dove)

Order 22. Psittaciformes

1. The parrots and parakeets have the size of a pigeon, although the African species is larger (up to 16 in./40 cm long) and the South American species is the largest (up to 18 in./45 cm); the smaller Lorikeets are also included in this group; mainly frugivorous, but also feeds on and damages large amount of crops and cereals.
2. The parakeets are parrots with long tail.
3. Beak stout, gradually narrower rostrally, sharp-edged and hooked at tip; upper beak movable on frontal bone of skull; cere present at base of the upper bill; tongue fleshy.
4. Toes are 2 in front and 2 behind; outer hind toe not reversible; feet adapted for grasping; plumage bright green, blue, yellow or red.

   e.g., *Psittacula krameri* (Roseringed Parakeet)

   Psittacula eupatria (Large Indian Parakeet)

   Amazona aestiva (Turquoise-fronted Parrot)

   Psittacus erithacus (African Grey Parrot)

   Loriculus vernalis (Lorikeet).

Order 23. Cuculiformes

1. The old world Cuckoos, the Hawk Cuckoos, the Great Spotted Cuckoos, the African Cuckoos, the Asian Koel, the Malkoha, the Coucal and the Road-runner of which both parasitic and some

non-parasitic species are distributed almost worldwide. The Great Spotted Cuckoo is about 19 in. (48 cm) long.

2. Known as songbirds due to their musical whistling notes.
3. Toes are 2 in front and 2 behind, outer hind toe reversible; feet not adapted for grasping; tail long, bill moderate.

e.g., *Eudynamys scolopacea* (Koel)
Clamator jacobinus (Pied-Crested Cuckoo; Papiya)
Coturnix coturnix (Migratory Quail)
Centropus sinensis (Coucal)
Clamator glandarius (Great Spotted Cuckoo).

Order 24. Strigiformes

1. The various owls and owlets are noctural predators and are worldwide in distribution; their size varies from the Common Scops owl (7.5 in./19 cm) and the little owl (8.5 in./21.6 cm) to the large Great Grey owl (22.5 in./57cm).
2. Large round head with large eyes forwardly directed and facial in position, surrounded by a monkey-like facial disc of radial feathers; ear-opening large.
3. Beak short; legs may be feathered or unfeathered; feet adapted for grasping; claws sharp.

e.g., *Tyto alba* (Barn Owl)
Speotyto cunicularia (Burrrowing Owl)
Bubo bubo (Indian Great Horned Owl)
Bubo zeylonensis (Brown Fish Owl)
Strix nebulosa (Great Grey Owl)
Athene brama (The Spotted Owlet).

Order 25. Caprimulgiformes

1. The nocturnal Nightjars and Whippoorwills are included in this order; size about 10 in. (25 cm); distributed in Europe, Asia and North-West Africa.
2. Beaks short; mouth wide, and in the Whippoorwills, margined with long bristlelike feathers; fly like owls without making sound.
3. Legs and feet small and not adapted for grasping; do not build nest.

e.g., *Caprimulgus asiaticus* (Common Indian Nightjar)
Caprimulgus europaeus (Eurasian Nightjar).

Order 26. Apodiformes

1. The Swifts and Humming birds are included in this order; these birds are perhaps most perfectly adapted to volant life; size very small, some humming birds attaining the smallest size—up to 2.4 in. (6 cm) long. The Giant Humming bird is the largest species of Trochilidae.
2. Legs and feet very short and small; the swift has a long and pointed wing; the humming bird has its feathers with slots so arranged that the feathers can beat against the body at right angle at tremendous speed (about 150-200 times/sec.), the flight being known as, 'hovering flight' i.e. it can remain suspended in air at the same place, by flapping its wings. This helicopter-like characteristic is unique to this bird.
2. Bill small and weak in Swifts which are insectivorous; the humming bird has a long pointed beak with a long, tubular tongue that helps in nectar-feeding.

e.g., *Apus affinis* (House Swift)
Patagona gigas (Giant Humming bird)

Order 27. Coliiformes

1. The Colies or Mousebirds of Africa are called so because of their brownish-grey plumage and their swift movements like mice through bushes; inhabit forest along rivers.
2. 1st and 4th toes reversible; expert in climbing trees; tail very long, the rectrices may be up to 10 in. (25 cm) long.

   e.g., *Colius striatus* (Speckled Mousebird)

Order 28. Trogoniformes

1. The Trogons have short but stout bill with bristles at base; they inhabit Asia, Africa, Central and North America.
2. Feet small and weak; plumage soft and brilliant; tail long and upper beak distinctly hooked downward.

   e.g., *Harpactes erythrocephalus* (Red-headed Trogon)

   Harpactes fasciatus (Southern Trogon).

Order 29. Coraciiformes

1. The Kingfishers, Bee-eaters, Rollers, Hoopoes and Hornbills are included in this order; their distribution is practically worldwide; several representative species of each group occur in India, Pakistan, Bangladesh, Sri Lanka & Burma (Myanmar).
2. Size usually varies from Myna to Pigeon, although the Giant Kingfisher *(Magaceryle maxima)* of Africa, may be up to 16 in. (40 cm) long and a subspecies of the Great Pied Hornbill *(Buceros bicornis)* is larger than a vulture and may be up to 47 in. (1.2 m) long.
3. Bill stout and strong in general, but straight in the Kingfishers, while it is longer, gradually curved downward and with a casque on the upper beak in the Hornbills.
4. 3rd and 4th toes fused at base.

   e.g., *Alcedo atthis* (Small Blue Kingfisher; Eurasian Kingfisher)

   Pelargopsis capensis (Brown headed Storkbilled Kingfisher)

   Halcyon smyrnensis (White-breasted Kingfisher)

   Tockus birostris (Common Grey Hornbill)

   Buceros bicornis (Great Pied Hornbill)

   Merops orientalis (Small Green Bee-eater)

   Upupa epops (Hoopoe)

Order 30. Piciformes

1. The Jacamars, Puffbirds, Barbets, Honeyguides, Toucans and Woodpeckers are included in this order; size variation in body length and in the length of the bill are quite different in different species; among the Woodpeckers, the Black Woodpecker of Europe and Asia *(Dryocopus martius)* is the largest (18 in. /45 cm long); the Blue-throated Honey guide (*Indicator indicator*) is the smallest (4 in.-8 in./10-20 cm long); the Toco Toucan of South America (*Ramphastos toco*) is the largest among these birds (24 in./60 cm long); the bill and the tongue are 9 in./22.5 cm and 6 in./15 cm long, respectively.)
2. Bill stout chisel-like (Woodpeckers) or thick, short, heavy and conical (Barbets); tongue roughened or with barbs near tips and protrusible; rectrices stiff with pointed tips.
3. Toes 2 in front and 2 (rarely 1) behind and are not reversible; cling to tree trunks, make holes in the wood and dig

out insects and their larvae to feed on.

e.g., *Dinopium benghalense* (Golden-backed Woodpecker)
Dendrocopos major (Great Spotted Woodpecker)
Dryocopus martius (Black Woodpecker)
Picoides tridactylus (Northern Three-toed Woodpecker)
Megalaima asiatica (Blue-throated Barbet)
Lybius bidentatus (Double-toothed Barbet)
Ramphastos toco (Toco Toucan).

Order 31. Passeriformes

Total sp. 5,739 Total Bird sp. 9,702

1. This order consists of the largest number of species of perching birds whose number is about three-fifth of the total number of bird species (9702—as per 1993 checklist); divided into 4 suborders (2, according to Monroe and Sibley, 1993) and about 60 families (69 fams; Storer, 1943; 49 fams; Mayr and Amadon, 1951; 46 fams; Monroe and Sibley, 1993). This order is formed of a spectacular assemblage of diverse group of birds of which the majority of families constitute the 'songbirds'.
2. Toes are 3 in front and 1 directed backward, adapted for perching; toes neither reversible nor united; inhabit all types of habitats, some strictly terrestrial, majority arboreal, some (e.g., Eurasian Dipper, *Cinclus*) diving in fresh water streams and birds like Flycatchers, Swallows and Drongos catch insects in flight.
3. Though comparatively smaller in size than that of the birds of many other orders, there is a remarkable size variation ranging from the large-sized (28 in./70 cm) African white-necked Raven to the North American Chickadee (Titmouse) which is less than 4 in. (10 cm) long.

   e.g., *Hirundo smithii* (Wiretailed Swallow)
   Alauda gulgula (Indian Small Skylark)
   Oriolus oriolus (Golden Oriole)
   Dicrurus adsimilis (Black Drongo)
   Acridotheres tristis (Indian Myna)
   Corvus corax (Common Raven)
   Corvus splendens (House crow)
   Pycnonotus cafer (Redvented Bulbul)
   Terpsiphone paradisi (Paradise Flycatcher)
   Copsychus saularis (Magpie Robin)
   Motacilla flava (Yellow Wagtail)
   Nectarinia asiatica (Purple Sunbird)
   Passer domesticus (House Sparrow)
   Ploceus philippinus (Baya Weaver Bird)

17

Evolutionary Position of *Archaeopteryx* and the Origin of Birds

To explore into the possible sources for the origin of birds, a thorough discussion on the evolutionary position of *Archaeopteryx* and of the contemporary pro-avian fossils is essential. About 150 years ago, when Prof. T. H. Huxley (1825-1895) commented that 'the birds are glorified reptiles', he was clear in his conviction that the birds originated from the reptiles. About 100 years later Edwin Colbert (1955, 1969) amplified the comment of Prof. Huxley by specifying the 'glorified reptiles' as 'glorified archosaurians' which not only gave rise to birds through *Archaepteryx*, but also the flying pterosaurs. On the other hand, the question of the origin of birds and that of its flight cannot be looked upon as two separate episodes. Since the1980s, many new fossils, some with more reptilian features and some more proavian, have been discovered. And, above all, evidences gathered in recent years from works in Biochemistry, Physiology, Immunology and Molecular Systematics have shed new light in avian genealogy.

Our knowledge about the fossils of Jurassic and Cretaceous birds is rather fragmentary. The earliest known and the oldest fossil bird is called *Archaeopteryx*, meaning 'ancient wing'. It dates back to the late Jurassic period, about 150 MY (King and McLelland, 1984) or 150-170 MY (Zweers et al, 1997).

17.1. DISCOVERY OF *ARCHAEOPTERYX*

Archaeopteryx was discovered by Andreas Wagner in 1861, in the lithographic slate of Solnhofen in Bavaria (Germany). This lizard-like fossil so clearly resembled the smaller bipedal dinosaurs that it would have been classified as a reptile, were it not for the unmistakable impressions of its feathers. Wagner described the fossil under the name *Gryphosaurus macrura*. Its present name, *Archaeopteryx lithographica*, was given to it by Hermann von Meyer (1863) who had earlier described a feather—probably of the same species—in Spain. The detailed structure was studied by Owen in 1863.

The fossil of a more common specimen was discovered by Daves in 1877 about 10 miles/16km away from the same locality. It was named *Archaeornis siemensi*. A third specimen was discovered in 1956. All the three specimens probably belong to the same genus, or even species, in which case, as per priority, the name *Archaeopteryx lithographica* was retained. In 1974, Ostrom examined a specimen collected and labelled (1885) as a pterosaur at the Teyler Museum, Holland,

and confirmed it to be an *Archaeopteryx*. Wellnhofer in 1974 identified the 5th specimen of *Archaeopteryx* at the Eischstadt Museum, Berlin.

17.2. IMPACT OF THE DISCOVERY OF *ARCHAEOPTERYX*

The discovery of *Archaeopteryx* (1861) only two years after the publication of the 'Origin of Species.......' by Charles Darwin (1809-1882) in 1859, presented a concrete proof in support of his theory of 'connecting link'. It became established as an evidence that the birds originated from the reptiles.

Although *Archaeopteryx* was an evidence without any doubt that it descended from the reptiles, scientists were divided in opinions as to the immediate ancestor or ancestors of *Archaeopteryx* . During about last 146 years since the discovery of *Archaeopteryx* many contemporary fossils have been discovered, and some of these fossils belonging to the group called Deinonychosauria (Ostrom, 1969) (Coelurosaur maniraptorans) may hold the key to pre-*Archaeopteryx* radiation. Ostrom's view (1978) has now a wide acceptance that the pre-*Archaeopteryx* progenitor was a coelurosaur maniraptoran called *Compsongnathus longipes* which was a 'slender, bipedal, fast-running, predatory reptile, perhaps about 1 metre long, or about the same size as a large domestic cat' (King and McLelland, 1984).

17.3. PROBABLE HABITAT

All the specimens of *Archaeopteryx* were found embedded in marine beach deposits, but according to Yalden (1971), there are strong reasons to believe that *Archaeopteryx* was a forest-dweller. Perhaps they were blown seawards and were preserved in fine-grained mud deposits. Yalden (1971) states that though the sternum was without keel (recent findings suggest that *Archaeopteryx* had possibly a keeled sternum), the short rounded wings of the bird enabled it to flutter, rather than fly, and make short gliding flights. Its clawed digits suggest its ability to climb trees or returning to tree branches. Thus the primitive bird could also escape predators.

17.4. SOME AVIAN DISTINCTIONS FROM DINOSAURS

The general characters of the class Aves have already been mentioned. The birds have many similarities with the reptiles, particularly with the dinosaurs. Such structural similarities between the birds and reptiles led Prof. T. H. Huxley to designate these two groups under the common name 'Sauropsida', assigning it to a superclass.

However, birds possess certain remarkable distinctions from the dinosaurs; these are modification of the forelimbs into wings for flight, development of a unique feature, the epidermal feathers in birds, replacing epidermal scales of reptiles and the endoskeleton rendered light yet rigid due to the occurrence of numerous air cells and extensions of air sacs within the bones.

17.5. STRUCTURAL PECULIARITIES OF *ARCHAEOPTERYX*

According to King and McLelland (1984), "the fossil remains of *Archaeopteryx* are a mosaic of reptilian and avain features. Among the reptilian characteristics were teeth, a long tail, a short sacrum with no more than six vertebrae, a hand (manus) with free metacarpal bones and claws on all three digits, and a foot with free metatarsal bones. Its avain features included feathers (primaries and secondaries on the wings, and contour feathers all over the body), paired clavicles joined to form a typical 'wish bone' (furcula), and a foot with the first digit opposed to the other three digits as in many modern perching birds".

The structural specialisations of *Archaeopteryx* may thus be enumerated:

1. *Archaeopteryx* was presumably a land bird of about the size of a Jungle Crow or Black-billed Magpie (about 18 in./45 cm long).
2. The most obvious character is a long, slender, tapering lizard-like tail supported by 18-23 free caudal vertebrae, and without a terminal pygostyle.
3. Body long, lizard-like, with about 50-56 vertebrae, and with two flexures—cervical and thoracic.
4. The most remarkable pro-avian feature is the presence of feathers provided with barbs, barbules and barbicels (hamuli), as in modern birds.
5. At least 12 hindermost vertebrae, each bear a pair of large, well-developed lateral rectrices or tail feathers arranged obliquely along the axis of the tail.
6. Skull typically avian with large, rounded braincase; single occipital condyle extended posteriorly.
7. Beaks short and blunt; jaws thick and strong and jaw bones heavier than those of modern birds.
8. Each jaw bone is provided with a series of short, conical, enamelled, crowned and bony thecodont teeth.
9. The upper jaw has 13 and the lower jaw 3 teeth on either side, implanted in sockets.
10. Nostrils definitely holorhinal and divided into two holes by an alinasal growth, as in many living birds.
11. Orbits large and apparently avian, having a sclerotic ring of bones and completed below by a bony arch.
12. Presence of a large antorbital foramen in front of the orbit.
13. Skull possibly **mesokinetic** (Bock, 1964; Wellnhofer, 1992); Bühler's (1985) description of *Archaeopteryx* suggests a streptostylic quadrate and a prokinetic skull.

    NOTE: Heilmann's (1926) description of *Archaeopteryx* skull is much more reptilian than the Wellnhofer and Bühler descriptions (Zweers et al, 1997).
14. Vertebral column formed of about 50-56 vertebrae; number of cervical vertebrae more variable than in other regions, usually 10-14; thoracic 10-11; lumbar 5; sacral 6 (fusued) and caudal 18-23 (usually 20).
15. Vertebrae amphicoelous, occasionally with a few amphiplatyan type; there was no indication of heterocoelous vertebrae.
16. Presence of a large mandibular foramen between the supraangular and dentary of the lower jaw.
17. Recent studies reveal that a keeled sternum was present, however, the keel was possibly a weak one.
18. The scapulae are slender, curved bones, eminently avian; the coracoids are small and partially visible; supracoracoid similar to that of the mesozoic reptiles.
19. In all probability, coracoid and scapula meet at right angle or even at acute angle.
20. Presence of a strong, wide U-shaped furcula, like that of a bird, formed by the fusion of the clavicles.

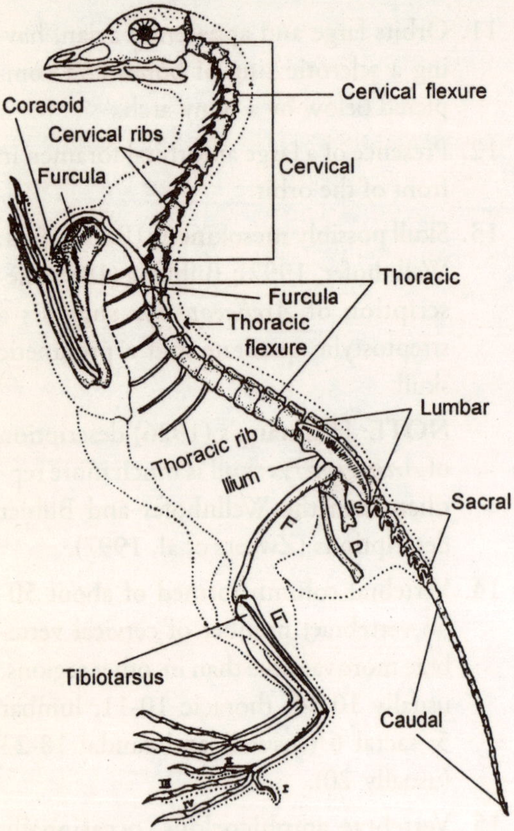

Fig. 17.1A: Skeleton of *Archaeopteryx*

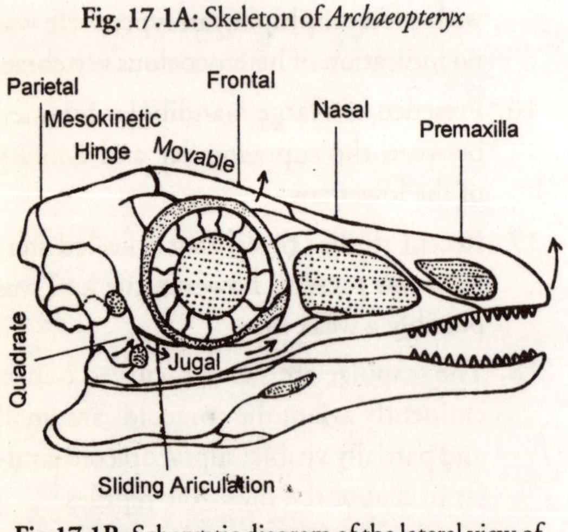

Fig.17.1B: Schematic diagram of the lateral view of the skull of *Archaeopteryx*, based on figures in Heilmann (1926), deBeer (1954) and Bock (1964). A mesokinetic hinge between parietal and frontal bones and a sliding articulation between the postfrontal and jugal bones (shows by thick lines) could possibly raise the entire snout

Box 17.1: Vertebral Column
1. Long, lizard-like with about 50-60 vertebrae, and with 2 flexures—cervical and thoracic.
2. Number of cervical vertebrae more variable than in other regions—usually 10-14; sacral only 5-6 (fused); caudal 18-23, without pygostyle.
3. Vertebrae amphicoelous, occasionally with a few amphiplatyan type.
4. Thoracic ribs without uncinate processes.
5. Possibly with a keeled sternum.

21. A hand (manus) with free metacarpal bones and claws on all three digits; foot with free metatarsal bones; first digit of foot opposed to other three digits.

17.6. EVOLUTIONARY POSITION OF *ARCHAEOPTERYX*

Since the discovery of the fossil remains of *Archaeopteryx*, the scientists all over the world turned their attention to this unique reptile-bird that possibly held the key to the ancestry of birds. But during the last 146 years since its discovery (1861) by Andreas Wagner, a precise conclusion to the debate has not yet been reached as to the possible source of the reptilian stock from which *Archaeopteryx* evolved and also about post-*Archaeopteryx* radiation. With the recent discovery of at least a dozen of pre-*Archaeopteryx* and post-*Archaeopteryx* fossils, some revealing more reptilian and some more avian features (either of the cranial or of the post-cranial segment or of both), it becomes all the more difficult indeed to construct a generally agreed genealogical tree of birds.

de Beer (1956), while discusing 'The evolution of the Ratites' referred to Owen (1866) who believed that the ratites are degenerate descendants of the carinate birds. Huxley (1867) proposed a system of classification of birds based on palate and was of the opinion that the dromaeognathous (= palaeognathous) palate characteristic of the ratites was the most primitive. Lowe (1928, 1935) also supported this view and stated that the neognathous palate, characteristic of the modern flying birds, is derived from the primitive palaeognathous palate. Lowe believed that the ancestors of the ratites never flew. So, according to him, as de Beer (1956) writes: "*Archaeopteryx*" could not have been in the line of ancestry of birds, but must have been an independent offshoot from the reptiles".

According to de Beer (1956), the structure of feathers and their arrangements on the wing are so similar that an independent evolution of *Archaeopteryx* from the reptiles—unrelated to the line of evolution that gave rise to the modern birds—is unbelievable. Simpson (1936), Steiner (1956) and a few others also believed that *Archaeopteryx* was ancestral to modern birds. de Beer (1956) believed that the palate of the ratites is not primitive but neotenous, and represents an early stage through which the palate of many carinates passes during development. Other neotenous features of the ratites are the plumage and the persistence of sutures between the bones of the skull.

According to King and McLelland (1979, 1984) (In: Form and Function in Birds, Vol. 1; Academic Press, London, N.Y., etc.) "While the therapsids were declining, a reptilian line, the Archosauria was beginning a radiation which dominated the vertebrate scene for 120 million years. The first group to appear in the early Triassic was the Thecodontia which, either directly or indirectly, gave rise to four other orders of Archosauria, i.e., Crocodila, Ornithischia, Saurischia and Pterosauria, and also to Birds". More recently, Zweers et al (1997) have more precisely summarised the reptilio-avian transitional relationships. They have quoted Gauthier (1986) who rejected the Thecodont-Bird (Heilmann, 1926), the Mammal-Bird and the Crocodile-Bird hypothesis in favour of Dinosaur-Bird hypothesis. Based on Gauthier (1986) and Benton (1985; 1990), Pough et al (1989) have summarised the reptilio-avian transitional stages as:

The three divergent lines of amniote taxa at the base are: Testudimorpha (turtles), Sauria/Diapsida (other reptiles and birds), and Synapsida (e.g., mammals). The diapsids include two major groups: the Archosauromorpha (including the Archosauria and Rhynshosauria) and the Lepidosauromorpha (including the Ichthyosauria, Plesiosauria and Lepidosauria). The Archosauria include the Phytosauria, Crocodilia, Pterosauria, and Dinosauria.

The Dinosaur stock includes the Aves within the Theropoda. The Dinosaur group is constituted of large, herbivorous, mostly quadruped Ornithischia (e.g., *Stegosaurus*) and smaller to large-sized Saurischia that comprised two suborders: The very large, quadruped, herbivorous, long-necked Sauropodomorpha (e.g., *Brontosaurus, Triceratops, Diplodocus*), and comparatively smaller Theropoda. The theropods were biped and carnivorous, and they included the large-sized carnosauria (e.g., *Tyrannosaurus*), the middle-sized ceratosauria and the small **Coelurosauria**.

The Coelurosaurians include the Ostrich-like Ornithomimidae (e.g., *Struthiomimus*) and the Maniraptora. Maniraptora includes the small-sized carnivorous Deinonychosauria and the Aves. Deinonychosauria include two taxa, the Dromaeosauridae (e.g., *Deinonychus, Dromaeosaurus, Velociraptor*) and the Troodontidae (e.g., *Troodon*), while some taxa, e.g., *Compsognathus*, are placed here *incertae sedis*. Gauthier (1986) mentioned that Ornithomimids and maniraptorans (including birds) are reported from upper Jurassic sediments: "The re-discovery of *Deinonychus* (Ostrom, 1969) and the re-examination of closely related specimens of *Velociraptor* and *Compsognathus* (Ostrom, 1978) revealed close relationship of Deinonychosauria, *Archaeopteryx* and all other birds. They are considered as the most derived maniraptorans". Benton (1990 a) concluded that "birds are derived from theropods close to Dromaeosauridae and that Dromaeosauridae, Troodontidae and Aves are left as an unresolved trichotomy for the present."

Currie (1987) suggests that Troodontidae are the closest outgroup of Aves, and Martin (1995) does not exclude close relationship with Ornithomimidae (Zweeers et al, 1997). Zweers and Vanden Berge (1997-98) mention that Novas and Puerta (1997) described a non-avian theropod dinosaur called 'half-bird' (*Unenlagia*) from the Lower Cretaceous (90MY) in Argentina, which—by cladistic analysis of the few postcranial elements fits between the Dromaeosauridae and *Archaeopteryx* (See Figs. 17.2, 17.3).

As regards structural details of the pelvis, forelimbs, hindlimbs and skeleton of Ornithischian and saurischian dinosaurs, King and McLelland (1979, 1984; In: Form and Function in Birds, Vol.1; Academic Press; London, N.Y., etc.) have referred to observations by Galton (1970), Bakker and Galton (1974) and specially Ostrom (1969, 1973, 1975, 1976, 1978, 1979) and state that though the Ornithischian pelvis is more bird-

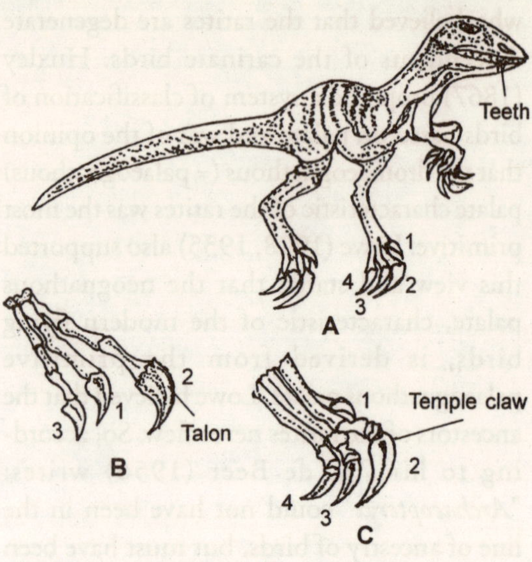

Fig.17.2: A. Deinonychus sp. (Lateral view); a dromaeosaurid mid-cretaecous dinosaur with three clawed fingers and four toes, extraordinary among the dinosaurs; B. hand with three clawed fingers of which the second forms sharply curved 'talon'; C. foot stout, heavy and with four pointed clawed toes of—which the second one has an extraordinarily large, curved and massive claw called 'temple claw' (or the 'terrible claw') that can be raised off the ground while running (Based on and modified from Cranfield, I ed., 2001; In: Dinosaurs and other prehistoric creatures; Greenwich edition, London)

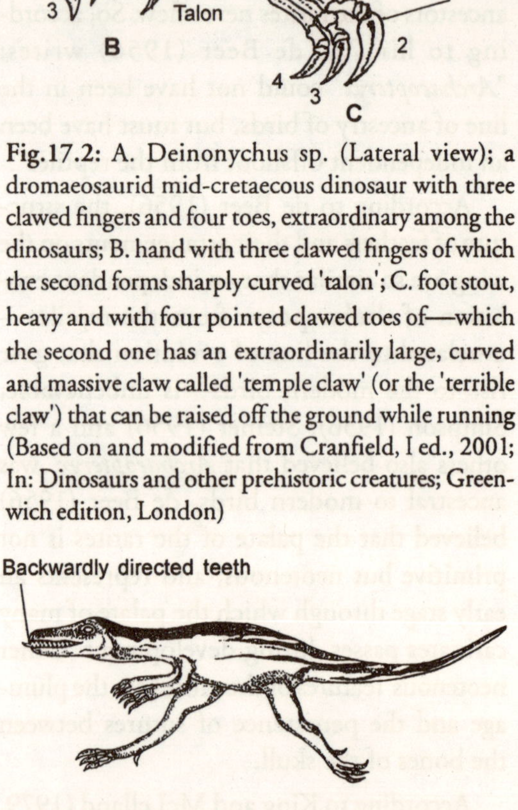

Fig. 17.3: *Compsognathus*, a coelurosaur from the Late Jurassic of Southern Europe (Length 70 cm-1.4 m)

like, this group, from various other considerations, is not considered to have given rise to birds; instead, the smaller coelurosaurian theropod dinosaur appears to be the basal stock that gave rise to *Archaeopteryx* and to the modern birds.

Ostrom (1985, 1991) in his later works has vary reasonably explained that the structural similarities observed between *Archaeopteryx* and coelurosaur dinosaur are so obvious that these cannot be the result of parallel evolution. In fact, according to Ostrom, such a type of small, coelurosaur carnivore (e.g., *Compsognathus*) represented the pre-*Archaeopteryx* stage. In the skull models of the ornithomimid *Struthiomimus* and in those of the Deinonychosaria (e.g., *Deinonychus*, Ostrom, 1969; *Compsognathus* Ostrom, 1978), the pre-bird kinetic features are such developed as to allow either pre-(Rhyncho-/Pro-) kinesis (e.g., *Struthiomimus*) or *mesokinesis* (e.g., *Deinonychus, Compsognathus*). So, from the kinetic point of view, as Zweers et al (1997) have put it, both ornithomimid-like and deinonychosaurian-like skull designs may fit into the possible pre-*Archaeopteryx* ancestral models.

The skulls of *Protoavis* (220 MY; Chatterjee, 1991, 1995) from the Triassic-Jurasic deposits and *Confuciusornis* (170; Hou et al, 1995) from late Jurassic of China have both been described as prokinetic. Zweers and Vanden Berge (1997/98) comment: (1) "This late Jurassic avian fossil, *Confuciusornis sanctus* may represent an undiscovered pre-*Archaeopteryx* radiation, or (2) the resemblance of this fossil to *Arhaeopteryx* may confirm an extremely rapid post-*Arhaeopteryx* evolution in birds as evidenced by its differences from early Cretaceous avian fossils, *Cathayornis* (70-105 MY), *Gobipteryx* (70-85 MY) and *Gansus* (140-150 MY)."

However, Hou et al (1995) specifically mention that *Confuciusornis* differs from *Arhaeopteryx* and agrees with *Gobipteryx*, a fossil of much later age. Dyke and Thorley (1998) have thoroughly discussed the inter-relationships of *Protoavis*, *Avimimus* and other Mesozoic birds. As regards development of flight in archaeic birds, it has been mentioned that a few fossils—like *Sinornis santensis* (140 MY; Enantiornithes), *Iberomesornis* (125-135 MY) and *Eoalulavis* (70-105 MY)—indicated much better development of flight than that developed in *Arhaeopteryx*.

Ostrom (1979) stated that *Arhaeopteryx* was a fast-running, cursorial, predatory biped, preying on insects by chasing. The coelurosaurs also possessed similar habit. Bellairs and Attridge (1975) also held nearly similar view.

King and McLelland (1979, 1984; In: Form and Function in Birds. Vol. 1. Academic Press, London, New York, etc.) have summarised the scheme of the phylogeny of reptiles, birds and mammals and 3 possible sources of the origin of *Arhaeopteryx*. These are: (1) *Arhaeopteryx* had its origin directly from the thecodont reptiles in the Mid-Triassic. This view has been held by Broom (1913), Heilmann (1926), Swinton (1960), Romer (1966) and de Beer (1975); (2) Arhaeopteryx arose from the Mid-Jurassic Coelurosauria (Theropoda), as proposed earlier by Huxley (1868) and on later re-examination strongly supported by Ostrom (1973, 1976, 1979, 1985), and (3) a possible source from the late Triassic Coelurosauria, as proposed by King and McLelland (1979; In: Form and Function in Birds, Vol. 1) and the time-scale for the period of origin supported by Young (1981).

According to Young (1981), *Archaeopteryx* probably arose from the larger predator coelurosaurian group through some form of '**Protoarchaeopteryx**'. It may be assumed that the narrow feathers on hand became longer. This offered much greater advantage to the bird in hunting insects. Gradually, with better development of the pectoral muscles,

Proto-*archaeopteryx* gave rise to *Archaeopteryx*. Thus the principal progressive stages in the origin of *Archaeopteryx* and Birds, according to Young (1981) and Ostrom (1976, 1978, 1979, 1985), may be depicted as:

Thecodont Pseudosuchia (e.g., *Saltoposuchus*) → Coelurosauria (Late Triassic)
↓
Proto-*Archaeopteryx*
↓
Archaeopteryx
↓
Birds

Young (1981)

Thecodont Pseudosuchia → Coelurosauria (e.g. *Compsognathus*) Mid-or Late Jurassic
↓
Pre-*Archaeopteryx*
↓
Archaeopteryx
↓
Birds

Ostrom (1976, 1978, 1979, 1985)

Bock (1964; In: Kinetics of the avian skull, J. Morph., 114) while discussing skull kinesis in birds—both ancient and modern—has traced the evolution of *Archaeopteryx* from the pseudosuchian reptiles and the 'Coelurosauria' through a stage called Pre-*Archaedpteryx*. This Pre-*Archaeopteryx* might have been represented by the Coelurosaurian, *Comsognathus* as suggested by Ostrom (1976) or Proto-*Archaedpteryx* as mentioned by Young (1981). Controversies exist about the probable pseudosuchian stock that gave rise to *Archaeopteryx*, either directly or through a pre-*Archaeopteryx* stage.

Some authors mention about the psuedosuchian *Euparkeria* (Colbert, 1969), some about *Saltoposuchus* as the basal thecodont stock that gave rise to *Archaeopteryx*, either directly or through coelurosaurian dinosaur like *Compsognathus* (Ostrom, 1976; Heilmann, 1926).

Again, there are controversies about the time-scale for the origin of *Archaeopteryx*. Some authors consider it to be Mid-Triassic (Heilmann, 1926; Swinton, 1960; Romer, 1966; de Beer, 1975), some Mid- or Late-Jurassic (Huxley, 1868; Ostrom, 1976), still some consider it to be Late Triassic (Young, 1981). The fossil records of birds are so scanty and incomplete that to consider *Archaeopteryx* as the immediate ancestor of modern birds is far from reality.

After the discovery of the fossil of *Archaeopteryx*, a post-*Archaeopteryx* fossil of an aquatic wader called *Gallornis* was discovered. Its appearance was estimated to be between 3 to 25 MY after that of *Archaeopteryx*. King and McLelland (1979, 1984) comment that within that short period of time, it was not possible for an wader to evolve from an archaic bird like *Archaeopteryx*. Nor was it possible for the early Cretaceous birds to have evolved directly from *Archaeopteryx*. Considering all these points, most authors believe that in the phylogeny of birds, *Archaeopteryx* was the source beyond doubt, but the evolution of modern birds progressed through several steps from *Archaeopteryx* through pre-modern birds (e.g., *Iberomesornis*, Enantiornithes, *Patogopteryx*, Hesperornithiformes, etc.).

However, it is certain that at the end of the Cretaceous (or rather at the KT transition period; Zweers and Vander Berge, 1997/98), there appeared quite a few aquatic birds. But, till then, the fully terrestrial birds were absent; rather extensive radiation of modern birds, both aquatic and terrestrial occurred during early Tertiary.

There are several hypotheses on the time-scale of the radiations of modern bids. The common and the more traditional one states that all the existing orders did develop before the Cretaceous-Tertiary boundary (KT boundary) through which they successfully passed and survived. Zweers and Vanden Berge (1997/98) comment that this does not explain as to how they could pass the KT boundary without being affected by environmental harzards and why the bird-like dinosaur taxa could not pass and became extinct.These authors also comment on the other extreme hypothesis (Feduccia, 1995) which states that only one representiative of one order was already present before the KT boundary.

King and McLelland (1979, 1984) state that about 35 MY ago, by the end of Eocene period, about 26 out of 32 orders of modern birds appeared, and then, after Pleistocene glaciation which resulted in mass extinction of birds, the present avian orders were established. However, according to Zweers, Vanden Berge and Berkhoudt (1997), most orders of modern birds were present between 60 and 75 MY ie., around KT boundary and early Tertiary.

Having arisen from the thecodont reptiles–either directly in mid-Triassic or through coelurosaur dinosaurs in the mid- or late Jurassic,—the birds show much greater affinities with the order crocodilia among the reptiles, and their relationships with the other groups of reptilies such as Lepidosauria, Chelonia and Rhynchocephalia are quite distant, as these groups diverged much earlier in late Permian age from the stem reptiles.

Regarding flight in birds, it may be asserted that though some short-distance flights in amateur fashion have been observed in some species through vertebrate series, flight in Pterosaur among the reptiles, in Bats among the mammals, and in all carinate birds (with a few rare exceptions) is recoginsed as true flight—although the nature and mechanism of their flight are different. Some authors assume that the primary functions of the feathers in birds were heat-regulation and heat-conservation and that the flight came in as the secondary innovation.

All modern reptiles are ectothermic, i.e., they depend on external environment for maintaining their body temperature and for meeting energy requirement for the body. However, quite a few species of dinosaurs were endothermic. They could produce, at least partially, some amount of internal heat to maintain their body temperature. Hence, the acquisition of endothermy by birds is claimed by many authors to have preceded the development of feathers, as, otherwise, feathers would have blocked the absorption of heat by the body through solar radiation.

As already mentioned, the question of the origin of flight is closely linked with that of the birds. It is really surprising that the feathers contained within them the inherent pre-adaptive qualities of flight.

17.7. ORIGIN OF FLIGHT IN BIRDS

There are several theories to explain the origin of flight in birds :

17.7.1. Cursorial Origin of Flight

The Hungarian palaeontologist Francis Baron Nopcsa (1907) advocated that the ancestral birds must have been cursors or fast runners. The primitive birds could run so fast on the ground that, at that time, their feathered fore-limbs were extended and moved, or, rather, flapped up and down. This led to some upward thrust of the body from the ground. According to Nopcsa, both these acts—fast running and moving the extended and feathered forelimb up and down—were designed for chasing and catching insects. Nopcsa believed that the birds evolved from the long-tailed, biped, cursorial reptiles which, by running and moviñg the forelimbs up and down, acquired flight.

17.7.2. Arboreal Origin of Flight

The Danish scientist Gerard Heilmann (1929), on the other hand, did not believe Nopcsa's theory. Heilmann believed that for purely cursorial animals, the grasping hind toe or the hallux would have a tendency to disappear and the forelimbs would have been reduced in size, rather than elongated. So Heilmann advocated that the birds and certain dinosaurs came from the same arboreal stock from which the dinosaurs early returned to a terrestrial mode of life. But for birds, the return was delayed long until after the assumption of flight.

In Heilmann's opinion, the pro-avian was a form intermediate in structure between certain Pseudosuchia and the Jurassic *Archaeopteryx*. It had a grasping feet which were in no way concerned with flight, a hand with three long slender fingers and an elongated reptilian tail. Heilmann thought that the feathers first evolved as a heat-retaining clothing, from which flight could be derived by further elongation of feathers, as an outcome of climbing and gliding. The pro-aves used arms and tail for support in their soaring leaps from tree to tree or to the earth.

Young (1981) comments in this connection, "Thus the two theories that the flight came 'from the ground up' and 'from the trees down' may both be true.

According to Heilmann (1926), the flight of primitive birds was in the form of gliding. In the intial stage, climbing trees and then gliding in air—these processes made true flight possible in later stage.

17.7.3. Diving Origin of Flight

Newman (1950) regarded the pro-aves as aquatic reptiles that used to take advantage of soaring in thermal upcurrents and diving in water to catch fishes. The most primitive orders of birds are divers, or at least aquatic, while the most specialised orders are arboreal. Thus Newman believed that flight in birds originted from the habit of soaring and diving in water.

17.7.4. Gliding Origin of Flight

de Beer (1954) thought that simple gliding preceded flapping, because *Archaeopteryx* had no keel and its pectoral muscles must have been very feeble.

This view may be regarded as an extension of the arboreal origin of flight.

To achieve sustained flight with the help of flapping of wings, the birds had to acquire many morphological adaptations such as limiting body-shape to the most suitable model, decreasing body weight, conversion of the forelimbs into wings for flight and modifying the wing skeleton suitably, modification of the sternum and greater development of the pectoralis muscles and, finally, suitable changes in the circulatory, respiratory and nervous systems.

17.8. SUMMARY OF DISCUSSION ON THE ORIGIN OF BIRDS

Although biologists are confronted with a genuine gap of evidences in constructing a generally agreed genealogical tree of birds, the evidences available at hand suggest that the birds evolved from the archosaurian diapsids, closely allied to the small pseudosuchian of the Triassic like *Euparkeria* (Colbert, 1969) or *Saltoposuchus* (Young, 1981). From some such bipedal pro-avian type, the evolution of birds might have taken place through a theropod dinosaurian group called the Coelurosauria.

Archaeopteryx, the most primitive bird discovered so far, originated from some form of coelusrosaurian, possible closely allied to *Compsognathus longipes* as conceived by Heilmann (1926) and, after 50 years, strongly supported by Ostrom (1976). According to Ostrom, the time of origin was late Jurassic. Young (1981) believed it to be late Triassic, and the origin of *Archaeopteryx* took place through some Proto-*Archaeopteryx* stage.More recent studies by Cracraft (1974, 1989), Ostrom (1976), Gauthier (1986), Sibley and Ahlquist (1990), Chiappe (1995a, 1995b), Chatterjee (1995), Hou et al (1995), Feduccia (1996), Benton (1997), Mindell (1997), Zweers and Vanden Berge (1997), Dyke and Thorley (1998) and many others reveal in-depth studies of many new fossils discovered before and after *Archaeopteryx*. These studies are not only based on palaeontology but also on cladistic analyses of osteological form characters and on molecular systematics. But the data-sets of these three disciplines are not always uniformly fitting one another.

However, such studies lead to few deductions :

1. *Archaeopteryx* evolved from some form of maniraptoran coelurosaur allied to Deinonychosauria which may be designated as the Pre-*Archaeopteryx* stage.
2. Most authors consider that the neognath birds are monophyletic in origin, assuming that feathers and true flight developed only once and that modern avian orders have been established in time and space through a series of post-*Archaeopteryx* stages.These stages may be represented by *Iberomesornis*, Enantiornithes, *Patagopteryx*, Hesperornithiformes, Ichthyornithiformes and Neornithes.

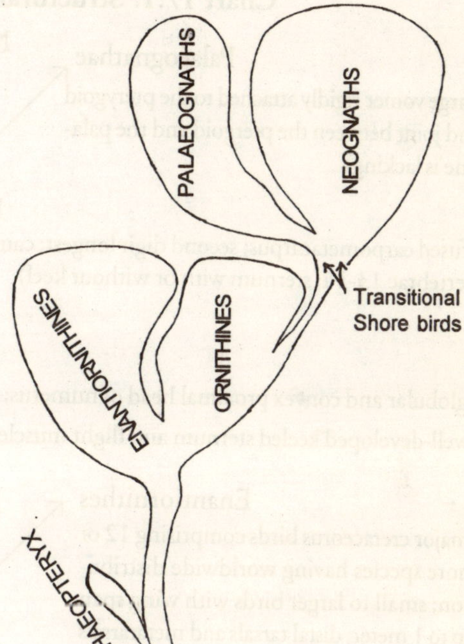

Fig.17.4: Radiations of birds during the Mesozoic and Caenozoic eras, based on Kardong (1998). According to this hypothesis both the major groups of birds (Palaeognaths and Neognaths) evolved from transitional shore birds

3. Recent studies in Molecular Systematics by Sibley and Ahlquist (1972-1992), using DNA sequences and DNA hybridisation techniques have revealed many new facts about the age and primitiveness of fossils as well as their relationships which, in many cases, are not in conformity with those established by

palaeonotological records. Cooper and Penny (1997), using differences in DNA sequencing from mitochondrial and nuclear genomes in modern birds, have provided strong evidences in favour of their hypothesis that the ratitomorph and gallimorph lineages diverged early, followed later by psittacimorphs, and these three ordinal groups are a sister clade to a major clade of many other ordinal systematic taxa.

As regards the origin of flight in birds, most authors are in favour of the arboreal origin. The primitive birds could climb trees and could come down from the trees through gliding; flapping of the wings came later.

Chart 17.1: Structural Distinction in Avian Evolution

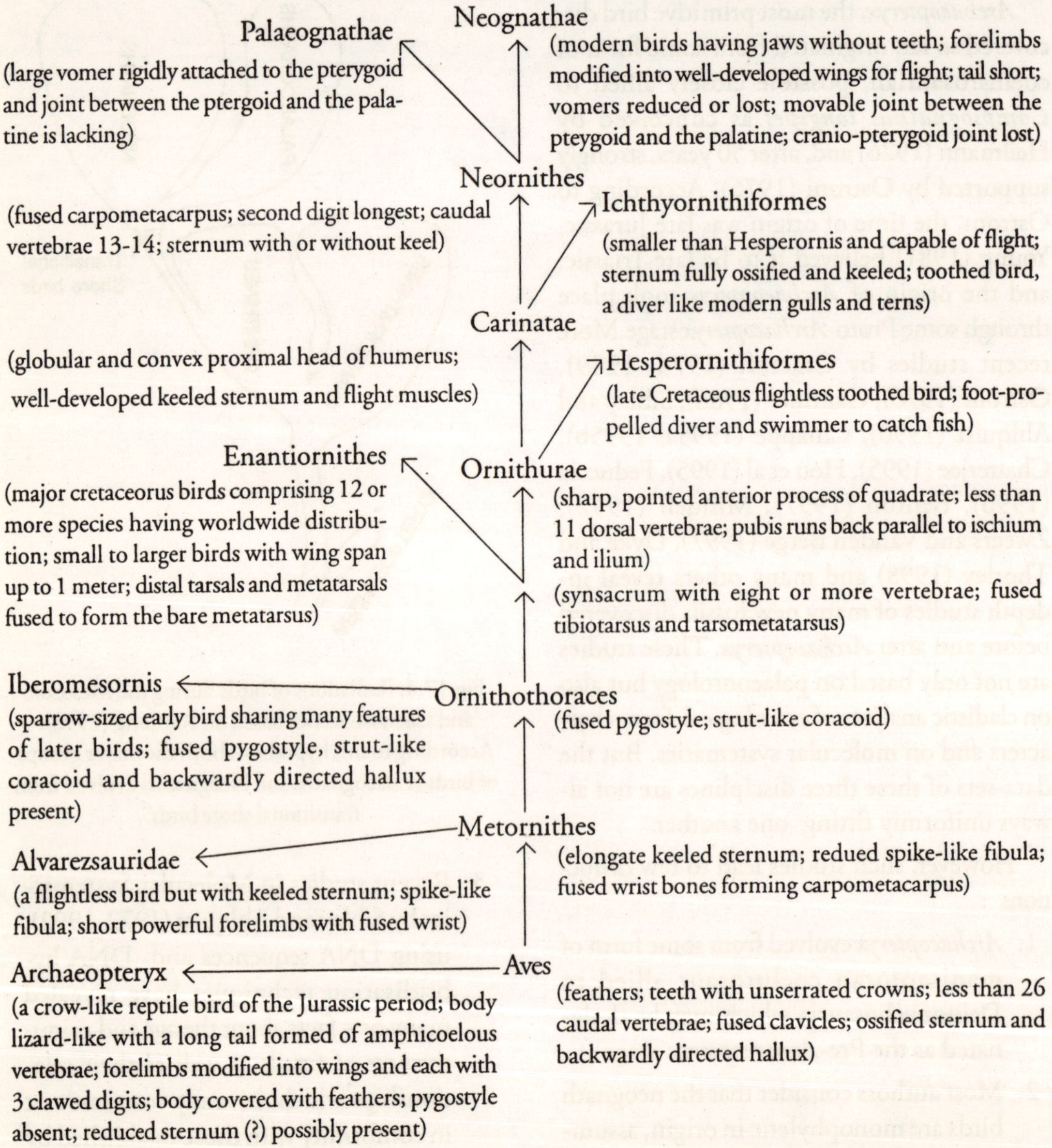

Chart 17.2: Summary of the Phylogeny of Saurians and Birds— Based on the Cladograms of Gauthier (1986), Cracraft (1986), Sanz et al (1995), Chiappe (1995) and Zweers et al (1997)

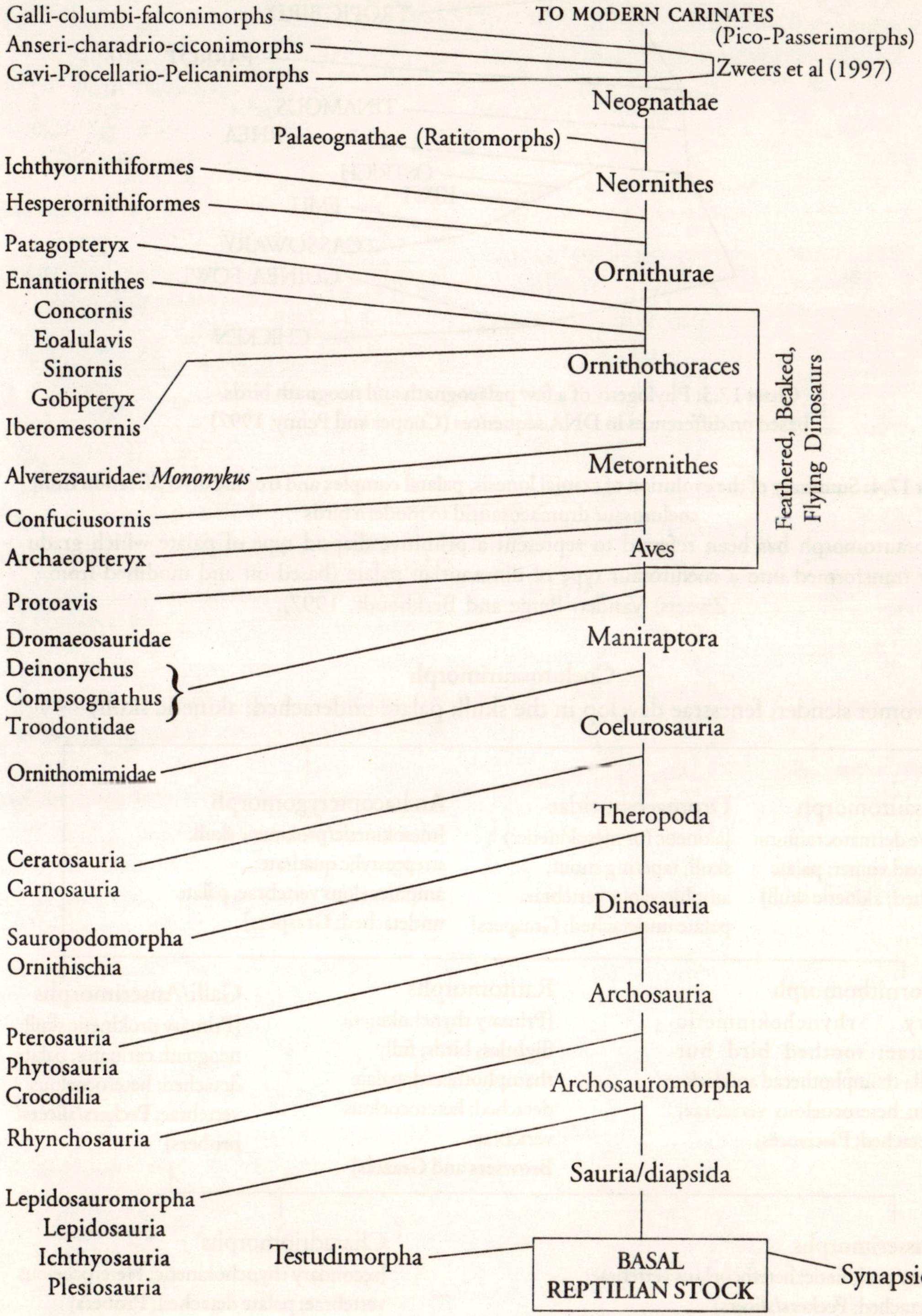

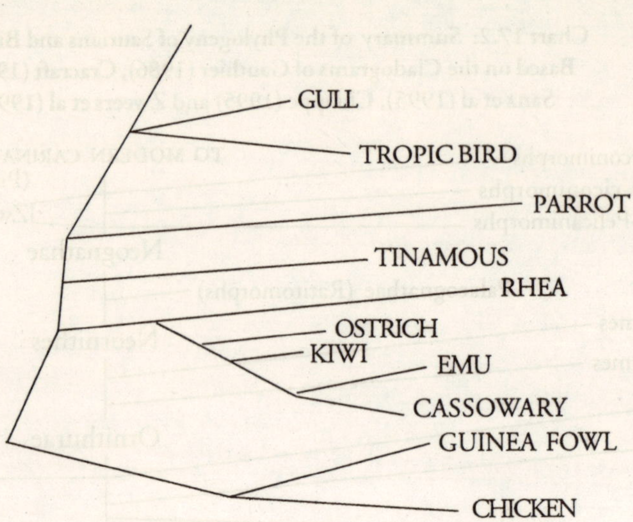

Chart 17.3: Phylogeny of a few palaeognath and neognath birds, based on differences in DNA sequences (Cooper and Penny, 1997)

Chart 17.4: Summary of the evolution of cranial kinesis, palatal complex and trophic diversification from coelurosaur dromaeosaurid to modern birds

Lepidosauromorph has been referred to represent a primitive diapsid type of palate which gradually transformed into a coelurosaur type of dinosaurian palate (based on and modified from Zweers, Vanden Berge and Berkhoudt, 1997).

Coelurosaurimorph
[vomer slender; fenestrae develop in the skull; palate undetached; akinetic skull]

Lepidosauromorph
[extensive dermatocranium; large, broad vomer; palate undetached; akinetic skull]

Dromaeosauridae
[akinetic (or mesokinetic?) skull; tapering snout; amphicoelous vertebrae; palate undetached; **Graspers**]

Archaeopterygomorph
[mesokinetic/prokinetic skull; streptostylic quadrate; amphicoelous vertebrae; palate undetached; **Graspers**]

Hesperornithomorph
[Primary rhynchokinmetic ornithurae; toothed bird but premaxilla rhamphothecad and lacking teeth; heterocoelous vertebrae; palate detached; **Piscivores**]

Ratitomorphs
[Primary rhynchokinetic flightless birds; fully rhamphothecad; palate detached; heterocoelous vertebrae;
Browsers and Grazers]

Galli/Anserimorphs
[Primary prokinetic skull; neognath carinates; palate detached; heterocoelous vertebrae; **Peckers/slicers/probers**]

Pico/Passerimorphs
[secondary prokinetic; heterocoelous vertebrae; palate detached; **Peckers/slicers**]

Charadriomorphs
[secondary rhynchokinetic; Heterocoelous vertebrae; palate detached; **Probers**]

Box 17.2: Major trophic groups of birds crossing KT-boundary

According to Zweers et al (1997), neognaths have been divided into 4 major trophic groups of which the first 3 groups—Gavi-Procellario-Pelicanimorphs, Ansero-Charadrio-Ciconimorphs and Galli-Columbi-Falconimorphs–successfully crossed the KT boundary and survived, while the 4th group, Pico-Passerimorphs,is supposed to be a later radiation following congenial change of climate and vegetation.

All the 4 major trophic groups of birds have been shown in the phylogenetic Chart 17.2

Chart 17.5: Possible Sequences in the Evolution of Avian Cranial kinesis
(Summary of Different View points)
Modified from Bock, 1964)

Coupled prokinesis
↑
Uncoupled prokinesis
↑
(Holorhinal Nostril:
Reduction of the Pol.)
↑
Coupled prokinesis
(Holorhinal Noostril)

Ratite Rhynchokinesis
(Holorhinal Nostril; Gap in Lateral Bar)
↑
Coupled prokinesis: Holorhinal Nostril
(MODERN BIRDS–ANCESTRAL STOCK)

Secondary Prokinesis
(Holorhinal Nostril)
↑
Secondary Prokinesis
(Schizorhinal Nostril)
↑
Charadriiform
Rhynchokinesis with
Schizorhinal Nostril

↑
Mesokinetic hinge disappears: Prokinetic hinge established with uncoupled prokinesis, holorhinal nostril: enlargement of the Brain and Eyes, as Flight developed
(PRE-MODERN BIRDS)
↑
Most reasonably mesokinetic hinge with streptostylic quadrate or possibly amphikinetic (i.e., both mesokinetic and Prokinetic)
(ARCHAEOPTERYX)
↑
Metakinetic hinge disappears: Mesokinetic hinge appears with streptostylic quadrate
(PRE-ARCHAEOPTERYX)
↑
Metakinetic or possibly Amphikinetic (i.e. both metakinetic and mesokinetic with the beginning of stereptostylic quadrate)
(PSEUDOSUCHIA)
↑
Metakinetic skull with Monimostylic Quadrate
(STEM GROUP OF THECODONT REPTILES)

Chart 17.6: Dendrogram Showing the Probable Changes in Kinesis and Associated Elements of the Head from the Pseudosuchian Ancestors of Birds to Modern birds (Adopted from Bock, 1964)

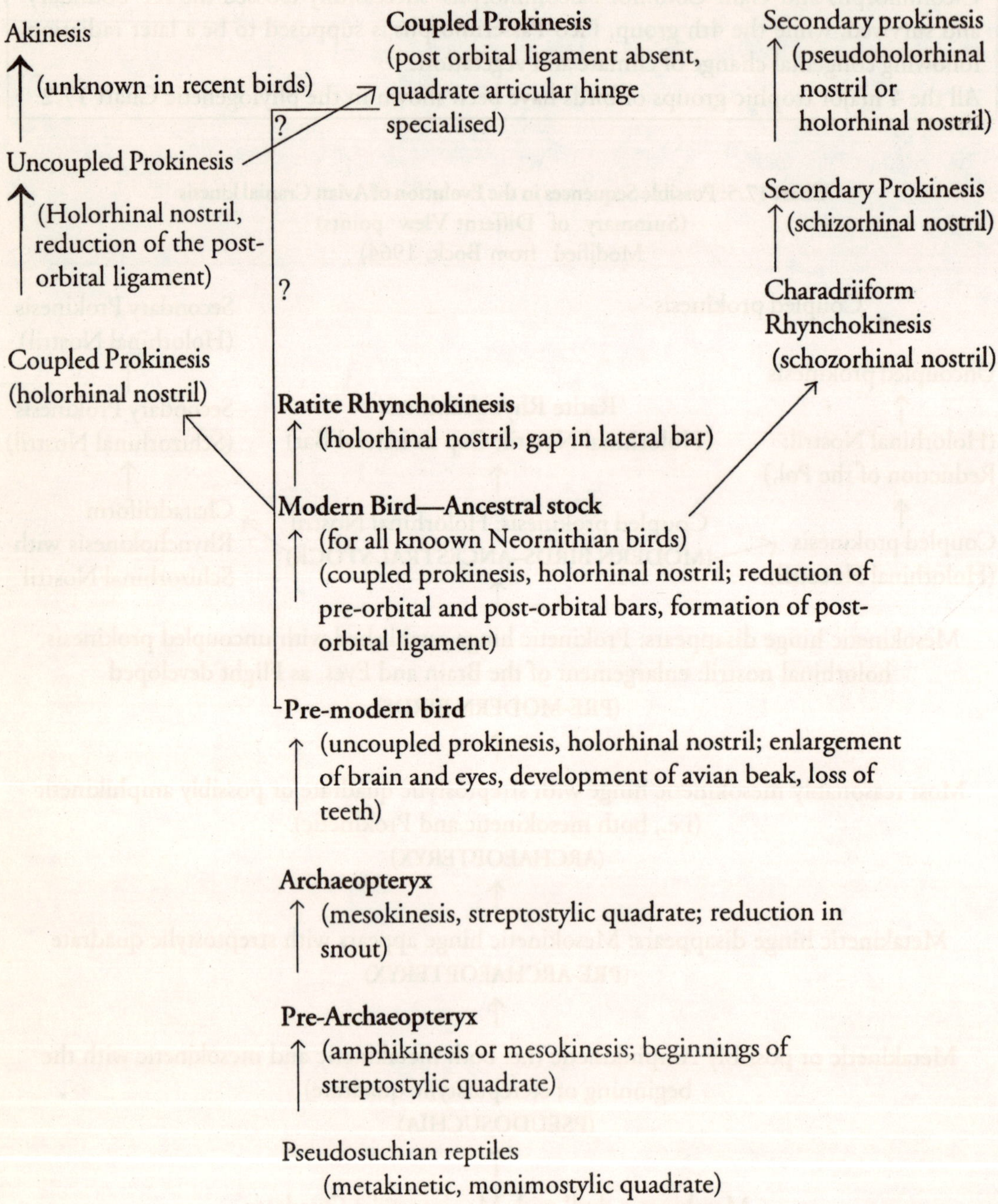

18

Exoskeletal Structures and their Modifications

The exoskeletal structures in birds are integumentary derivatives and are exclusively peculiar characteristics of the class. The exoskeletal structures are so called as they form a hard outer covering of the body, being extensively modified in structure and function as are the endoskeletal parts. The exoskeletal structures in birds are derived from epidermis as in the reptiles and they comprise the **beaks, claws, scales and feathers.**

Among them, the feathers are unique and extraordinary modifications which have singularly isolated the birds from the rest of the vertebrates. The development of feathers in birds—replacing reptilian scales—was a marvellous innovation of nature. Each of the exoskeletal structures mentioned above has undergone wide range of variation in structure and function—in response to habitation, feeding habits and other changes in the environment.

The skin in birds—unlike in other vertebrates—calls for a special mention. It is dry, loosely held and has the corneal layer of epidermis reaching a climax of development. The skin comprises several specializations of which, in birds, the beaks, claws, scales and feathers are collectivity known as exoskeletal structures.

18.1. THE HORNY BILL OR THE BEAKS

Ever since the reptilian forelimbs became modified into wings in birds, the versatile functions of the forelimbs had to be taken over by a remarkable rostral prolongation of the snout—the beaks,formed by the premaxillary and mandibular extensions together with their horny sheaths, the **rhampohotheca**. Thus the upper and lower horny sheaths are known, respectively, as the **maxillary rhamphotheca (= rhinotheca)** and the **mandibular rhamphotheca** (= **gnathotheca**). The mid-dorsal ridge of the maxillary rhamphotheca is called the '**culmen**' while the mid-ventral ridge along the mandibular rhamphotheca is called the '**gonys**'.

The lower beak is formed by the joining of the two mandibular rami rostrally and the two rami enclose a space forming the floor of the oropharynx whose caudal part is known as the gular region. The cutting edge on either side of the maxillary and mandibular rhamphothecae is called in '**tomium**'. The maxillary and the mandibular tomia match one another during closure of the bill.

The bill of birds is a unique multi-purpose epidermal structure of variable thickness, rendered hard in all birds due to deposition of keratin. The bill is usually formed of a single sheath, but, in some birds like albatrosses, pelicans, etc., this sheath is divided into a number of prominent sections. In some birds, the maxillary rhamphotheca at the base of the bill is modified into large ornamental protruberances–the frontal shield of coots and the magnificent casque of hornbills.

18.1.1. Histology of the Bill

Histologically, the actual part of the bill (i.e., the entire sheath covering the jaw bones) may be compared to the skin in which the stratum corneum of epidermis has reached the climax of modifications.

The keratinised layer becomes very thick and cells of this layer contain free calcium phosphate and crystals of hydroxyapatite, in addition to the abundant keratin and keratin-bound phospholipids and calcium. Although the bill is mainly derived from the epidermis, it is not totally unrelated to the dermis of the skin, as the bill-sheath is attached to the periosteum of the jaw bones.

King and McLelland (1984) mention that the entire bill is not always very thick and hard, as most charadriiform birds (waders) have a rather soft and leathery bill, and, in ducks and geese, the bill is hard only at the tip, forming the so-called maxillary and mandibular nails. Each nail along with its dermis forms a complex sensory organ. The nail is perforated by canals (originating from tips of the dermal papillae) which open through minute pores on the inner, ventral surface of each nail, and, thus, the complex sensory organ formed within each nail is called the **'bill tip organ'**.

The bill tip organ responds to tactile stimuli capable of sensory discrimination between food and other particles, and is mainly formed of a type of mechanoreceptors called the **'Herbst's corpuscles.** These are the most widely distributed mechanoreceptor cells in the avian body, occurring at the bases of feather follicles, and along the wing and pelvic bones, as well as in muscles, tendons and joints.

Another type of sensory corpuscles known as **'Grandry's corpuscles'** are distributed in the dermis at the sides and tip of the anseriform as well as grain-eating passeriform bill. Each Grandry's corpuscle consists of one or two granular cells with large number of axonal endings. Structurally, these are very similar to Merkel cells. The nerve endings of the upper bill of birds are usually supplied by the maxillary and ophthalmic divisions of the trigeminal nerve and those of the lower bill by the mandibular division.

18.1.2. Morphology of the Bill

The study of the bird-bill has always been a fascinating subject–not only from the point of view of functional morphology, but also in providing some evidence to the phylogeny of different groups. In the adaptations for obtaining food, the characteristic shape of the beaks and their epidermal investment play a significant role. The beaks of birds vary in shape, size and structure–according to varying food-habits.

However, beak characteristics are not always dependable in ascertaining phylogenetic relationships. Thus, Bertin (1967) observes that even unrelated species with the same type of diet may have the same kind of beak, as in herons and storks.

In recent times, avian anatomists lay great emphasis on the complicated bill functions which they explain not merely by the structure of the bill, but on the principles of biomechanics. Nevertheless, the shape, size and structure of the bill do provide a clue to the nature of diet of the birds. The functional significance of the bill shape has been elaborately discussed by Bock (1966).

As already mentioned, the beak characteristics may be correlated to specific food-habits of birds. This takes into account the principal food-habits. Accordingly, the beak characteristics may be described, based on undermentioned principal food-habits:

1. **Seed-eating beaks:** Such beaks are short, stout and conical i.e., wide at base and then gradually narrowing. Some birds have strong beaks with enough power to crush seeds, e.g., Pigeons, Sparrows, Finches, Grosbeaks, Crossbills, etc.
2. **Slender cutting beaks for omnivorous birds:** The Crows, Ravens, Jungle Crows, Eurasian Jackdaws, Tree-pies, etc. have a thick, strong and elongated bill with hard tomial edges for cutting. These birds are generally omnivorous.
3. **Fruit-eating beaks:** A large number of birds are frugivorous (fruit-eating)–some exclusively and some having fruits as their principal diet. The birds like Parrots and Parakeets, Cockatoos, Macaws, Hornbills, Toucans, Trogons, Green Pigeons and Imperial Pigeons are fruit-eating, but their bill shape and size are so distinctly variable that these may be associated with the type of food and feeding behaviour of these birds.

Examples: The Parrots and Parakeets have a stout, short bill, quite broad at base and the upper beak curves downward over the lower one rostrally. The Great Sulphur-crested Cockatoo of Australia and New Guinea are larger psittacine birds (about 20 in./50 cm long) which possess a rather stout and strong blackish bill that appears broadly triangular. The upper beak curves downward over the lower, farther forward. In Parrots, Parakeets and Cockatoos, the upper beak has a prominently sharp hooked tip. The Hornbills are mainly frugivorous, but they also feed on lizards, mice and small birds. The Great Indian Hornbill (*Buceros bicornis*) and the Malabar Pied Hornbill *(Anthracoceros coronatus)* possess much longer, elephant tusk-like gradually narrowing upper and lower beaks–equal in size and, gradually, curved downward. The Toucans (Piciformes), distributed from Mexico to South America, have a distinctly large, tough and strong but light bill having diagonal serrations on the outer edges. The Common Green Pigeons and the Imperial Pigeons have a stouter, stronger and longer bill than that of the common Pigeons and Doves. In the former, the gape of the bill and the gullet are enormously extensible. The wide gape of the beaks helps these fruit-eating birds in grasping and swallowing even large fruits.

Thus it is observed that the birds with similar food-habit may have different bill characteristics.

4. **Insectivorous beaks:** The beaks of insectivorous birds have also undergone wide range of variations; though, in general, the beaks of these birds are thinly built, less specialised and lighter. The author has studied in some details the morphology of the beaks of many insectivorous birds, including those of the Old World insect-eaters (Bhattacharyya, 1987).

Some important characteristics of the bill in a few insectivorous birds studied by the author are mentioned:

i. **Tickell's Blue Flycatcher** *(Muscicapa tickellie)*: The bill is nearly straight as the upper and lower mandibles close in a line which more or less coincides with the axis of the head. The bill length is slightly less than the cranial length. The tip of the upper beak is slightly longer than the lower, and has a small nick in the tomium just behind the tip. This bird has a fairly strong bill and a reasonably wide bill-base beset with rictal bristles.

ii. **Jungle Babbler** *(Turdoides striatus)*: The bill is much longer than that of the fly-catcher and is moderately thick and broad. The anterior half of the upper beak has a gradual downward curve which, at the tip, completely closes the tip of the lower. Malhotra (1970) srates that the gradual increase of the bill-length over the cranial length indicates greater affinity towards more insectivorous diet.

iii. **Tailor Bird** *(Orthotomus sutorius)*: The bill is narrowly acute, thin and its length slightly exceeds that of the cranium. The external nares are large, oval and more backward in position, so that the bill-length anterior to the nares is about 2/3 of the total bill-length. This bill-form is better adapted for insect-eating.

iv. **Magpie-Robin** *(Copsychus saularis)*: The bill axis has a gradual downward curve from the naso-frontal hinge and the bill-tip is slightly hooked. The upper beak anterior to nares is considerably long and the bill-length exceeds the cranial length. Such a bill is suitably adapted to insect-catching and, to a small extent, grasping fruits and seeds.

v. **Yellow Wagtail** *(Motacilla flava)*: The bill is narrow and pointed, appearing somewhat similar to that of *Orthotomus*. The bill axis is almost in a line with the cranial axis. The bill length is nearly the same as the cranial length. This bill-form is suitably adapted for catching small insects.

vi. **Indian Pipit** *(Anthus novaeseelandiae)*: The bill is shorter and broader than that of *Motacilla* and has a slightly gradual downward curve. The bill length is slightly shorter than the cranial length. This bill is better adapted for feeding on ground insects.

vii. **Black Drongo** *(Dicrurus adsimilis)*: Both the upper and lower beaks are thicker and stouter than those of the other species studied. The upper beak has a gradual downward curve up to the tip which ends in a prominent, sharp hook. The bill length anterior to nares is 2/3 of the total bill length. This characteristic, together with the small size of the nares, greater width at bill base and the prominent hook at the bill tip indicate quicker action of the bill for catching insect food in air and grasping it with a moderate compressive force.

viii. **Green Bee-eater** *(Merops orientalis)*: The bird has a long pointed bill of which the upper beak has a very gradual downward curve from the base to the tip which completely closes that of the lower mandible. *Merops* has a long bill, double the length of the cranium, and its width at the base is considerably narrow. The bill length anterior to nares is more than 2/3 of the total bill length, an adaptation perfectly suited for insectivorous diet.

Besides the above birds, the swallows, swifts, goat-suckers, nightjars, hoopoes, vireos and the tree-creepers are mainly insectivorous birds, but their food-types and feeding behaviour are different in different species.

5. **Tearing and piercing beaks of the birds of prey:** The falconiform and strigiform birds—like the vultures, eagles, hawks, kestrels, merlins, buzzards, kites, shrikes, owls—are carrion-feeding and flesh-eating. Some of them also feed on large insects and other food-types. The beaks of falconiform birds are either short or medium-sized, rhamphothecal investments thickly cornified and hardened and a downwardly directed, sharp and pointed hooked tip of the upper beak helps in tearing and piercing flesh from the carrion or from any other food-prey freshly killed (e.g., frogs, lizards, rats, mice and smaller birds). The falcon has a sharp toothlike process just behind the hooked upper beak. This tooth-like process becomes very effective in tearing flesh.

6. **Anseriform bill for water- and mud-straining:** The anseriform bill—possessed by a large variety of birds like ducks, geese, teals, swans, gadwalls, pochards, shovellers and flamingos—is a highly characteristic bill form that strains aquatic insects, water weeds, small crustaceans and molluscs from water and mud. Some are surface-feeders and some dive quite deep in water.

The anseriform bill is broad, flat and its margins are provided with cornified lamellae arranged in transverse rows, forming an efficient straining apparatus that strains insects and other food from water or mud. The lamellae act as the sieving apparatus through which large amount of water taken inside the mouth is expelled, leaving only the food-preys inside the mouth.

In flamingos the bill-form is slightly different, as its rostral half is sharply bent downward. This helps in scooping the mud-bottom in shallow waters and then straining its food that comprises crustaceans, worms, larva of insects, seeds of marsh plants, and organic ooze. The lamellae of the beaks and the fleshy tongue perform similar functions as in the typical anseriform bill.

7. **Long slender bills for probing into water and mud bottom:** A large number of charadriiform birds like the jacanas, oyster-catchers, plovers, curlews, terns, godwits, snipes, sandpipers, stilts, avocets, stints, lapwings have their beaks slender and long, either straight or curved for probing into certain depth of water, or for reaching the mud-bottom in search of crustaceans, molluscs, insects and their larvae, as wells as small fishes.

The curlews (*Numenius* sp.), sandpipers, (*Tringa* sp.) woodcocks (*Scolopax* sp.) and the snipes (*Gallinago* sp.) have each a characteristic bill, either long and straight up to tip or straight and curved downward rostrally.

It is to be noted that most charadriiform birds possess rhynchokinetic upper jaw in which the flexible zones in the upper jaw may be located rostrally, thereby causing up and down movement of the rostral part of the upper jaw only, and not of the entire jaw. This facilitates movement of the beak-tips within a small area of soil or mud in search of food.

The avocet (*Recurvirostra* sp.) has its slender bill curiously turned upward—'The curved part skimming the semi-liquid mud with a back and forth rotatory or churning motion, washing out the food particles' (Salim Ali, 1987).

Among the charadriiform birds, the Skimmer (*Rynchops* sp.), known as the Scissorbill, has 'a peculiar laterally flattened, knife-bladelike, orange-yellow bill' in which the upper beak is shorter than the lower. The lower mandible is kept immersed in water, ploughing the water surface, while the upper mandible is kept open for snap-closing on any fish touching the beaks. The Atlantic Puffin of the Arctic coasts of Northern Europe and America has its summer beak red, blue and yellow and of a triangular and laterally compressed shape, while in the winter, the bird sheds part of the beak, losing its bright colour. The Spoonbill (*Platalea* sp., Ciconiiformes, Threskiornithidae) has a distinctive, large, flat, black and yellow spoon-shaped bill. The beaks are held partly open for probing into mud bottom with the help of the lower beak in search of tadpoles, frogs, molluscs, insects and vegetable matter.

The large Shoebill (*Balaeniceps rex*; Ciconiiformes; Balaenicipitidae) is about 4 feet (1.2 metre) long with a very distinctive shoe-like deep and wide beak for feeding on fishes, amphibians and molluscs.

8. **Beaks for swooping, catching and grasping fishes:** The Darter or Snake-bird (*Anhinga* sp.), the Herons, some Storks and the Grey Pelicans—which feed on fish as their staple food—have their beaks long, gradually narrowing, but stout. The snake-bird has a long, flexible neck and a long bill that may be thrust like a javelin to chase the fish and grasp it below the water surface, the neck acting as a spring to thrust the bill forward.

The Pelicans, on the other hand, drive the fishes in a group towards the shore or shallow water, then scoop up the fishes independently into the remarkable pouch hanging from the lower mandible. Some storks use their long, beaks as efficient tools for catching fishes, frogs, reptiles, crabs, molluscs and large insects. The Little Cormorant *(Phalacrocorax niger)* is exclusively a fish-eater, the fish being chased and caught under water. This bird is an expert diver and underwater swimmer. The Kingfishers mainly feed on fishes, although tadpoles, lizards and aquatic insects are also taken as food. Their beaks are quite long, straight and stout, broad at base and then gradually tapering rostrally.

The Kingfishers usually sit hiding among foliage of tree-branches overhanging water and swoop and grasp fish from the surface water with tremendous speed. The grasping power of the bill is endowed by the strong mandibular and pterygoid muscles. The Pied Kingfisher hovers for quite some time in air before dashing on the quary at top speed, catching and grasping it with the help of the beaks.

Among the Terns and Gulls, the Caspian Tern (*sterna caspia*) has a comparatively long, stout and strong bill. The Caspian Tern mainly feeds on fishes. Both Caspian Terns and Arctic Terns are long distance migratory birds. Many other fish-eating Terns and Gulls, as also the Loons (*Gavia* sp.), possess moderately slender bill, suitably adapted for either swooping on and catching the fish from nearer the surface water or plunging deeper (e.g., Loon) in water to catch and grasp the fish. However, beside fish, most Terns and Gulls also feed on molluscs and other invertebrates.

The Grey-headed Fishing Eagle has a moderately long but very strong and stout bill; the upper beak-tip is hooked downward. This bird swoops on to the fish and catches it from the surface water.

Fig. 18.1: A few representative beaks for different food-habits. A. Pigeon (Granivorous); B. Fly catcher (insect eating); C. Crossbill (seed-eating, extracting from the kernel of hard coniferous seed); D. Flowerpecker (long-billed honey eater); E. Woodpecker (long, chisel-shaped beaks for wood probing and insect-eating); F. Pelican (fish-hunting; with mandibular gular pouch); G. Egyptian vulture (sharply-hooked flesh-eating beaks); H. Falcon (with tooth in the upper jaw for shearing flesh); I. Crow (with cutting beaks); J. Jungle Crow (with cutting beaks); K. Parrot (fruit-eating beaks); L. Hornbill (fruit-eating beaks); M. Crested Serpent Eagle (flesh-eating beaks); N. Kingfisher (fish-hunting and gripping beaks); O. Coromorant (fish hunting and gripping beaks); P. Spoonbill (mud-probing beaks); Q. Avocet (mud-skimming up turned bill)

9. **Drilling, drumming and wood-chiselling beaks:** The woodpeckers have a remarkably thick, strong and gradually narrowing elongated bill which, at the rostral end, appears sharp and chisel-shaped. The woodpeckers drill and bore through the bark of the wood, striking it with great force accompanied by amazingly backward and forward jerks of the head.

   The intermediate series of the cervical vertebrae are each provided with a pair of **Pr. caroticus** rostro-ventrally. In the woodpeckers and in a few other birds, this paired processes become ankylosed, forming a complete osseous canal through which passes the internal carotid artery. The ankylosis of the Pr. caroticus helps the woodpecker in throwing its head forward with great force in the act of drilling, drumming and chiselling wood.

10. **Beaks adapted for nectar-feeding:** Yet there are a variety of small birds whose beaks, tongue and flight have been marvellously adapted for probing flower-tubes and feeding principally on flower nectar.

The Australian Honeyeaters (Meliphagidae), the Humming birds of South America (Trochilidae) and the Sunbirds (Nectarinidae) of our country have the unique adaptation of hovering flight to remain stationary for quite some time at the level of the flowers by which these birds—with the help of their long, pointed and downwardly curved bill and a long, tubular tongue—probe into the flower tubes and suck in nectar. The tongue, being covered with bristles, and very often bifid at the tip, can be rolled up from the sides into a tubular structure—a very suitable adaptation for sucking in nectar.

The Flowerpeckers (Dicaeidae) and the White-eyes (Zosteropidae), having comparatively shorter, slender bill, do also largely subsist on flower nectar, besides feeding on insects, spiders and other invertebrates (Dicaeidae), as well as fleshy pulp of fruits and berries (Zosteropidae).

Among the nectar-feeding birds, the Humming birds are the smallest, sometimes attaining the size of barely 2 inches (5 cm) or even less. The Giant Humming bird *(Patagona gigas)* of South America may attain larger size up to 9 inches (23 cm) or more and its long, slender and straight beaks may measure even up to 4.5 inches (11.5 cm).

The Sicklebill (*Epimachus* sp., Fam.Corvidae; subfam.Corvinae; Tribe Paradisaeini) of New Guinea has a slender bill curved downward. This bird, as well as the Magpie-Robin (Turdidae), the Green Bulbul (Pycnonotidae), the Blackbird (Turdidae) and quite a few other species of our country feed mainly on the insects, spiders, other invertebrates, some fruits and berries, but they also feed on flower nectar.

Thus it is observed that the modifications of beaks and tongue in exclusively (or nearly so) nectar-feeding habit have been much more pronounced than in others with mixed feeding habits.

The hard horny tissue of the bill is normally lost by wear and is continuously replaced. King and McLelland (1984) report that in caged birds, where normal wear may not occur, the bill may have to be clipped—the process being known as 'Beak trimming'—to prevent feather pecking and cannibalism.

18.1.3. Bill Functions

As already stated, the versatile functions of the reptilian forelimbs have been largely compensated by the avian bill. No doubt the beaks of birds are principally used as food-getting device (pecking, or, by other means–grasping and manipulation). Their other functions are exhaustive collection of materials and arranging them in specific designs for the purpose of nest-building, preening and cleaning the feathers with the secretion of the oil gland (preen gland) which also makes the feathers waterproof, and using them for offensive and defensive purposes as and when needed.

18.2. CLAWS

The embryogeny of an avian claw is not different from that of a reptilian one. In birds, the terminal phalanx of each digit of the foot is enclosed by a horny claw. It is formed of a very hard and strongly keratinised dorsal plate that builds up the dorsal ridge and lateral walls of the claw and a softer ventral plate that forms the sole of the claw. As the dorsal plate grows faster than the ventral plate, the claw curves downward in varying degrees.

In the pigeons, as well as in the great majority of birds, the digits of the wings are devoid of claws. However, Van Tyne and Berger (1976) describe that vestigial claws occur in the 'hands' of many birds. These wing claws are horny sheaths that articulate with the distal end of the phalanx of one of the digits.

In *Opisthocomus* sp. (Order Galliformes) and in many adult ratite birds, the digits of wing are provided with claws. The large digital claws were also present in *Archaeopteryx*. Wing claws are usually present in digit I, but in *Opisthocomus* and in newly-hatched anseriform birds, they are present in digit II. Van Tyne and Berger (1976) report that the vestigial claws have also been found in the young of the African Finfoot *(Podica senegalensis)*, as well as in the adult of several passerine birds, White-necked Crow (*Corvus leucognaphalus*, and Red-winged Blackbird *(Agelaius phoeniceus)*.

In response to the enormous varieties of the environment, food-habits and modes of locomotion, appropriate modifications have taken place in the structure of the feet, toes and claws in birds.

In shorebirds, waders and aquatic birds, the toes and their claws have been variously modified. Webs between the toes, either fully covering the toes (eg., anseriform birds) or partially (e.g., some storks, avocets, etc.) are of common occurrence. The partial webs form 2-5 lobes in each toe of the grebes (Podicipediformes) and coots (Gruiformes). The lobate toes are peculiar characteristics of these birds. The claws in these birds are short to medium-sized, whereas, in leaf-walking birds like pheasant-tailed Jacana and Bronze-winged Jacana, the toes and claws are light-built and exceptionally long. The pelicans and the cormorants have webs between 2nd, 3rd and 4th toes, each ending in a short claw. The anseriform birds, for the purpose of swimming and paddling in water, have their toes fully webbed but with short claws. Most galliform birds have medium-sized strong claws for scratching soil as well as running on the ground.

The **raptors** such as the falconiform and strigiform birds have thick and strongly developed toes provided with very strong, hard and sharply curved claws called talons. These help in grasping and anchoring the fast-moving preys.

18.2.1. Spurs

In most Phasianidae (order Galliformes) including the domestic fowls and turkeys, some pheasants, francolins and patridges a spur occurs as a bony projection on the caudomedial surface of the tarso-metatarsal region. It is particularly well-developed in males and reduced in females. In fact, the age of the male domestic fowl can be calculated on the basis of the growth of the tarsal spur which lengthens about 0.4 in. (1 cm) per year and reaches a maximum of 2.4 in. (6 cm)only. In some phasianidae, more than one spur may be present (e.g., Indian game fowl) on each leg (See Fig. 20.1.H).

Yet another type of spur developing from the carpo-metacarpal region and known as the wing-spur occurs in some birds like the Spur-winged goose, the Spur-winged Lapwing, Jacanas, some plovers and in two species of ducks (Van Tyne and Berger, 1976). Tarso-metatarsal spur differs from the wing-spur in that the former has a projecting bony core with an external investment of horny material derived from the epidermal skin, whereas the latter is entirely formed of claw-like horny material. Rand (1954) and Weller (1968) report that the outermost horny layer of the wing-spur is subject to annual moult.

Functions: The epidermal sheath of the spur may be very hard, sharp and pointed. The birds possessing spurs use them in fighting with one another. In fact the spurs are used more as offensive organs than defensive. The fighting among galliform birds, including domestic fowls, is a spectacular entertainment for the villagers.

18.3. SCALES

The epidermal scales of reptiles were replaced by feathers in birds in most parts of their body except at the lower parts of their legs and feet (tarso-metatarsal region) and at the base of the beak where the scales persist. The non-feathered skin of the foot is termed as podotheca. King and McLelland (1984) define that the scales are raised areas of highly keratinised epidermis separated by folds of less keratinised skin.

The scales are usually arranged in overlapping manner as in snakes and lizards. In some species, however, feathers are observed between the scales in the region of podotheca. The scale patterns are of considerable interest in taxonomic studies. In most passerines, only a small number of scales are present at the bases of the toes, the foot in such case being known as holothecal or "booted". Perfect overlapping of scales is observed in the Kingfishers and Woodpeckers. In Pallas's Sandgrouse, Ptarmigan, Owls and quite a few other birds, the foot is almost entierly covered by feathers.

The scales are of little functional significance except that they provide some additional rigidity and protection to feet and toes.

18.4. FEATHERS

Perhaps the most magnanimous gift of nature to the class of birds is their feathers (also called plumage) which perform multiple functions, besides their main role in flight. And the ability to fly is to be considered as the sum total effect of several other factors that are largely controlled by feathers. Such factors are lightness of the body, buoyancy, thermoregulation, etc. The skin of birds is very thin and has no glands, except the uropygial or preen gland situated near the tail. Its oily secretion is smeared over the feathers by the beaks, thus rendering them waterproof.

'In general, there are three main types of feathers, filoplumes, or hair-feathers; plumulae, or down feathers, and plumae, or contour feathers' (Weichert and Presch,

1978). This is considered to be a broad classification of feather types in which the flight feathers have been included under contour feathers and several other types of feathers have been considered as modifications of either the filophemes or the contour feathers. Bologna (1995; In: Macdonald Encyclopedia of Birds of the World) has distinguished one more type of feathers, the semiplumes. Stettenheim (1972); Pough, Heiser and McFarland (1990, 1996) have classified feathers into five types: **Contour feathers** (including **coverts** and **flight feathers**), **Filoplumes, down feathers, semiplumes** and **bristles.**

Young (1981) comments that in adult birds, the semiplumes are the down feathers whereas the authors mentioned above consider the semiplumes to be halfway between the contour feathers and the down feathers. Most of these authors also consider that the bristles are modifications of the filoplumes. However, to describe the feathers in details, the flight feathers, the bristles and other types have been described here separately.

18.4.1. Develpoment of Feather: A General Account

The study of the development of a feather in general strongly indicates its similarity (and hence its homology) with the epidermal origin of scales in lizards and snakes.

Initially, the germ cells developing a typical down feather appear as papillae of dermal cells (dermal papilla) pushing the overlapping epidermis up ahead of it. The projection of the papilla is rather cylindrical, and not flat as in the development of a reptilian scale. However, the process of development of a reptilian scale and that of a feather is basically similar. Around the base of a feather papilla, an annular groove appears. This groove later forms the feather follicle. The malpighian layer of the distal part of the feather germ proliferates in such a way as to form a number of vertical radiating ridges, the barbs of the shaft. In fact, these vertical ridges—which extend through the dermal pulp arise from a collar of cells of the malpighian layer.

The outer horny layer of the epidermis forms a temporary sheath, the periderm, which is later thrown off as the feather grows and expands. The feather germ with its peridermal sheath grows fast and it projects from the feather follicle outside the skin. As the feather grows, the dermal pulp and the germinal layer of epidermis gradually recede. The periderm at the apex of the growing feather splits and the distal ends of barbs peep outside through the region of the breaking periderm sheath.

As the periderm gradually perishes the barbs spread out. The bases of the barbs by now separate from one another, become cornified and remain attached to the quill. A small papilla remains at the base of the feather. This gives rise to another feather later on.

The development of a contour feather is basically similar to that of the down feather as described above. Here the process is more complex. Because, in the development of a contour feather, the growth of the feather becomes differential on one side of the collar of the germinal layer. Further, the barbs also arise from the sides of the rachis, instead of arising directly from the quill as in the down feather.

Feathers arise from the complex beta keratin layer of cells of the epidermis that gives rise to scales in lizards.

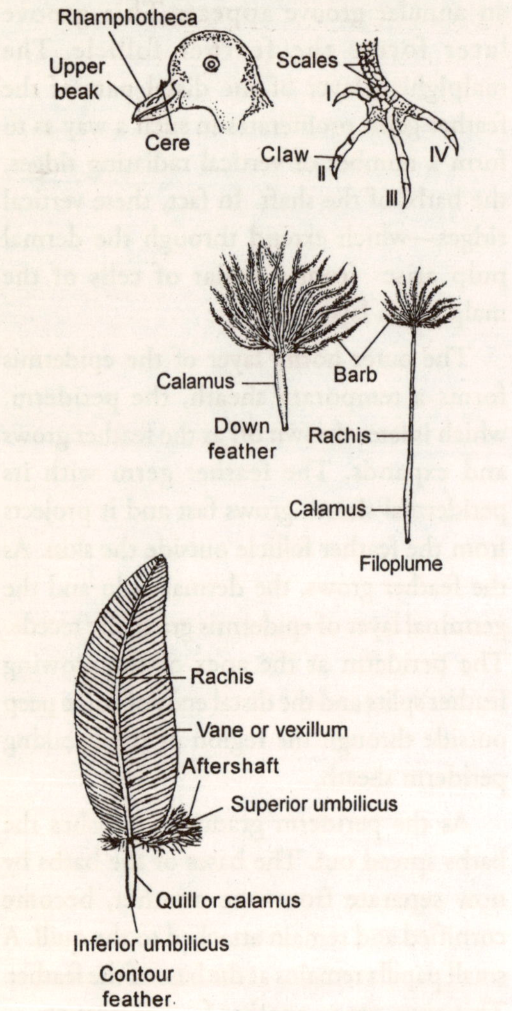

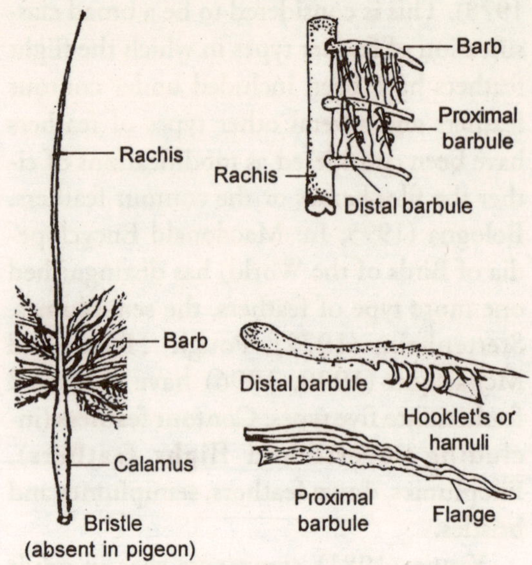

Fig.18.2: Beaks, claws, scales and feathers in a Pigeon (as an example of a generalised bird)

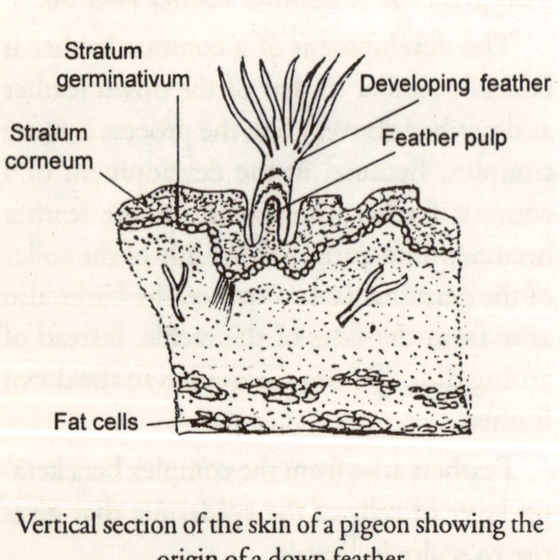

Vertical section of the skin of a pigeon showing the origin of a down feather

18.4.2. Structure of Different Types of Feathers

Based on structural details, the feathers may be classified into the following types:

1. **Plumulae or Down feathers:** After plucking off the larger covering feathers from the body surface, the adult birds are found to possess a kind of small, soft woolly feathers called 'down feathers'. Birds, at a very early stage of their development, are covered with a sort of very soft, temporary **'nestling down'** feathers which provide excellent insulation to the body.

Structurally, a down feather has a short, hollow quill, reduced shaft and long flexible barbs with short barbules and without barbicels (hooklets). Such numerous flexible barbs arising from the top of the quill give the appearence of a woolly tuft. The barbs bear very short, slender barbules along their edges.

The **nestling down feathers** are replaced by contour feathers in the adult birds. In most adult birds, the down feathers lying beneath the layer of contour feathers are known as **'powder down'** feathers, because, in this case, the tops of down feathers break down into a powdery substance. The powder down feathers are especially abundant in many aquatic birds like herons, ducks, swans, etc. and in many birds of prey.

The powder down feathers form a thick, as well as warm insulating layer that helps in controlling body temperature and warmth in incubation.

2. **Filoplumes or Hair feathers:** In a plucked bird, a number of minute, hair-like delicate feathers of unknown function are found distributed over the body along with down feathers. These are called filoplumes or hair-feathers or pin-feathers. The structure of a filoplume is rather simple, as it consists of a short calamus and a long threadlike rachis or shaft which bears a few weak barbs and barbules at its distal end. There are no hooklets. The base of the calamus lies in the feather follicle of the skin. The filoplumes are found scattered over entire surface of the body beneath the contour feathers.

The filophemes attain unusual length in peacocks.

3. **Bristles:** It is believed that a modification of the filoplumes—called rictal bristles—are possessed by some birds at their rictus i.e., around junction of the upper and lower beaks. Each bristle has a short calamus and a long rachis which bears a few barbs and barbules at its base i.e., at the junction of the calamus and the rachis. The rest of the rachis becomes stiff, completely devoid of barbs and it gradually narrows towards the distal end. In Muscicapidae (e.g., Flycatchers) and Caprimulgidae (Nightjars, Whip-poor-will, etc.), the rictal bristles are quite prominent. These help the birds in catching insects-preys in air.

4. **Plumae or Contour feathers:** The large majority of feathers that provide general covering of a bird's body and maintain its outline shape or contour are called contour feathers. The larger flight feathers of the wing and tail and the smaller wing and tail coverts are regarded as specialised type of contour feathers.

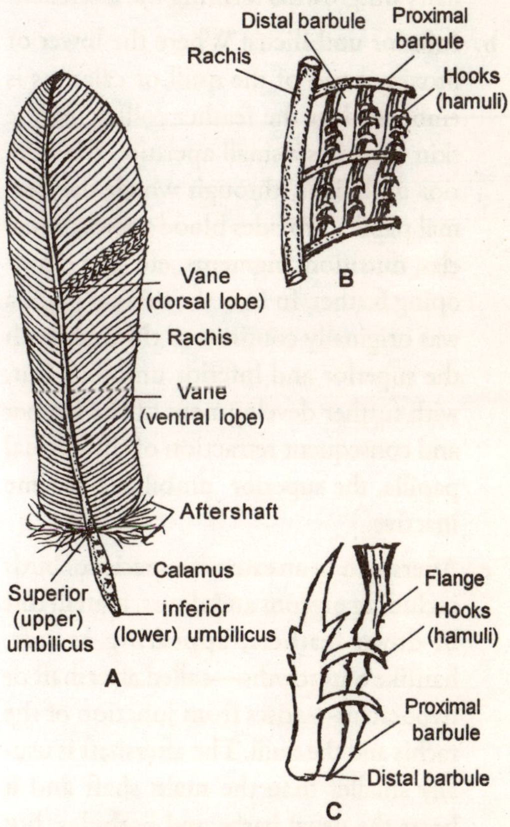

Fig. 18.3: A generalised flight feather.
A. Parts of a flight feather; B. A portion of the vane showing interlocking arrangement of the barbules; C. Enlarged view of a distal and a proximal barbule

A typical contour feather consists of two main portions: a hollow, almost transparent stalk, the **quill** or **calamus** embedded in the skin and an expanded, leaf-like distal portion, the **vane** or **vexillum**, supported by the axial rachis which is an extension of the calamus.

i. Quill or Calamus

The hollow and slightly swollen basal part of the feather devoid of barbs is called the calamus. It consists of these structures:

a. **Superior umbilicus:** It is a minute aperture situated ventrally on the quill, at its junction with the rachis. This aperture is mostly covered by a tuft of delicate hairy outgrowths forming the aftershaft.
b. **Inferior umbilicus:** Where the lower or proximal end of the quill or calamus is embedded in the feather follicle of the skin there lies a small aperture, the inferior umbilicus, through which the dermal papilla provides blood vessels, muscles, nutrition, pigments, etc. to a developing feather. In fact, the dermal papilla was originally continuous through both the superior and inferior umbilici, but, with further development of the feather and consequent retraction of the dermal papilla, the superior umbilicus became inactive.
c. **Aftershaft:** In an extensive species of birds including pigeons and doves, a small tuft of down feathers, appearing as soft, hairlike outgrowths—called aftershaft or hyporachis—arises from junction of the rachis and the quill. The aftershaft is usually smaller than the main shaft and it bears the usual barbs and barbules, but the hooklets are generally absent. In some ratite birds, such as emu, cassowary and in the extinct *Dinornis* the aftershaft is very well-developed and is as long as the main shaft. This gives the appearance of a double feather.

ii. Vane or Vexillum

It is the leaf-like expanded part of the feather, supported by a midrib-like solid longitudinal axis, the rachis. The rachis divides the vane into two unequal lateral halves and from the base i.e., from its proximal quill-end, the rachis gradually narrows down towards its distal tip. From the superior umbilicus to the distal tip of the rachis, a mid-ventral groove runs along the whole length of the rachis. This is known as the **umbilical groove**. In transverse section, the central core of the rachis is found to be compactly filled up with a mass of pith cells.

The vane is formed of a series of closely-set, parallel thread-like structures, the **barbs** which arise somewhat obliquely rostro-lateral from the two lateral sides of the rachis. The barbs become gradually shorter in length distally. Each barb in its turn gives rise on either side of it (ie., on both proximal and distal sides) to a row of more delicate, oblique filaments, the **barbules**. Thus the barbules, according to their site of origin on the barb, are of two types: the **distal barbules** directed towards tip of the feather and the **proximal barbules** directed towards base of the feather.

The rostral half of each distal barbule bears along its lower edge a number of curved, minute **hooklets** (also called **barbicels** or **hamuli**), while rostral two-third along upper edge of the proximal barbule curls inwardly forming a **groove or flange**. As the barbules are obliquely directed forward or backward, they overlap causing anchorage of the hooklets of the distal barbule with the flanges of the proximal barbules of the preceding adjacent barb. Thus an efficient interlocking arrangement among the barbules provides a flat, wide and unbroken surface of the vane which is helpful for displacement of air in flight. However, at or near base of the vane, many distal barbules may lack in hooklets. If the barbs split and the interlocking arrangement is damaged, the bird is capable of repairing the damaged part of feather through preening with the help of its beaks and restore original position of the barbs.

The ratite birds generally lack in hooklets in their barbules, and, hence, are incapable of sustained flight.

5. **Flight feathers:** The large feathers of bird wing and tail directly taking part in flight are known as flight feathers. These are specialised contour feathers to which they bear structural similarlity and from which they can be distinguished by their longer and prominent hookelts. The remiges and rectrices are the flight feathers of the wing and tail, respectively.

i. Remiges or Wing-quills

These are large flight feathers of the wing and usually have a definite number, such as 23 (with exceptions in a number of species), which are attached along the hinder border of each wing. Of the 23 wing-quills, the 11 attached to the hand (7 metacarpals and 4 digitals) are called '**Primaries**' and the 12 attached to the tendon and muscles of the ulna (forearm) are called '**Secondaries**' or '**Cubitals**'. A small tuft of feathers, attached to the pollex or thumb, is called '**ala spuria**' (**alula or bastard wing**'). By controlling small air-currents, the alula helps the rectrices in their landing manoeuvre. In all the remiges, the portion of feather inner or posterior to the vane is broader than that of the outer or anterior to the vane. The alula consists of 3-4 pollical remiges.

Both the primary and secondary remiges are variable in number in different birds, the primaries ranging from 9-12 and the secondaries vary from 9 (in small passeriform birds) to 40 (in albatrosses).

ii. Rectrices or tail-quills

The rectrices or the tail-quills are very important in steering through air and in landing on a perch or on the ground. In piegon, there are 12 large rectrices arranged semicircularly around the uropygium. The rectrices are also quite variable in different birds, being very short or absent in some ground-dwelling birds as tinamous and quails, but very large in size in the birds which fly speedily, as in the swallows. In these birds, the outer rectrices are enlarged for steering. The attractive colour patterns displayed by a peacock through fanning out the rectrices is an example of one of the special uses of tail coverts and tail quills.

Landing in birds is also an intricate process, particularly when a bird suddenly comes to rest on a perch. In such an event, the tail quills are lowered and fanned outwardly a process that acts as a brake.

iii. Coverts

These are smaller wing feathers located at the base of the wing-quills and tail-quills and are usually arranged in one to several rows, lying parallel to and in the direction of the flight feathers. The coverts are present on both the upper and under surfaces of the wing and tail feathers, being known, respectively, as the **upper and under wing coverts** and **upper and under tail coverts**. The wing coverts, usually arranged in several rows, are larger distally and smaller proximally. Accordingly, they have been named as greater wing coverts, median wing coverts, and lesser wing coverts. The coverts may also be mentioned as primary and secondary wing coverts.

The coverts close the interstices between the adjacent flight feathers, thus providing a continuous surface that controls buoyancy of air in flight and also strengthen the bases of the flight feathers.

6. **Semiplumes:** A reference to these feathers has already been given at the beginning of the description of feathers. Although Young (1981) comments that 'the semiplumes are downy feathers of the adult, lying beneath the contour feathers and giving warmth and shape', most authors (Stettenheim, 1972; Bologna, 1995; Pough, Heiser and McFarland, 1990, 1996) recognise them as a distinct type–halfway between the contour feathers and down feathers. Structurally, the semiplume has a completely plumulaceous vane and a long rachis that is longer than the longest barb of a down feather.

FUNCTIONS OF FEATHERS

The feathers are not only the most diagnostic features in birds, but they also perform some remarkable functions in the life of birds.

Some essential functions of feathers:

1. Just outside the body of a carinate bird, as in the pigeon, the feathers form a smooth texture of light yet rigid, flexible and dry coating.
2. In the pigeons and in all flying birds, the wing quills and the tail quills—due to their lightness, rigidity, and flexibility and with the attributes like keratinised composition, and **barbs and barbules with hooklets (hamuli)**—function as a superb aerofoil that maintains buyoancy in flight.
3. Heat conservation and temperature regulation are also very important functions of feathers which, being non-conductor of heat, retain the high degree of body temperature (average 40°C/104°F characteristic of birds.
4. The feathers, being smeared with the secretion of preen gland, become waterproof; feathers also protect the soft skin underneath from environmental hazards and frictional injury.
5. The diverse colouration of feathers in birds performs several important functions; it acts as protective colouration to some birds which, being matched with the environment, protect them from the attack of predators; the brighter colouration in the male species helps to attract the female during breeding season when such colours become most conspicuous in the crown, throat, breast and in the wings and tail.
6. The feathers are widely used by the birds themselves in nest-building and by human beings in the preparation of various usable items as well as articles of decoration.

19

Pterylosis

The feathers in all carinate birds (except in penguins) are not distributed uniformly all over the body of a bird; there are certain distinct areas of feather tracts called '**pterylae**' which may be differentiated from the featherless tracts called **apteria**. The apteria, however, may contain some sparsely distributed down feathers and filoplumes.

The distribution and arrangements of feathers over the skin of a bird is called **pterylosis** which is variable in different groups of birds. In external view, the entire body of a bird appears to be covered with feathers which—by their longer size and overlapping arrangement—cover up all the apteria. The pterylae and the apteria in many other birds are much more distinct than in the pigeon. Thus the variable form and arrangement of pterylosis in different birds provide important clues in the classification of birds. Clench (1970) reports that pterylosis within the individuals of the same species remains practically the same, but considerable differences exist between the pterylae of adults and those of nestlings and juveniles of a species. Pterylosis can be best observed through **X-ray pterylography** of the tracts.

The nomenclature of pterylae and apteria in different birds has been mainly based on morphological locations of the body. The pteryla capitalis (also known as cephalic pteryla or head tract) are in the head region, pteryla cervicalis in the neck, pteryla scapulo humeralis in the shoulder and humeral regions, pteryla dorsalis (also known as pteryla spinalis or spinal tract) along the entire dorsal region, pteryla ventralis along the ventral region, the femoral and crural pterylae along the thigh and shank, respectively. These are some of the feathered tracts in the body of a bird. The nearly bare skin in the apteria is protected from external injury by the feathers lying close to the apteria.

In the flightless ratite birds, apteria are usually found only in the young, the adult birds possess a uniform covering of feathers. In the ratitae, the feathers are downy and the barbs and barbules are without hooklets. In the penguins, the wing feathers are degenerate.

The patterns of the pterylosis of a pigeon and a generalised passerine bird have been shown here for an easy identification of the pterylae and apteria.

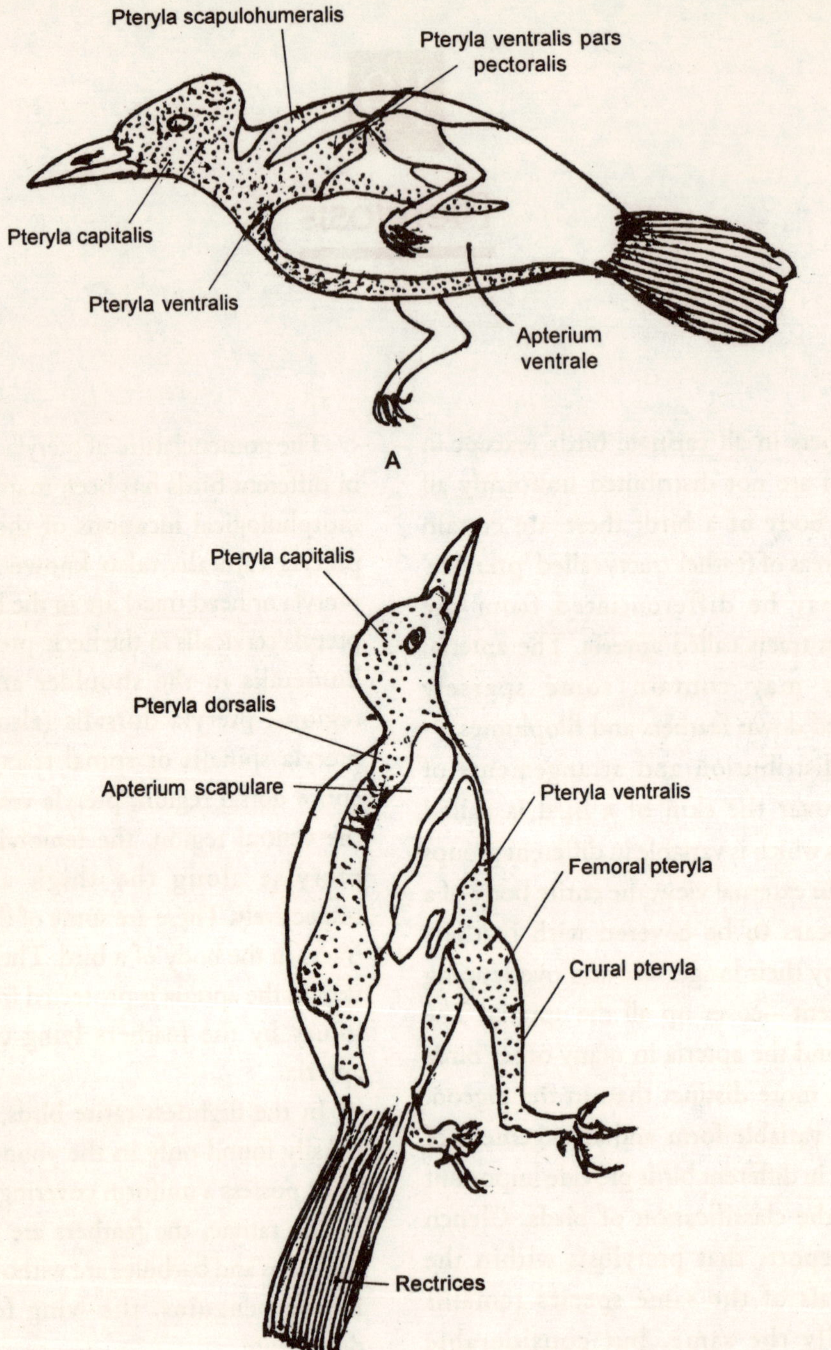

Fig. 19.1: Pterylae and apteria in a generalised passerine bird. A. Latero ventral view; B. Dorso Lateral view (Based on and modified from NAA, 1993)

20

Modifications of Feet and Toes

While describing the claws it has already been mentioned that the feet and toes in birds have been variously modified in nature in response to their habitats, different modes of locomotion and feeding habits. In the course of evolution of birds from the possible coelurosaur-like dinosaurian stock, the rudimentary 5th digit present in most dinosaurs appears to have been completely lost in birds—through 60 millions years of evolution. Thus the feet in birds possess four toes of which the first or the hallux is, usually, directed backwards. However, reduction in the number of toes is quite common among the ratite birds which have been perfectly adapted to running on the ground. Among the carinate birds, diverse modifications have taken place in feet and toes adapted to life on land, trees, water and aquatic vegetation which provide an enormous versatility in food niches.

The modifications of feet and toes are briefly described under the following heads:

20.1. RATITE FEET AND TOES FOR FAST RUNNING ON GROUND

Most flightless ratite birds have their feet quite tall and stout and usually with three toes directed forward [e.g., *Rhea*, *Dromaius* (Emu), *Casuarius* (Cassowary), *Eudromia* (Tinamou)].

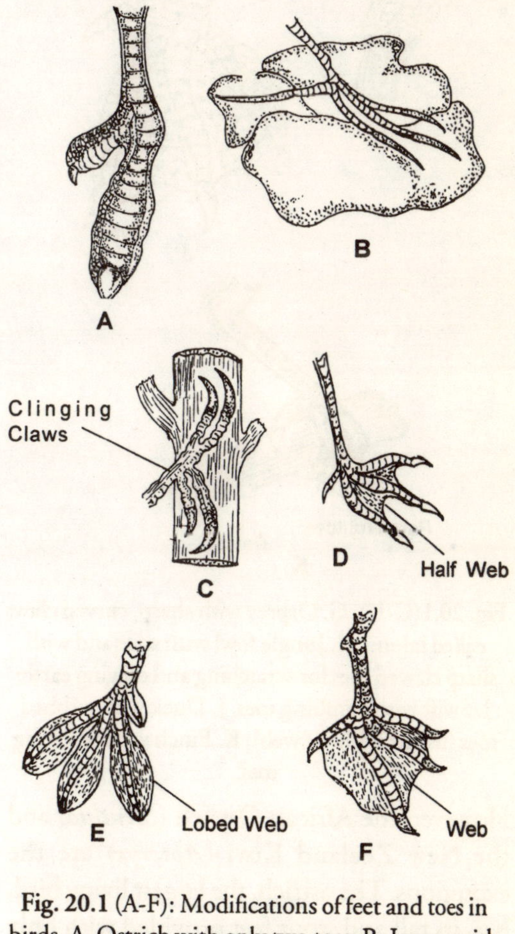

Fig. 20.1 (A-F): Modifications of feet and toes in birds. A. Ostrich with only two toes; B. Jacana with long light toes for walking on aquatic leaves; C. Woodpecker with climbing and clinging toes; D. Avocet with toes partially webbed (half-web); E. Grebe having toes with lobed webs; F. Cormorant having toes fully webbed

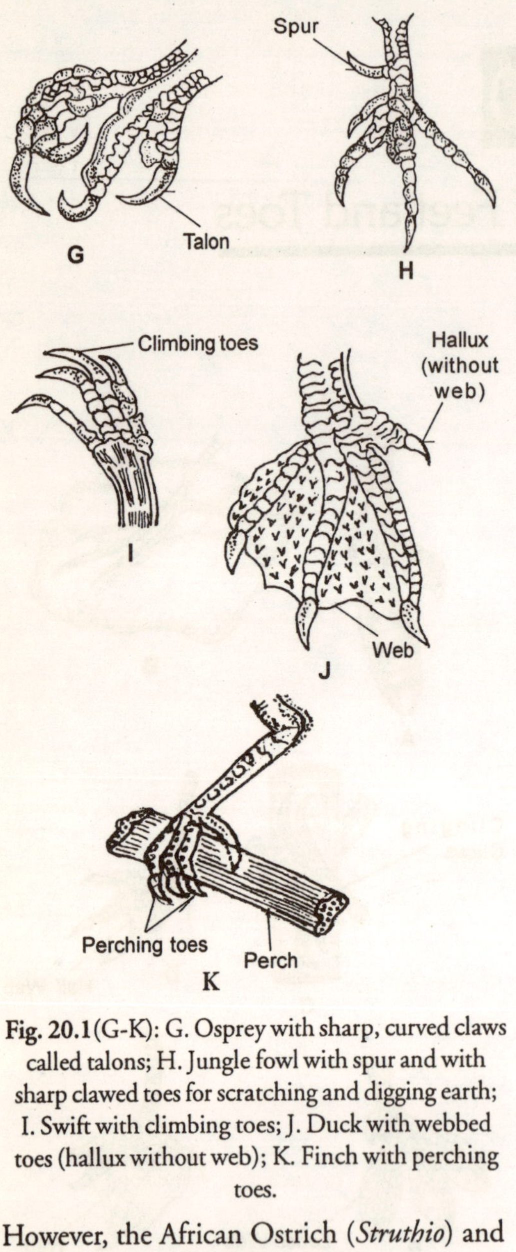

Fig. 20.1(G-K): G. Osprey with sharp, curved claws called talons; H. Jungle fowl with spur and with sharp clawed toes for scratching and digging earth; I. Swift with climbing toes; J. Duck with webbed toes (hallux without web); K. Finch with perching toes.

However, the African Ostrich (*Struthio*) and the New Zealand Kiwi (*Apteryx*) are the exceptions. The ostrich, the largest living bird, has, its tall and stout feet provided with only two toes—the enlarged middle or the 3rd toe having a flat nail at its tip and comparatively smaller outer or the 4th toe having a claw. Both the 1st and 2nd toes are absent. (Fig. 20.1.A). In the Kiwi, the feet are comparatively shorter and provided with all the 4 toes, the 1st being much reduced.

The following are the different modifications of feet and toes in carinate birds.

20.2. SCRATCHING AND RUNNING TOES IN GALLINACEOUS BIRDS

In different species of jungle fowls, domestic fowls, partridges, quails and other gallinaceous birds, all the four toes are provided with sharp claws which are used for both running on the ground and scratching and digging earth to make small holes. Further, the males of certain species of fowls develop a sharp blade-like **spur** on one side of the foot above the toes (in a line with the hallux). The spur is used for both offensive and defensive purposes (Fig. 20.1.H).

20.3. TOES ADAPTED FOR PERCHING ON TREE-BRANCHES

The great majority of passerine birds (and some from other Orders too) are perfectly adapted to perching on tree branches. A typical perching foot of our common passerine birds (e.g., *Passer*, *Corvus*, *Copsychus*, *Dicrurus*, *Oriolus*, etc.) has certain distinctions. It is characterised by four toes of which the first one or the hallux is directed backward so as to grasp the tree branch from behind, while the three other toes, directed forward, bend inwardly–to grasp the tree branch firmly. All the toes are slender and provided with curved claws. The hallux is thick, strong and apposable which ensures a rigid grasp of the tree branch. After a period of flight the bird sits on a perch for resting or just hops on a tree branch from one place to other. This kind of movement is known as **perching movement.**

In this intricate mechanism of perching, the muscles and tendons of the feet play the most significant role. In perching birds the muscles of the feet are very well developed. Attachments of these muscles are mostly performed by long, cylindrical but quite strong tendons that either flex the toes or extend them. As such, the perching muscles are divided into two major groups–the flexors, and the extensors. These muscles help birds not only in perching but also in walking on land. These muscles are briefly described:

20.3.1. Flexor Muscles

i. **M. fibularis (peroneus) longus:** The muscle arises from the proximal end of tibia near its junction with the fibula and stretches along the fibula and, distally, its long tendon combines with that of M.flexor perforatus digiti III. The tendons are effective flexors of the toes (Fig. 20.1.K).

ii. **M. fibularis (peroneus) brevis:** This muscle, shorter in length than the above-mentioned 'longus' muscle, arises from the tibial joint at the knee and is attached by a tendon at the inner aspect of the tarso-metatarsus bone. This muscle maintains flexibility of the tarso-metatarsal region.

iii. **Mm.flexores perforantes et perforati digiti II et III and Mm. flexores perforati digiti II, III et IV:** Such groups of flexor muscles have been named differently, based on their location, nature of tendons and their actions in flexing the toes. Among the passerine birds–as also in the pigeons and many other birds–the foot has the prominent feature that its second and third digits are each provided with three flexor tendons that flex the toes around a tree-branch during perching of a bird.

iv. **M. flexor hallucis longus and M.flexor hallucis brevis:** These are the two flexor muscles of the hallux which, unlike in other charadriiform birds, is much elongated in the leaf-walking Jacanas. Andrei Zinoviev (2002; IOC, Beijing, China), while studying the evolution of foot in Jacanas, reports that M. flexor hallucis longus, the main flexor of the hallux, has lost its connection with the hallux, while that of M. flexor hallucis brevis persists, changes its insertion distally and functions as the only flexor muscle of the hallux. Zinoviev also asserts that the extraordinary elongation of the hallux in Jacanas is a secondary acquisition.

20.3.2. Extensor Muscles

Just as the flexor muscles are required for anchorage in perching on a tree branch, the extensor muscles are equally important in dislodging the anchorage while taking off the tree branch.

A few such extensor muscles are:

v. **M. gastrocnemius:** This is a large, powerful and strong calf muscle of the tibiostarsus. The muscle is formed of three parts: pars lateralis, pars intermedia and pars medialis–arising, respectively, from the lateral condyle of the femur, medial aspect of the tibia and from the ilium. The long tendons of these muscles run along the inner aspect of the tarsometatarsus and then combine with the flexor tendons of the peroneus muscles at the bases of the phalanges. This muscle performs a dual role, acting both as extensor and flexor muscles.

vi. **M. extensor digitorum longus:** This muscle arises from the cnemial crest of the tibiotarus and its branched tendons are inserted at the bases of the second, third and fourth toes. Contraction of this muscle extends the above-mentioned toes, thus loosening their grip on the tree branch before the bird takes off.

vii. **Mm. extensor hallucis longus and extensor hallucis brevis:** Both the muscles arise from the tibiotarsus and their tendons are inserted at the base of the first toe i.e., hallux. Both the muscles act as extensors of the hallux.

As the bird perches on the tree branch, its knees bend inward, resulting in a pull on the tendons of the flexor muscles. This, in turn, folds the toes which strongly grasp the tree branch. With such an anchorage of the toes, the bird may rest for several hours and does not fall–even in sleep!

20.4. FEET AND TOES IN BIRDS OF PREY

Raptors predating on live animals like fishes, lizards, snakes, small birds and mammals and feeding on the carrion of dead animals are provided with prominent morphological adaptations which include robust feet and strongly curved and sharply hooked claws called **talons** as in vultures, eagles, hawks, kites; large, fleshly bulbs called **tylari** present on the undersurface of the toes as in the sparrow hawk of Europe and Asia (Fam: Accipitridae); rough skin on the undersurface of lower feet and toes provided with **sharp teeth-like horny spines** as in **Ospreys**; **strong feet and toes** as in **Goshawk** (Fam: Accipitridae) that hunts and kills mostly birds and also some lizards, hares and mice on the ground. These are few examples among many falconiform birds which have diverse modes of hunting preys with the help of their strong feet, large, sharp and curved claws, acuity of vision and speed in flight while swooping on the prey (Fig. 20.1.G).

20.5. FEET AND TOES ADAPTED FOR CLIMBING AND CLINGING

The feet and toes of birds which climb trees or vertical wall have been suitably modified for grasping or clinging to such surfaces. In climbing birds, the 4th digit is usually directed backwards for grasping. But for more efficient grasping and climbing a vertical surface, both the 4th and 1st digits of many birds are modified to form a forcep-like structure with curved claws. Such an arrangement—called **zygodactylous feet**—is observed in birds like woodpeckers, cuckoos and parrots. In trogons, the hallux and the 2nd toe turn backwards—an arrangement known as **heterodactylous feet.** The arrangement in toes may be variously modified. In most coraciiform birds (e.g., Kingfishers, Bee-eaters), either all the front toes or the middle and outer toes may be united at their bases up to certain length. In certain apodiform (e.g., Swifts, Humming birds) and coliiform birds (Mouse birds), all the four toes may be directed forward—a condition that facilitates climbing vertical surfaces—this arrangement of toes being known as **Pamprodactyl feet** (Fig. 20.1.C).

20.6. MODIFICATIONS OF FEET AND TOES FOR WADING (I.E., WALKING ON AQUATIC VEGETATION AND WETLAND)

In a few ciconiiform (e.g., Herons and Storks) and in most charadriiform birds (e.g., Jacanas, Plovers, Snipes), the feet become narrow, lightweight with elongated toes bearing thin and long claws which are perfect adaptations for movement on leafy aquatic vegetation or in marshes and muddy wetlands. Usually, the hallux is directed backward while the three other toes are directed forward and spread apart in leaf-walking birds, so that the pressure on the leaf is uniformly distributed and balanced, due to radical expansion of the area of contact between foot and the floating substrate. This prevents sinking of the bird down the floating leaf. The expansion of the area of contact of the foot on the floating leaf is achieved by extraordinary elongation of all the toes and claws.

In truly leaf-walking birds, the web between the toes is absent, but in some wading birds, webs may develop to a small extent. In Water Rails (Fam: Rallidae; Order Gruiformes) the legs are provided with long toes and claws. Webs are absent (Fig. 20.1.B).

20.7. MODIFICATIONS OF FEET AND TOES FOR SWIMMING AND DIVING

A large number of birds are very good fliers as well as swimmers. In most aquatic birds belonging to the Orders Sphenisciformes (Penguins), Gaviiformes (Loons), Podicipediformes (Grebs), Procellariiformes (Petrels), Pelecaniformes (Pelicans, Cormorants), Ciconiiformes (some Storks), Phoenicopteriformes (Flamingos), Anseriformes (Ducks and Geese) and in some charadriiform birds (e.g., Avocet : partially webbed) the toes are webbed in varying degrees, either reduced and restricted to bases of the toes or extended along half or even the entire length of the toes. The webs are membraneous extensions of the cuticle extended between the adjacent toes in different degrees. The webs are used for efficient paddling through water in swimming birds. Webs are especially well-developed in ducks and geese. However, there are quite a number of aquatic birds that are very good swimmers even in the absence of webs.

In ducks and geese, gulls and, to a lesser extent, in some terns, the webs are fully extended between the three forwardly directed toes, leaving the backwardly directed hallux free, whereas, in the cormorants, all the four toes are fully webbed between the claws (Fig. 20.1.D, E, F, J).

21

Endoskeleton and Muscles

Of all body systems that have contributed to the unique flight adaptations in birds, the skeleto-muscular system holds the key position. This is because the endoskeleton in birds has acquired such marvellous distinctions as lightness, flexibility yet rigidity of the bony framework due to frequent fusion of bony segments, pneumaticity of bones due to the profuse air-spaces within them, and excellent buoyancy of the body endowed by the airsacs and their extensions. The entire bony framework has the mechanical advantage of a hollow girder from which are suspended the rib-basket, the sternum and the keel, the pectoral and pelvic girdles, and the skeletons of the wings and feet.

21.1. ENDOSKELETON

The endoskeleton in birds offers a very wide range of variations in conformity with the divergence in flight, food and feeding habits, habits and habitats, and various other environmental factors. However, the endoskeletal system in birds can be described under few broad heads:

ENDOSKELETON IN BIRDS

- AXIAL
  - Skull
    - Cranium
    - Sense capsules
    - Jaws
    - Hyoid apparatus
  - Vertebral column
  - Sternum and ribs
- APPENDICULAR
  - Pectoral girdle and wing skeleton
  - Pelvic girdle and hind limb bones

21.1.1. Axial Skeleton

Endoskeletal components as the skull, vertebral column, ribs and sternum form the axial skeleton.

21.1.2. Appendicular Skeleton

The bones of the pectoral girdle and forelimbs and of the pelvic girdle and hindlimbs which support the appendages of the body (such as the forelimbs and hindlimbs) and connect to the axial skeleton are known as appendicular skeleton.

As already mentioned, enormous variations exist in the form-function complexes of the endoskeleton in different birds; yet a uniformity is also maintained in the diagnostic features of a carinate bird. The salient features in the endoskeleton of carinate birds are described here briefly.

21.2. SKULL

The skull is a composite structure that includes the cranium proper, the sense capsules and the jaws and hyoid apparatus. As per recent nomenclature (NAA, 1993), the jaws, Ossa faciei comprising Quadrate, pterygoid, palatine, maxilla, premaxilla, nasal, Jugal, Quadratojugal, etc. and the Ossa mandibulae are designated as 'Facies' and the hyoid apparatus as the *'Apparatus hypobranchialis' (A. hyolingualis)*.

The avian skull, in general, possesses certain distinctive features:

i. The skull is rostrally elongated but centrally and caudally more or less rounded; skull compact but lightweight due to the presence of pneumatic bones.

ii. Rostral extension of the skull forms the characteristic beaks (upper jaw and lower jaw) or bill of birds; jaws toothless in modern birds.

iii. In the skull of adult birds, the bones are so completely fused that sutures are obliterated.

iv. Large orbits are very distinctive features of the avian skull; the two eyes may be equal to or even greater than the weight of the brain; a median interorbital septum separates the two orbits in almost all birds; Bellairs and Jenkin (1960) report that Kiwi (*Apteryx*) lacks such a septum. Eyes in most birds have limited mobility and those of owls are immovably fixed in large sclerotic rings.

v. The enlarged nasal capsules intervene between the orbits, much in the way as in most mammals.

vi. The large Foramen magnum is an opening at the base of the skull. It transmits the spinal cord and its meninges and is bounded by the supraoccipital, exoccoipital and basioccipital bones.

NOTE: For sketches on skull and other skeletal parts see Chapter 30.

vii. A prominent, single occipital condyle is mainly formed by the Os basioccipitale; minor lateral contributions are made by the exoccipitals on either side.

viii. Proc. paroccipitalis (Proc. exoccipitalis; Richards and Bock, 1973; Johnson, 1984) forming caudal wall of cavum tympanicum projects ventrolaterally in modern birds. This process is especially well-developed in *Gavia*, *Pelecanus*, *Anser* and *Caprimulgus*.

ix. In certain birds, the temporal region of the skull may be smoothly convex and, in the great majority of birds, lateral temporal fossa, either shallow or deep, may develop distinctly. The fossa serves as the origin site of some important jaw muscles (e.g., M. adductor mandibulae externus superficialis pars temporalis).

x. Birds possess characteristic autostylic jaw suspension in which the movable quadrate-pterygold-palatine complex is attached to the cranium in front and by a basipterygoid process behind. This condition appeared much early from the modern lung fishes.

xi. Majority of flying birds exhibit variable degrees of **upper jaw kinesis** in which the maxillae and premaxillae of the upper jaw undergo some upward and downward movement either as a whole at the craniofacial hiatus (also known as the **nasofrontal hinge**) or bend up and down at distinct '**flexion zones**' (zona elasticum craniofacialis) of the upper jaw. The flexion zones are also variable in different birds. The craniofacial zones, either single or double, develop particularly in **rhynchohinetic skull** with a **schizorhinal nostril** (e.g., Gulls, Plovers). The upper jaw in these birds bends at the flexion zones located either rostrally and/or caudally. In the majority of flying birds, the skull is prokinetic with a holorhinal nostril where whole of the upper jaw moves as a unit at the distinctly formed craniofacial hiatus or a flexion zone forming a hinge-joint known as the nasofrontal hinge. Other kinds of kinesis, such as **mesokinesis** and **amphikinesis**, also seem to have appeared in course of avian evolution. The upper jaw has virtually no movement in the ratite birds (e.g., Ostrich, Emu, Cassowary, Kiwi, etc.) and in the hawfinch (*Coecothraustes*) where the skull type is known as akinetic.

xii. Based on the relations between vomer and maxillopalatines, T. H. Huxley (1868) divided avian skull into four principal types:

a. **Dromaeognathous**—with large and broad vomer; palatines not connected with the rostrum and maxillopalatines from both sides remain connected with the vomer. e.g., ratite birds (*Struthio*, *Casuarius*).

b. **Schizognathous**—vomer small and pointed in front or may be absent; maxillopalatines are not connected with one another or with the vomer. e.g., *Columba*, *Gallus*.

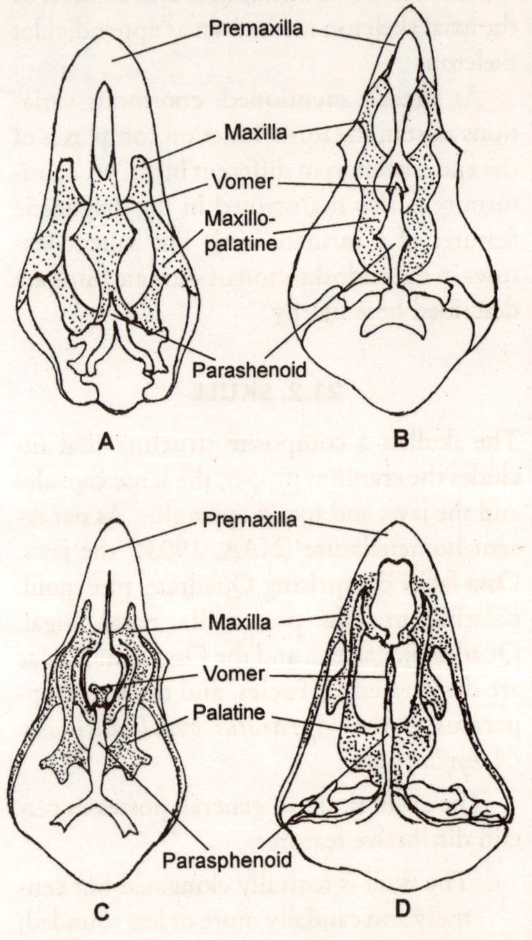

Fig. 21.1: Four principal skull types based on relation between vomer and maxillo-palatine.
A. Dromaeognathous skull (*Struthio*)
B. Schizognathous skull (*Columba*)
C. Aegithognathous skull (*Corvus*)
D. *Desmognathous skull* (*Anser*)

c. Aegithognathous—schizognathoyus type of the skull, but the vomer, instead of becoming pointed in front, appears as a cut-end or blunt. e.g., *Corvus.*

d. Desmognathous—vomer reduced, and ventral to the vomer along midline, the maxillopalatines from both sides join to form a large sponge-like plate. e.g., *Anas, Anser.*

NOTE : Srl. nos. (b), (c) and (d) type of the palates are present in flying birds.

21.2.1. Cranium

The cranium proper is fully ossified, lacking sutures, and is formed of paired occipitals, parietals and frontals from behind forwards. The dome of the skull called calvaria is somewhat broader caudally. The occipital region of the skull is typically formed of a crescentic supraoccipital dorsally, a pair of exoccipitals laterally and a single, distinct ventral basioccipital which forms occipital condyle in birds. The three occipital parts enclose a large cavity (somewhat basally) through which the medulla of the brain and its meninges become continuous with the spinal cord. In the nuchal region of the skull dorsal to the foramen magnum, the median convexity of the supraoccipital and the parietal is called '**Prominentia cerebellaris**' which overlies the cerebellum.

Two distinct cristae (elevated ridges) develop in the nuchal region of the cranium—one along middorsal line of the supraoccipital called '**Crista nuchalis sagittalis**' and the other transversely along hinder border of the parietal bone on either side called '**Crista nuchalis transversa**'. The latter crista separates the muchal plane of the supraoccipital from the rostrally located calvaria of the parietal and squamosal bones.

The parietals form the main broader dome or calvaria of the cranium, fused so neatly along the midline that the sutures are obliterated. The caudolateral cranium on either side of the parietal forms the temporal region, each with distinct temporal fossa. The temporal fossa is an oval or leaf-like elongated depression (either shallow or deep) which may extend dorsally up to the parietal or cauldodorsally up to the squamosal. In gulls, terns and herons and in a few other birds, the temporal fossa is wide and forms a deep concavity into the cranium. The fossa is bounded dorsally and laterally by a well-marked '**Crista temporalis**'. The temporal fossa provides extensive site for the origin of M. adductor mandibulae externus superficialis pars temporalis.

The roof of the rostral part of the cranial box is formed of a pair of frontals, lying in the front of the parietals. The frontals in most birds may be flat or dorsally convex, but a shallow, longitudinal depression occurs in certain birds (e.g., *Ardea, Anser*).

The caudolateral end of the frontal bone on either side projects downward, forming usually the caudoventral border of the orbit. In most carinate birds and in the ratite bird, *Rhea*, this projection is known as the '**Postorbital process**' (**Proc. postorbitalis**) which is largely formed by the laterosphenoidal bone. The base of the postorbital process is partly contributed by the squamosal in some galliform birds and by the frontal in Ostrich. The zygomatic process of the squamosal is situated below the postorbital process in certain birds (e.g., psittaciform and galliform birds). The sidewalls and ventral parts of the frontal region of the cranial box are formed by the laterosphenoid (syn: Os pleurosphenoidale; Os orbitosphenoidale; Os alisphenoidale) bone, while the caudoventral base is formed either by the parasphenoidal rostrum or by the basisphenoid. The frontal region is somewhat narrowed in front due to the presence of large orbits which are separated by an ossified **interorbital septum**. The vertical septum is formed mainly by the sphenoidal bones, but is also contributed by the mesethmoid, ectethmoid and the lachrymal bones. The mesethmoid bone forms much of the rostral interorbital septum.

21.2.2. Sense Capsules

The auditory, optic and olfactory sense organs are enclosed by and protected within the bony capsules formed early in the development of the cranium.

The **auditory capsule** housing the middle and internal ear are largely ossified and formed by the pro-otic bone. The small opisthotic of the embryo unites much early with the exoccipital and the epiotic with the supraoccipital. The tympanic cavity (**Cavum tympanicum**) of the middle ear is a shallow, open concavaity in the dried skull. Owing to the presence of the tympanic membrane, only the ventral part of the bony concavity is the tympanic cavity proper while the caudodorsal part is '**Meatus acusticus externus**'. The bony wall of the Meatus is formed by the parasphenoid wing (*Ala parasphenoidalis*), lateral margin of the paroccipital process and **Crista tympanica** of the body of the quadrate bone.

Columella (Syn: **Columella auris**; NAA, 1993) is the only auditory ossicle in the middle ear of a bird. It is a rod-shaped bone derived from the hyoid arch and it extends from the tympanic membrane to the vestibular window. At the tympanic or distal end, there is a tripartite extracolumellar cartilage and at the proximal or vestibular end there is an expanded basis or 'foot-plate' occluding the vestibular window. The tripartite distal end of the columella fixed to the tympanic membrane is referred to as '**Cartilago extracolumellaris**' (NAA, 1993) which is also mentioned by the synonym '**Pars extrastapedialis**' of the Columella (H. J. Müller, 1963, for a detailed synonymy for parts of the avian columella). **Fenestra vestibularis** (Syn: **Fenestra ovalis**) opens into the osseous labyrinth or the vestibule and is closely apposed by the foot-plate or base of the columella, while the **Fenestra cochlearis** (**Syn: Fenestra rotunda**) opens within the opisthotic bone that is closed in life by the tympanic membrane. The Eustachian canal communicates between the tympanic cavity and the mouth.

The **optic capsules** or **orbits** in birds are quite large and spacious, and are bounded mainly by cranial bones. In most birds, the floor of the orbit is not bony but it consists mainly of jaw muscles, except in woodcocks and snipes where the orbit is almost completely enclosed by bone (Hofer, 1955). In birds, the two large orbits are separated from one another by a thin, longitudinally placed vertical **interorbital septum** formed mainly by the mesethmoid and parasphenoidal bones. The interorbital septum has a few unpaired gaps and in the caudal wall of the orbit, there occur a few paired gaps through which the orbit communicates with the cranial cavity. These gaps are referred to as '**Fonticuli interorbitales**'. The **optic foramen** (**Foramen opticum**) is located near the boundary between the caudal margin of the interorbital septum and the caudal wall of the orbit. In carinate birds, the optic foramen is a single opening in the cranial cavity, but it may be divided into two foramina, one in each orbit, by the thin interorbital septum. Zusi (1978) reports that, in psittaciform birds and in cardueline finches, these two foramina are widely separated from one another by a thick septum. In many other birds, the optic foramina may be continuous with other foramina.

A large lachrymal bone with usually a downwardly directed process forms the rostral boundary of the orbit (caudal margin of the antorbital fenestra in all archosaurs, including birds). A vertical transverse plate of bone, the **ectethmoid (Os ectethmoidale)** forms part of the rostral wall of the orbit, separating it from the nasal cavity. In a number of birds the lachrymal is fused with the ectethmoid, forming a **lachrymal-ectethmoid complex** (Cracraft, 1968), as in some charadriiforms (Johnson, 1984).

The **olfactory capsule** is rather poorly developed in birds, as their organs of smell are less significant. The shape of the nasal apertures (external nares)—whether **schizorhinal** or **holorhinal**—is related to upper jaw kinesis (Hofer, 1955; Bock, 1964; Yudin, 1965; Biihler (1981 and Zusi, 1984). The aperture leads to a nasal chamber on either side, enclosed mainly by the nasal and premaxillary bones. The median column of the nasal bone forming dorsal border of the external nares is called **'Pila supranasalis'**. The nasal is a sort of tri-radiate bone having its caudal process meeting the frontal and lachrymal bones and the two rostral processes—one superior and the other inferior–meet the maxillae and the caudal process of the premaxillae, thus covering a part of the nasal chamber caudally. The two nasal cavities are separated by a medial **nasal septum**. The rostral part of the septum and the nasal conchae in the caudal part of the nasal cavity may ossify in different degrees in different birds. Otherwise, these remain more or less cartilagenous. The ossification may extensively take place in birds like albatross (*Diomedea*), some parrots, birds of prey (*Buteo, Strix*), herons and ibis, some pelecaniforms, trochilids, some coraciiforms, and passeriforms.

The **vomers** are paired elements that, in adults, very often fuse into a single median structure, articulating with the parasphenoid rostrum and/or pars choanalis of the palatine at the base of the nasal chamber. The vomer is not present in all birds, very weakly devloped in galliform birds and absent in a few (e.g., Pigeon). The vomer widely varies in shape and size in different birds and in passerine birds, it extends into ethmoid tissue–a condition unique to this group. In palaeognaths, the vomer has articular surfaces for pterygoid, palatine, maxillary and premaxillary bones, besides the parasphenoid rostrum. In palaeognaths, the vomer and pterygoids exclude contact between the paired palatines.

21.2.3. Jaws

Avian jaws—both upper and lower together with the hyoid apparatus—are parts of the visceral skeleton. In modern birds, the toothless upper and lower jaws being ensheathed by the thick and cornified rhamphothecal covering constitute the upper and lower beaks, respectively. The ornithologists often refer to them as 'bill' of birds. Both the jaws are rostral prolongations of the skull and by their characteristic shape, size and location compensate for—to a large extent—the functions of the forelimbs which are transformed into wings:

(i) **Upper jaw : Os premaxillare, Os maxillare, Os nasale** and their processes are the bony components of the upper jaw. The premaxillae are united into a single large tri-radiate bone which forms practically the whole of the upper beak. The maxillae, on the other hand, are smaller and have their rostral ends produced inwards into spongy **maxillo-palatine processes.** The slender, rod-shaped caudal end of the maxilla is continued backwards by an equally slender **Jugal** and **quadratojugal** bones to the quadrate. Hence the jugal arch (**Arcus jugalis**) consists of three elements ankylosed with one another: Proc. jugalis of Os maxillare, Os jugale proper, and Os quadratojugale.

In the skull characteristics of birds, it has already been mentioned that in all modern prokinetic—and in many neognathine rhynchokinetic—skull, the upper jaw is kinetic and it moves either wholly at the **craniofacial hiatus (prokinetic)** or partly at the **elastic flexon zones** in the upper jaw (**rhynchokinetic**). The craniofacial hiatus is the interval between the rostral edge of the interobital septum and the caudal edge of the osseous nasal septum (where present). The hiatus may be completed by a septum of cartilage that accomplishes a hinge-joint at this site or, more usually, a craniofacial flexon zone may be present here. Bock (1964), Butendieck (1980; as cited in NAA, 1993), Bühler (1981) and Bühler et al (1988) have vividly described the nature of **jaw kinesis** in different birds. It should be remembered in this connection that the upward thrust on the upper jaw is first initiated by the forward movement of the quadrate which is turned upward and forward due to depression of the mandible. The force of protraction of the quadrate being aided by the action of the **M. protractor pterygoidei et quadrati** is transmitted through the pterygoid-palatine-maxillopalatine complex to the upper jaw which is thus raised either at the cranio-facial hinge joint or at the flexion zones. The upward movement of the mandible restores initial position of the quadrate and this—together with the actions of the retractor muscles of the upper jaw and the quad rate—accomplishes closure of the beaks.

The inner series of bones of the upper jaw and also a part of the visceral skeleton are represented by an efficient kinetic bridge between the upper jaw and the cranio-mandibular end. These bones are:

(ii) Os quadratum: Avian quadrate bone offers variable shapes and sizes in different taxa. The quadrate bone is somewhat tringular having thicker cranial and mandibular arms and a comparatively narrower, rostral orbital arm. At the cranial end, **Processus oticus** of the quadrate almost immovably articulates with the cranium usually by two capitula separated by an incisura*. In *Hesperornis*, ratites and some neognath birds (Witmer, 1990), the intercapitular incisura is indistinct and the two capitula merge, thus apperaring single-headed. At the mandibular end, the mandibular end process of the quadrate (**Proc. mandibularis** in most avian taxa possesses three condyles (Bock, 1960) **Condylus medialis, Condylus lateralis** and **Condylus caudalis**—fitting into the respective cotyla at the caudal mandible (mostly articular bone). The quadrato-mandibular joints are movable.

The rostral arm of the quadrate forms the **orbital process (Proc. orbitalis quadrati)** whose tip bears ventrally an oval surface, providing origin-site for the quadrate retractor muscle. Just outer to the condylus medialis of quadrate, condylus pterygoideus is formed for ar-

* These two capitula (or condyles) of the Proc. oticus of the quadrate articulates independently of the other; so two separate joint cavities—namely, Artc. Quadrato-otica and Artc. quadrato-squamosa—are present (e.g., *Columba*). In many other taxa, the two condyles of the quadrate and their joint cavities on the otic and squamosal bones are enclosed within one common articular capsule (e.g., *Gallus, Corvus*).

ticulation with the head of the pterygoid bone.

(iii) Os pterygoideum: The flat, rod-shaped pterygoid bone is somewhat obliquely disposed between the quadrate and the palatine as well as parasphenoid rostrum. The quadratic condyle has a rather convex rounded articular facet to which articulates the single cotyla of the head or proximal end of the pterygoid. The **'foot of the pterygoid' (Pes pterygoidei**; Johnson, 1984), i.e., the expanded rostral end of the pterygoid bone, has articular surfaces at the caudal end of the palatine and the parasphenoid rostrum. In ratites, **procellariiforms, anseriforms, galliforms, many** charadriiform and cathartid vultures; in some caprimulgiforms, strigiforms and trogoniforms, a process called **'Proc. basipterygoideus'** occurs on each side of the sphenoidal rostrum (**Rostrum parasphenoidale**) for articulation with the pterygoid bones.

(iv) Os palatinum (Os pterygopalatinum): Four morphological features of the palatine bone are common in most birds: (1) **Pars choanalis**—forms boundary of the fossa choanalis which is the paired, shallow depression between the medial and ventral crests of Pars choanalis; (2) **Pars lateralis** (Lamella caudolateralis)—an expanded caudolateral **plate**, lateral and ventral to pars choanalis and **serving** as the wide origin-site for the M. pterygoideus; (3) **Proc. maxillaris,**–the maxillary process of the palatine, connected to **the** maxilla; it is usually a slender bar which **in** all neognath birds has a bending zone (Zona flexoria palatina); (4) **Proc. pterygoideus** which connects the palatine with the pterygoid bone and, in some birds, may include a fused portion of the pterygoid bone.

Hofer (1945), Richards and Bock (1979) and others have discussed different shapes and relationships of palatine bones in different avian taxa. Ziswiler (1985) has given a summary on the types of avian palates.

It has already been mentioned in the description of the upper jaw that the complex of jugal arch which forms the outer series of bones of the visceral skeleton appears as a sling of the inner series of bones forming arch of the bow. The jugal process of the maxilla is flexible, whereas the small bar-shaped quadratojugal has a condyle that fits into a cotyla on the lateral aspect of the quadrate. Usually this joint is synovial in most birds, but in *Sturnus* (De Kock, 1995) and *Caprimulgus* (Bühler, 1970) the joint is syndesmotic.

(v) Lower jaw (Ossa mandibulae): The characteristic lower jaw in birds, derived from the second branchial arch, consists of two rami—the **'Ramus mandibulae'**—the two rami—being united rostrally along midline to a pointed, apical symphysis called **'Symphysis mandibulae'** (**Rostrum mandibulae**; new term; NAA, 1993). The length of the symphysial region along midline from apex downward is variable in different birds (from about 0.08 in./2 mm to more than 0.4 in./1 cm usually). The joint between the two symphysial segments is actually a synchondrosis in early development, but due to overegrowth of the ossa dentalia (Dentary), the cartilagenous symphysis becomes ossified, leading to a synostosis joint in adults of most birds.

Each ramus mandibulae, may be subdivided into three parts: **(i) Pars symphysialis,** which is the rostral symphysial segment uniting with the opposite ramus at the Symphysis mandibularies; **(ii) Pars intermedia** which extends caudally up to the caudal intramandibular flexion zone where the oval **Fenestra rostralis mandibulae** may be present (e.g., *Columba*); **(iii) Pars caudalis** which extends from the caudal flexion zone up to the end of the retroarticular process. Morphologically it is the most important segment of the ramus, as it serves for the attachment of several jaw muscles, provides facets for articulation with the quadrate and contains **fenestra caudalis mandibulae** (where dual fenestrae occur on either side of the ramus as in Charadriiformes; Gruiformes, Psittaciformes and Strigiformes). However, both the mandibular fenestrae may be absent in certain taxa of birds.

Each ramus of the mandible, from the view-point of origin, consists of seven separate bones. Although the nomenclature of these bones has not been the same as used by different authors at different times, Müller's terminology and his synonymy have been followed here. These bones are :

(i) Os dentale (Syn.: **dentary; dento splenial; mentomandibulare**): This is the principal bony component of each mandibular ramus. It articulates with the supraangular and splenial bones at the junction of intermediate and caudal parts of the ramus.

Angulus mandibulae is the name given to the angular curvature of the ramus mid-dorsally wherefrom the ramus curves ventrally. The angle marks the caudal extent of the mandibular rhamphothecal sheath. The angle is well-pronounced in charadriiform, Caprimulgiforms, Falconiforms, Columbiforms and Passerines.

(ii) Os articulare: Lying in the caudal segment of the ramus, this bone bears several important structures such as **Fossa articularis quadratica** that forms facets for the articulation with the quadratic condyles—medial, lateral and caudal condyles; **Proc. (Processus) medialis mandibulae** and **Proc. lateralis mandibulae** which are processes of mainly Os articulare in the caudalis part of the mandible. Of the two processes, Proc. medialis mandibulae is much larger and stronger with its tip curved dorsally. The lateral process forms a part of the cotyla for the articulation of the lateral condyle of the quadrate. Lig. postorbitale—which plays an important role in the kinesis of the upper jaw–is strongly attached to Proc. lateralis mandibulae or to a tubercle slightly rostral to this process. The shallow concavity at the caudal surface of Proc. mandibulae medialis is called **Fossa caudalis**—it provides insertion-site for the M. depressor mandibulae.

(iii) Os prearticulare: It is a slender ventral bar lying rostral to articular and extending up to caudoventral limit of the dentary.

(iv) **Os angulare**: This bone lies ventral to articular forming ventral angle of the pars caudalis of the ramus and it projects caudally as the **Proc. retroarticularis** (Syn.: **Proc. angularis posterior**) which is mainly formed by the Os angulare behind Fossa caudalis and Proc. lateralis mandibulae. Proc. retro articularis is weakly developed in most birds, though quite prominent in Galliformes, Ciconiiformes, Psittaciformes, Anseriforms and in Phoenicopteriformes.

(v) Os Supraangulare (Syn: **Os surangulare**): It is a small, elongated bone that lies dorsal to the Os angulare, covering dorsally up to rostral limit of the Angulus mandibulae of the ramus.

Proc. Coronoideus (Syn: **Proc. pseudo coronoideus**): Os coronoideum is not ordinarily present in birds, but the Proc. coronoideus is present as a process on the dorsal margin of pars caudalis of the mandibular ramus to which is attached the thick and strong aponeurosis of M. adductor mandibulae externus superficialis pars rostralis (temporalis). In some birds the process may coincide with the Angulus mandibulae (Johnson, 1984). In anseriforms, this process is located on the lateral surface of the mandible (Zweers, 1974).

(vi) Os spleniale: Between the caudodorsal and caudoventral extensions of the dentary lies a thin, platelike bony element called the Os spleniale. It is located just ventral to Fossa aditus canalis mandibulae.

(vii) Os mentomandibulare: It is the fused, paired cartilagenous mentomandibular elements that are later replaced by bones that ankylose at the mandibular symphysis.

21.2.4. Hyoid Apparatus (Syn: Hypobranchial apparatus, Hyolingual apparatus)

The hyoid apparatus in birds is mainly derived from the hyoid arch and also from other caudal branchial arches. The nomenclature of the hyoid elements and their synonymy has been adopted from that by Müller (1963), McLelland (1968) and King and McLelland (1984). Zweers (1,974, 1982) described the hyoid apparatus in *Anas* and *Columba*, Bhattacharyya (1980, 1989, 1998) and vanGennip (1986) also followed the same terminology of hyoid apparatus in *Columba* and columbids.

The hyoid apparatus in a typical carinate bird consists of

i. A single median Paraglossum (Syn: **Entoglossum**) of the shape of an arrowhead (e.g, *Columba*) or a flat and wide paraglossum with a central foramen (e.g., Psittaciforms) or a paired paraglossalia united rostrally by a cartilagenous or bony isthmus. In *Psittacus* (Homberger, 1986), each of the paired paraglossals is bifurcate rostrally; the paraglossum in most birds has a short, spine-like caudolateral process ventrally.

ii. A single median basihyale (Syn.: **basibranchiale rostrale**; basihyoideum; Pars basihyalis copulae; Copula I-Muller, 1963) derived from the hyoid arch. In parrots, (Homberger, 1986), the enlarged caudolateral sides of the basihyale give rise to **Proc. parahyalis**, one on either side, which unite mid-dorsally forming the paraglossalis arch.

iii. **Urohyale (Syn. : Basibranchiale caudale,** Urohyoideum, Pars urohyalis copulae; Copula II-Muller, 1963): A slender, flat, bar-shaped caudomedial extension of basihyale. The basihyale and urohyale are separate in young birds, but they are fused in the adults.

iv. A pair of longer **Cratobranchiales** and a pair of shorter **Epibranchiales**, joined to one another articulate on either side of the broad caudolateral basihyale. When an additional, smaller cartilagenous segment is attached to the epibranchial, it is called Epibranchiale II, while the longer segment is called Epibranchial I.

Note: See Chapter 30 for sketches.

21.2.5. Vertebral Column (Columna vertebralis)

Avian vertebral column is a superb combination of flexiblity, rigidity and lightness. The total number of vertebrae—as well as the number of vertebrae in different regions, the shape, size and structural components of vertebrae—vary in different avian taxa. The total number of vertebrae varies from 39 (in some Passeriforms) to 64 (considering Pygostyle as a single vertebra) or 67-68 (considering 4-5 pygostyle vertebrae) in some Swans and ratites.

Flexures of the vertebral column: Due to the frequent fusion of vertebrae in the thoracic, lumbar, sacral and caudal regions, flexures in the vertebral column are observed mainly in the cervical and in the free caudal regions.

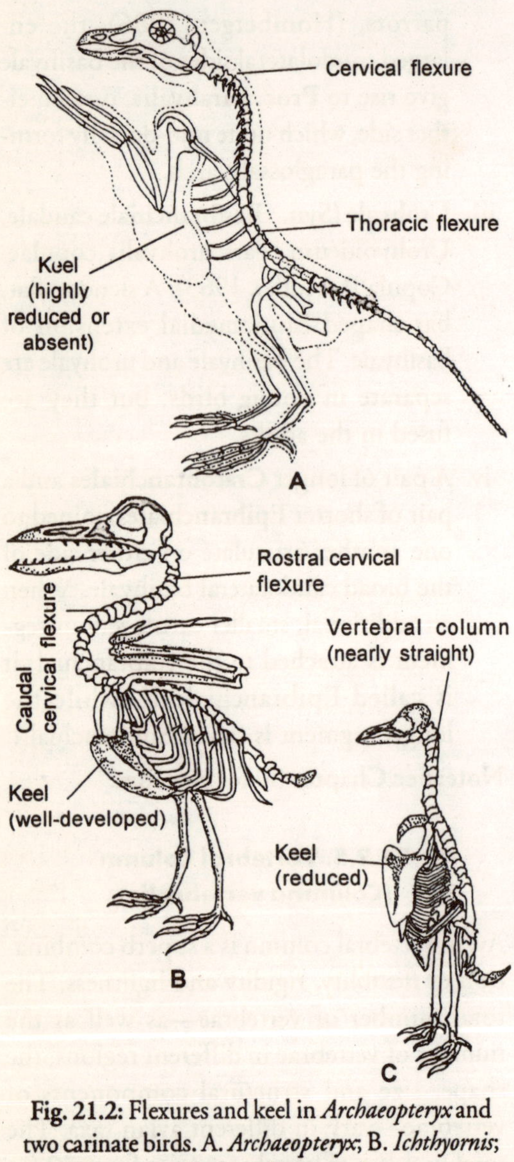

Fig. 21.2: Flexures and keel in *Archaeopteryx* and two carinate birds. A. *Archaeopteryx*; B. *Ichthyornis*; C. *Eudyptes* (Penguins)

Two **cervical flexures**–one rostral and one caudal–are quite common in most birds. In Penguins, where the hindlimbs have been shifted far backwardly to help the bird assume a rather standing position, the spinal column is more or less straight with slight curvature. In *Archaeopteryx*, the rostral cervical and thoracic, and in *Ichthyornis*, the rostral and caudal cervical flexures are quite prominent.

(i) Cervical vertebrae: In general, there are 13 or 14 to 16 cervical vertebrae in birds, but this number may be reduced to 11 as in certain parrots, or even 9 as in some songbirds and may increase up to 25, as in some swans.

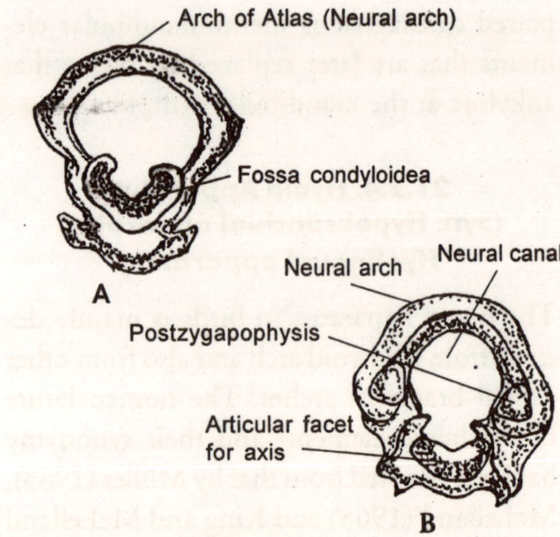

Atlas of a galliform bird; A. frontal view; B. caudal view

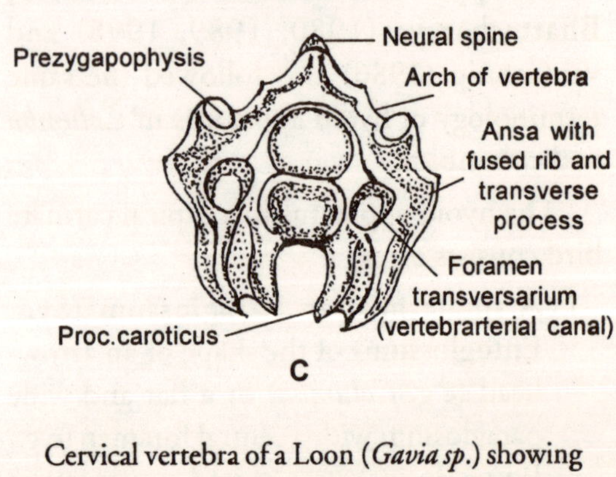

Cervical vertebra of a Loon (*Gavia sp.*) showing Proc. caroticus

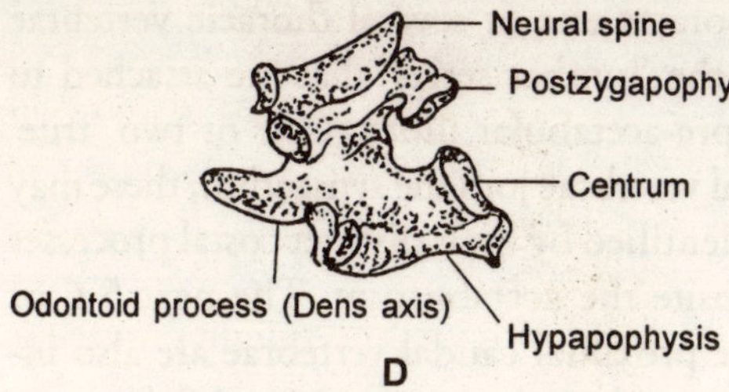

D

Axis of a galliform bird; lateral view

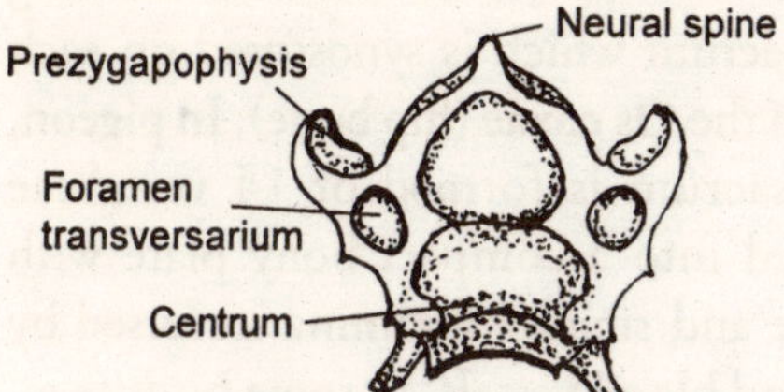

E

Cervical vertebra of a pigeon; frontal view

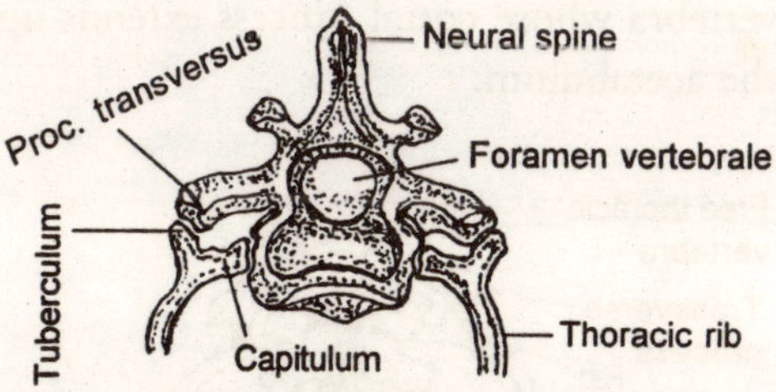

F

Thoracic vertebra of a Gull; frontal view

Fig. 21.3: (A-F): Some distinctive features in different vertebrae

All the cervical and free thoracic vertebrae are heterocoelous, i.e., saddle-shaped–a unique feature of the centrum in birds, so that rostrally the centrum appears concave from side to side and convex from above downwards, and caudally the situation is reverse. The only exception is observed in the extinct toothed birds *Ichthyornis* where the cervical and thoracic vertebrae were amphicoelous. In hetrocoelous vertebrae, the spaces between adjacent centra are traversed by a meniscus (a vertical plate of cartilage) through which a suspensory ligament connects the two vertebrae. The atlas and the rest of cervical vertebrae have their rudimentary ribs completely fused to the short transverse processes forming **'Ansa costotransversaria'.** These serve for attachment of tendons of lateral cervical musculature. The axis (second cervical vertebra; **Epistropheus**) is also specialised to have a **'Dens axis' (Proc. odontoideus)**, two zygapophysial processes and a well-developed hypapophysis. The intermediate series of the cervical vertebrae are each provided with a pair of **Proc. caroticus** rostro ventrally and all the cervical vertebrae have, each, a pair of **Proc. costalis** laterally and a pair of **Foramina transversaria**, one on either side at the base of the transverse process, respectively.

In *Pelecanus*, *Ardea* and *Dendrocopos* (Woodpecker), the paired Proc. caroticus become ankylosed, forming a complete osseus canal. This feature helps the bird in throwing its head forward, or in drilling and drumming as in woodpeckers. The series of transverse foramina form a continuous canal—one on either side of the column of cervical vertebrae. This canal is known as **Canalis vertebrarterialis** through which pass the ascending vertebral artery and vein.

The atlas is apneumatic in all the birds and the axis in many birds. The laterally compressed, platelike ventral process, the **hypapophysis,** is present on either side of the rostral and caudal cervical vertebrae, as well as in the thoracic vertebrae. These crests are very well-developed in the vertebrae of Penguins, *Gavia*, Kingfishers and in some ducks, as an adaptation for the underwater use of the neck. Hypapophyses provide origin-sites for the powerful flexor muscles of the neck.

(ii) Thoracic vertebrae (Vertebrae thoracicae): The first thoracic vertebra has been defined as the cranialmost vertebra with a complete set of ribs which articulates directly or indirectly with the sternum. The rest of the thoracic vertebrae (2-5 or 6; Storer, 1982) except the last are coalesced in adult birds into a single compact unit called the '**Notarium**' (**Os dorsale**). This Notarium is not fused with the synsacrum but may form one or two synovial joints with the synsacrum. The notarium is present in at least 17 families of birds, occasional in several other families (NAA, 1993) : e.g., tinamous, *Pelecanus*, ibises and spoonbills, galliforms, columbiforms, all podicipediforms and most falconids. Martin (1987) reports that the Mesozoic birds, *Archaeopteryx* and *Gobipteryx*, possess several "fused anterior dorsal (thoracic) vertebrae". Each thoracic rib has double-headed vertebral articulations (synovial joint for **tuberculum** and synchondrosis for **capitulum**) and **Proc. uncinatum** from the caudal border of each vertebral rib. The length of uncinate process is proportional to the effectiveness of the external intercostal muscle to set the rib into motion. Screamers and megapodids lack uncinate processes.

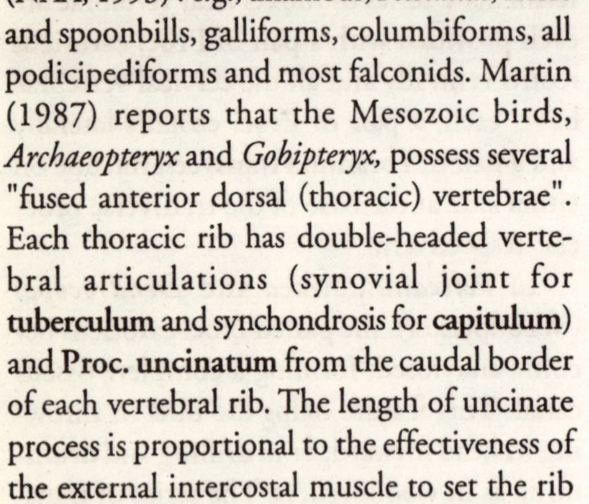

Fig. 21.4: A portion of the fused notarium of a pigeon (lateral view)

(iii) Synsacrum (Syn: Os lumbosacrale; Os pelvicum): The synsacrum is a rigid unit consisting of ankylosed vertebrae in mature birds. The pre-acetabular part of the synsacrum incorporates one or several thoracic vertebrae and the 'lumbar series' that are attached to the pre-acetabular ilium. One or two 'true' sacral vertebrae join the synsacrum; these may be identified by their distinct costal processes opposite the acetabutrum. The next 5-6 or more proximal caudal vertebrae are also incorporated in the composition of the post-acetabular synsacrum. Interspecific variation occurs in the number of vertebrae forming the synsacrum which is synostosed on each side with the **Os coxae** (hip bone). In pigeon, the synsacrum is formed of 14 vertebrae ankylosed into a compact bony plate with fenestrae and smaller foramina traversed by nerves and blood vessels. In some birds (e.g., *Meleagris*; wild turkey; Galliformes), '**vertebra acetabularis**' is the name given to a sacral vertebra whose costal process extends up to the acetabulum.

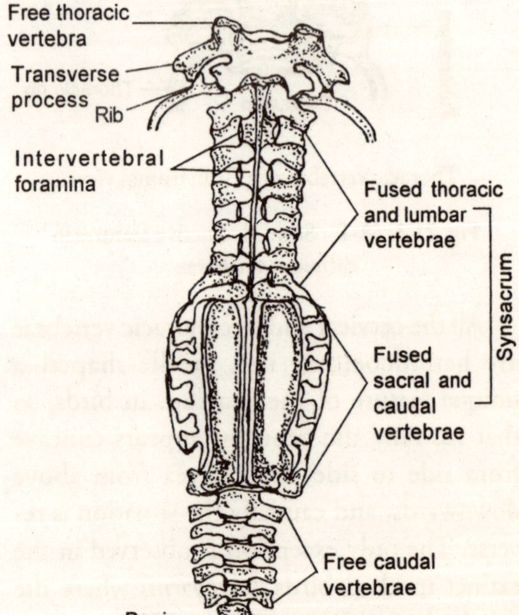

Fig. 21.5: Simplified dorsal view of the synsacrum of a pigeon

(iv) Caudal vertebrae: The number of caudal vertebrae is also variable in different species. However, the three segments or regions of the caudal vertebrae have been ankylosed in their rostral and caudal parts to form the synsacrum and the pygostyle, respectively, only the middle region comprising 5-7 caudal vertebrae (5 in *Meleagris*; 6 in *Columba*) remaining free. Of the 5-7 pre-acetabular caudal vertebrae fused to the synsacrum, the transverse processes, except in the last, are formed of two elements–the dorsal element completely fused to the synsacrum and the ventral element appears as a rod-shaped rib. The intermediate series of free, caudal vertebrae are amphicoelous, each with a pair of short transverse processes.

The last segment is a compound upturned **Os pygostylus** (Pygostyle; Ploughshare bone), usually consisting of 5-6 vertebrae formed into three parts, the **'Basis pygostyli'**, the **'Corpus pygostyli'** and the narrow **'Apex pygostyli'**.

The pygostyle of kingfisher has rudimentary transverse processes and that of woodpecker has a **'discus pygostyli'** at its apex for attachment of very well-developed depressor muscles of the tail.

21.2.6. Sternum and Ribs (See Chapter 30 for sketches)

(i) Sternum: Avian sternum is a broad bony plate which, in carinate birds, is produced along its ventro-median line into a triangular, vertical ridge or crest called the **keel** or carina. Fürbringer (1888) dstinguished sternum into a cranial part, the **'costosternum'** to which the ribs are attached, and a caudal part, the **'xiphosternum'** (= **metasternum**). Rostrally, the sternum has the **Rostrum sterni** comprising the **spina interna** and **spina externa, the Sulcus ar.iculatio coracoideus,** the **Incisura intercostalis, Sulcus carinae, Apex carinae** and **Pila carinae**. Laterally, the sternum has a pair of **Trabecula lateralis**, a pair of **Trabecula intermedia** and a pair of large **Fenestra lateralis**, and, caudally, the **Trabecula mediana** with the **Fenestra medialis** on either side of it.

In *Strigops* (Psittaciformes), the carina is lacking. Ratites generally lack a well-developed distinct carina.

(ii) Ribs: Last one or two cervical vertebrae have double-headed ribs with their distal ends free. The rest of the cervical vertebrae (except the 1st and 2nd) have vestigial ribs which appear as short, backwardly directed transverse processes. 5-6 pairs of true thoracic ribs consist of vertebral and sternal elements, the latter fully ossified—unlike in the mammals—and they articulate at the **Margo costalis sterni** with distinct notches for articulation. Each thoracic rib is provided with a dorso caudally directed **'Proc. uncinatus'** (**Uncinate process**), attached to the caudal border of the next vertebral rib.

Screamers (Anseriformes) and Megapodids (Craciformes = Galliformes) lack uncinate processes. The extinct toothed bird, *Ichthyornis,* possessed uncinate process.

21.2.7. Joints and Ligaments

Most of the joints of the vertebral column are simple diarthroses with synovial joints. Occasionally these joints are accompained by different **articular** and **linkage** ligaments. The joints may be symphyses as found between the free caudal vertebra in *Columba* where the joints are furnished by elastico-fibro cartilagenous discs. The joint between the last free caudal vertebra and the pygostyle is a synovial joint of the nature of a hinge-joint.

Diarthroses of the nature of ball-and-socket joints are found in such places as the shoulder and the hip. The rotatory joint between the atlas and axis or the gliding joints elsewhere are also diarthroses. **Amphiarthroses** have no joint cavity. Pelvic symphysis is an example of this type of joint. In the caudal thoracic region of some taxa–and in the synsacral region universally—the joints become ankylosed forming **synostoses.** The zygapophysial articulation (Artc. Zygapophysialis) is a synovial joint. The junction between each vertebral and sternal rib is a persistent cartilagenous joint. The joint between the capitulum and the centrum is cartilagenous, whereas that between the tuberculum and the transverse process is a synovial joint.

Most of the joints between vertebral bodies are synovial joints with intraarticular menisci.

Lig. elasticum obliquum and the **Lig. elasticum transversum** of one side are tensed when the vertebral column is flexed to the contralateral side. The strong paired **Lig. collaterale** connects vertebral bodies and is continuous with the intervertebral articular capsule.

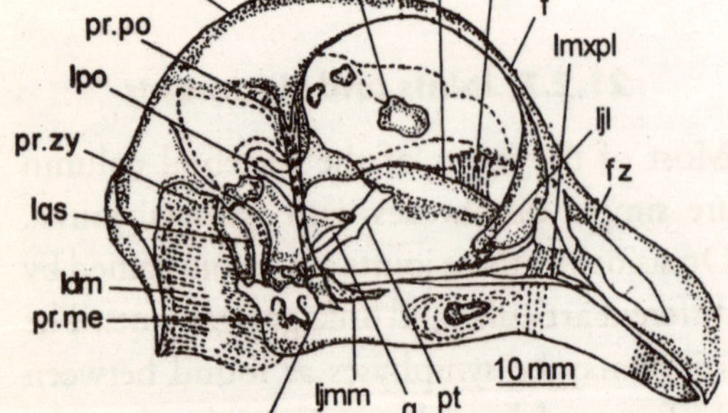

Fig. 21.6: Ligaments. ldm, Lig. depressor mandibule; ljml, Lig. jugomandibulare lateral; ljmm, Lig. Jugomandibulare mediale; lpo, Lig. postorbitale; lqs, Lig. quadrato-squamosale; lmspl, Lig. mesthmopalatinum; lmxpl, Lig. maxillopalatinum

Besides the above ligaments, Baumel (1985) has confirmed the presence of supporting apparatus of **long and short suspensory ligaments** that pass from the vertebrae to the roots of the Brachial plexus. The function of these ligaments is to protect the spinal nerve rootlets and the spinal cord by shunting the tensile force from the plexus roots via the suspensory ligaments to the vertebral column which absorbs the force in place of the spinal cord.

(Also see Chapter 30, Fig. 30.15—for ligaments of the vertebral column.)

21.3. APPENDICULAR SKELETON

[See Chapter 30 for sketches]

The shoulder girdle and the pelvic girdle which support the forelimbs (modified as wings in carinate birds) and hindlimbs and the bones of the forelimbs and hindlimbs are described as appendicular skeleton.

21.3.1. Pectoral Girdle and Wing Skeleton

In carinate birds the pectoral girdle offers some structural uniformity in that it consists of a **sabre-shaped scapula** whose pointed blade is directed caudally reaching up to the pelvis, a thick **pillar-shaped coracoid** and a **V-shaped furcula** (Merry thought bone; wish bone), and the scapula-coracoid angle is usually less than 90° degrees (in the ratites the angle is usually obtuse).

The middle or main body of the sub-cylindrical and then flat, sword-shaped scapula is the **corpus scapulae**, its pointed caudal end (pelvic end) being known as **Extremitas caudalis**. On one side of the blunt knob-like cranial end lies the pointed Acromion which is located near the glenoid process of the scapula. Rich (1985) reports that in some birds of the passerine suborder Menurae, the acromion is bifurcate, having two blunt knob-like processes. The glenoid process of the scapula forms surface for articulation with the head of the humerus and, in certain birds, this process also articulates directly with the procoracoid process of the coracoid. In some birds (e.g., *Ardea, Larus*), the cranial surface of the glenoid process of the scapula forms a convex tubercular process, called the **Tuberculum coracoideum**, which fits into a cup-shaped surface of the coracoid, both thus forming the **coraco-scapular joint**. The glenoid processes of the scapula and coracoid bearing articular facets for the head of the humerus join together to form the **glenoid cavity (Cavitas glenoidalis)** which performs articulation with the head of the humerus.

Dorsally, in most birds, the scapula, coracoid and the clavicle at their junction contribute to the formation of a '**Foramen triosseum**' which Fürbringer (1888) designated as '**Canalis supracoracoideus**' and in recent nomenclature (NAA, 1993) this passage has been designated as '**Canalis triosseus**'. This canal transmits the tendon of **M. supracoracoideus** (= **M. pectoralis minor**)

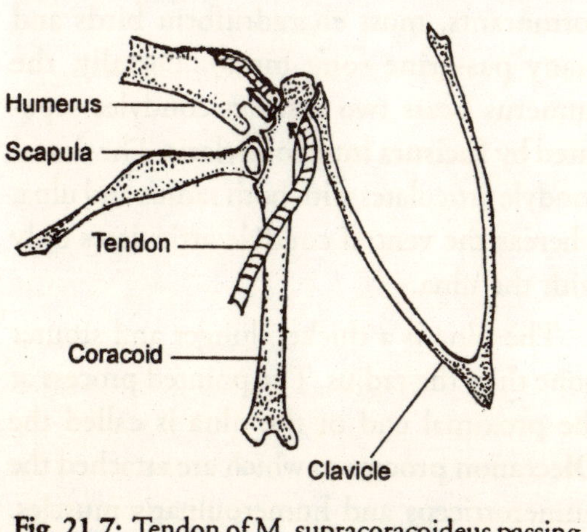

Fig. 21.7: Tendon of M. supracoracoideus passing through Canalis triosseus (lateral view) in *Gallus*. (Based on sketch by Yasuda, 2002)

which—by the pully-action of its tendon through the canal—serves as the major elevator muscle of the wing. In some birds (e.g., touracos; Berger, 1960), this canal is produced by only two bones, the procoracoid process of the coracoid and the scapula, there being no contribution from the clavicle. The canal may be formed only by the coracoid bone as in Musophagidae, Meropidae, Upupidae, Bucerotidae and in *Columba livia*.

The **coracoid (Os coracoideum)** is a thick, straight and stout pillar-like bone that articulates at one end to the coraco-scapular joint and then extends slightly caudoventral to articulate by its broad, knob-like caudal end at the elongated sulcus of the rostral sternum. The coracoids offer several structural details at both the ends. At the rostral end, a projection of the medial border of the coracoid is called '**Proc. procoracoideus**' which forms at its base a prominently curved **Sulcus supracoracoideus** through which the tendon of M. supracoracoideus passes.

Proc. procoracoideus has two more facets—a small cotyla scapularis which receives the Tuberaculum coracoideum of the scapula producing the coraco-scapular joints, and rostral to the cotyla lies the articular facet for the humerus. The sternal end of the coracoid is broad and produced into **Proc. lateralis** and **Angulus medialis**, enclosing an elongated articulating surface for the sternum (**Facies artc. sternalis**). This articulating surface with the sternum is usually divided into a dorsal and ventral facet (Facies externa and Facies interna, respectively). In *Ardea, Columba* and *Corvus* these two facets are not continuous.

'The coracoids, especially, serve as struts to hold the humerus away from the body and the sternum' (George and Berger, 1966).

Clavicula (Furcula): The clavicles are a pair of slender rod-shaped bones extended between the dorsal shoulder end and the ventral sternal end, 'Furcula' (Wishbone; Merry thought bone) is the name given to the paired clavicular rami ankylosed at their ventral ends and, hence, called 'V'shaped furcula, meaning a 'fork'. In most birds, the two rami of the furcula are ankylosed into a blade, rod or knob-like structure called the **'Hypocleideum' (Apophysis furculae; NAA, 1993)** which, in most birds, gets attached, either directly or indirectly, to the apex carinae of the sternum. Along the midventral line of ankylosis of the two furcular rami, a projection may be formed known as the **Proc. interclavicularis.** The dorsal expanded end of each clavicular ramus is called **Extremitas omalis claviculae** or the **Epicleidium.**

In certain birds, the dorsal end of each clavicular ramus forms a well-developed **Proc. acromialis** to articulate with the cranial tip of the scapula and **Proc. acrocoracoideus** with the uppermost narrower end of the coracoid. In *Diomedia, Ardea* and *Aquila*, Proc. acromialis is very well-developed, whereas in *Alcedo, Merops, Sturnus* and few others, both the processes are well-developed. The clavicles are fused to both the sternum and the coracoids in hoatzin (*Opisthocomus*). The clavicles are rudimentary or absent in the ratites, some parrots, pigeons and barbets.

NOTE : (See Chapter 30.13 for sketch)

The wing skeleton or the forelimb-bones in carinate birds comprise the humerus, radius, ulna, two carpals, carpometacarpus and, usually, three digits. The head of the proximal end of the humerus with its articular surface is called **Caput humerus.** The head of the humerus articulates in the shallow **glenoid cavity (Cavitus glenoidalis)** formed by glenoid process of both the scapula and coracoid. The scapula-coraco-humeral articulation is a ball-and-socket articulation which allows free rotation of the humerus. On either side of the caput humerus there is a prominent tubercle—the **Tuberculum dorsale**—to which is attached the tendon of the M. supracoracoideus; and the **Tuberculum ventrale,** to which are attached several shorter muscles of the scapula and coracoid. The ventral tubercle is especially well-developed in the ratite birds.

Below these tubercles, two crests, one on either side of the head of humerus, are quite distinct: **Crista deltopectoralis** below the dorsal tubercle and **Crista bicipitalis** below the ventral tubercle. On either side of the former crista are attached the M. pectoralis and M. deltoideus major, whereas the latter crista provides site for the origin aponeurosis of M. biceps brachii.

On the caudal surface of the head of humerus, a **pneumatic foramen** (or a number of foramina), which is the entrance to the air chamber, lies in the depressed pneumatic fossa. The pneumatic formen is lacking in many birds (e.g., Penguins, loons, grebes, cormorants, most charadriiform birds and many passerine song birds). Distally, the humerus bears two distinct condyles separated by **Incisura intercondylaris.** The dorsal condyle articulates with both radius and ulna, whereas the ventral condyle articulates only with the ulna.

The ulna is a thicker, longer and stouter bone than the radius. The pointed process at the proximal end of the ulna is called the **Olecranon process** to which are attached the humerotriceps and humeroulnaris muscles. The olecranon process is lacking and is replaced by a sesamoid bone in some birds (e.g.,

Micropus, Aptenodytes). The **Cotyla dorsalis** and the **Cotyla ventralis** are the articular surfaces of the ulna for the dorsal and ventral condyles of the humerus. Besides these, the **Incisura radialis** at the proximal ulna is a concave facet for the articulation of the radius. The ulna also provides, along its caudodorsal surface, a number of (11, 12 or more) **'Papillae remigales'**, arranged in a linear row, for the attachment of the ligaments of the follicles of the secondary remiges. At the distal end, the ulna bears two condyles—the **Condylus dorsalis** and the **Condylus ventralis**—with a groove between them known as **Trochlea carpalis**. There is a also a prominent carpal tubercle (**Tuberculum carpale**) adjacent to the ventral condyle.

Only two carpal bones, **Os carpi radiale** (Os scapholunare) and **Os carpi ulnare** (Os cuneiform) persists between the forearm and hand in adult birds, the other three embryonic carpal elements fuse with the compound bone **Carpometacarpus**. The hand or manus in birds typically consists of three metacarpal bones and three digits. The metacarpals are fused at both proximal and distal ends called **Symphysis metacarpalis proximalis** and **S. m. distalis**. The symphyses are, in fact, **synchondroses** which, when ankylosed, are transformed into **synostoses**. Of the three digits, usually Digit I or the Allular digit (Pollex or thumb) has one phalanx, Digit II or the Digitus major has two phalanges and Digit III or the Digitus minor always consists of a single phalanx. However, in a number of avian orders, the allular digit has two phalanges and the second digit in a few birds may have three phalanges (e.g., *Gavia*). Digital primaries are attached to the dorsal surface of the carpometacarpus. In flying birds, the number of primary remiges varies from 9-12, but that of secondary remiges are widely variable (from 6 in some Humming birds to 32 in Wandering Albatross). George and Berger (1966) report that a variable number of feathers (2 in Humming birds to at least 7 in some Turacos) may be attached to the pollex forming the **alula** or the **'bastard wing'**. The alula is the feathered forefinger which–by acting as a slot–can regulate air-flow in flight. (**See Chapter 30, Fig. 30.2.**)

21.3.2. Pelvic Girdle and Bones of the Hindlimbs

The large pelvic girdle, also known as **Os coxae** or the hip bone, is formed on either side by three elements—the **Ilium, Ischium** and the **Pubis**. The pelvis is formed by joining of the two halves of the pelvic girdle with the synsacrum.

Corpus ilii is the strongly developed central part of the ilium, cranial and dorsal to the Acetabulum from which its pre- and post-acetabular alae (wings) emanate (NAA, 1993). The dorsal or the dorsomedial border of the pre-acetabular ilium is called **'Crista iliaca dorsalis '**. The lateral ledge of the post-acetabular ilium marks the boundary between the dorsal and lateral surfaces of the ilium (absent or indistinct) in a few birds. It is called **'Crista dorsolateralis ilii'**. The ischium is a broad bony plate lying caudal to and also covering the ventral part of the **Foramen iliaischiadicum** which is a large oval aperture (for the transmission of ischiadic nerves and vessels), situated just caudal to the acetabulum. The ilium and ischium articulate at the acetabulum in all birds. In young carinate birds, the two bones are joined by cartilage which, in older birds, ossify, caudal to the Fensetra ilioischiadica. The terminal caudal process of the ischium (**Proc. terminalis ischii**) is often pointed and its lower border articulates with the pubis.

Foramen obturator is an oval opening situated caudoventral to the acetabulum. It is the detached cranial part of the Fenestra ischiopubica (NAA, 1993). **Fenestra ischiopubica** (Syn.: **Foramen obturatorium, pars caudalis; Foramen oblongum**) is usually an elongated slit-like gap between the rod-shaped pubis and the ventral margin of the ischium. This gap, however, may be of various shapes and sometimes very wide in some birds. **Antitrochanter** is a projection mainly formed by the ischium and bears an articular surface that is in contact with the neck and trochanter of the femur.

The long shaft of the pubis (Scapus pubis) lies closely parallel to the ventral border of the ischium, (the two bones being) separated by the obturator foramen and the ischiopubic fenestra. Usually, the pubis extends caudally beyond the terminal ischium, and, in some birds, the ends of the pubes of the both sides curve inward and lie close to one another.

The hindlimb bones on each side consist of the femur, tibiotarsus, fibula, tarsometatarsus and 2-4 digits. The long body or shaft of the femur is usually circular in cross-section, except in atypical femora (e.g., *Gavia*) where the femur is laterally compressed and truncated. On one side of the proximal end of the femur lies the elevated and expanded part of the neck—the '**Trochanter**'; on the other side lies the rounded head of the femur that fits into the acetabulum. The articular condyles at the distal end of the femur are known as Condylus medialis and condylus lateralis—also known as condylus internus and condylus externus, respectively. Also at the distal end of the femur lie the **sulcus intercondylaris** and **Sulcus patellaris** for articulation with the tibiotarsus and patella, respectively.

The proximal tarsal bones fuse with the tibia to from a tibiotarsus, just as the distal tarsals fuse with three (II, III & IV) metatarsals along with the hypotarsus to form a single compound tarsometatarsus. However, according to recent workers (Mc Gowan, 1985; Müller and Alberch, 1990), only a single distal tarsal bone can be identified, as against up to four bones previously described. The Metatarsal I does not ankylose; it joins the tarsometatarsus by a ligament. Patella is a sesamoid bone in the common tendon of Femorotibiales and iliotibialis muscles. The tendons have been ossified to form the patella. A patellar crest **(Crista patellaris)** connects the proximal ends of the two cnemial crests of the tibiotarsus. Ballmann (1969 a) reports that the proximal two-third of the tibiotarsus is mostly three-sided with cranial, medial and caudal surfaces. At the distal end of the tobiotarsus are located the medial and lateral condyles for articulation with the tarsometatarsus and, between these two condyles, lies the trochlea as a wide furrow serving as articular surface for the **cartilago tibialis**. The fibula does not ossify in most ratites and tinamous.

The digits of the feet in different birds have been described in details under 'Modifications of feet and toes in birds'. However, some relevant points are:

Most birds have four digits or toes; some, of course, have only three (e.g., Rhea, Cassowary, Emu among the ratite birds, Auks, most American Plovers, some Kingfishers and 3-toed woodpeckers). **The Ostrich has only two toes.** The hallux is lacking in most ratites.

The general phalangeal formula in 4-toed birds is: Hallux, two digits; second digit, three phalanges; third digit, four phalanges; and fourth digit has five phalanges. The Ostrich has two digits and the hallux is elevated in certain terrestrial birds (e.g., Cranes). (See Chapter 30, Fig.14.)

21.4. MUSCLES

Romer and Parson (1986) comment that from locomotion to the circulation of blood—or even in the highest functioning of a human brain—all are either caused by or associated with muscular activity. Such activity is also of primary importance in the production of body heat.

21.4.1. Classification of Muscles

Avian muscles, like other amniote vertebrates, have been variously classified based on their histology, embryology, locatoin and functional differentiation.

From histological and embryological viewpoints, muscles are distinguished into two major types and a third type—the cardiac muscles—which form a special category found in the heart and are not present in any other organ. The two major types are: **Striated and Smooth non-striated** muscles.

i. **Striated muscles:** The huge bulk of skeletal muscles initiating movements of skeletal parts are formed of striated fibres. These are voluntary muscles derived mainly from the embryonic myotomes. They are somatic and arise from mesodermal somites—directly or indirectly (exception: branchiomeric muscles). The muscle fibres are large cylindrical multinucleate cells which range in length from 0.04 in./1 mm to several inches/ cms. The fibers run parallel and are arranged in bundles or fasciculae as the connective tissues bind them together. These muscle-fibres show distinct cross-striations at regular intervals.

ii. **Smooth muscles:** Non-skeletal smooth muscles (as the cells are without cross-banding) are found in the walls of hollow viscera, tubes and vessels, eyeball,dermis and various other sites. Smooth muscle-fibres are simpler and smaller than the striated fibres. They are spindle-shaped and uninucleate and may remain either scattered or (very often) arranged into bundles. They are derived from the embryonic mesenchyme, and, so, primarily mesodermal in origin. These cells develop in association with connec-

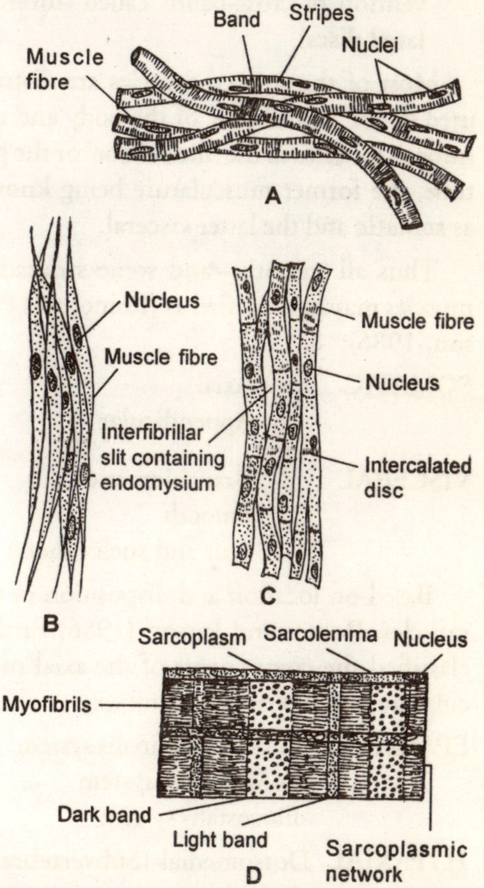

Fig. 21.8: Types of muscle fibres longitudinally disposed. A. striated or striped fibres; B. non-striated or unstriped fibres; C. cardiac fibre with intercalated discs; D. longitudinal section of striped fibres showing some details

tive tissues. The smooth muscle cells arise in connection with the connective tissues and they are typically involuntary, visceral and without cross-banding.

iii. **Cardiac muscles:** It is a special type of striated visceral musculature constituting the myocardium of the heart, and are not found elsewhere. Except similarity in cross-banding, cardiac muscles differ from the striated fibres. These muscles form a continuous network of dividing and rejoining strands which are separated into shorter units, through intervention of cross-bands called **'intercalated discs'**.

Most of the striated muscles are distributed in the **'outer tube'** of the body and the smooth muscles in the **'inner tube'** or the gut tube, the former musculature being known as **somatic** and the latter **visceral**.

Thus all smooth—and some striated—muscles may be classified as (Romer and Parson, 1986):

SOMATIC	Axial Appendicular
VISCERAL	Branchiomeric Smooth (gut and such others)

Based on location and disposition of the muscles, Romer and Parson (1986) further classified the components of the axial musculature of the tetrapod trunk as:

EPAXIAL	Transversospinalis system Longissimus system Iliocostalis system	
HYPAXIAL	Dorsomedial (Subvertebralis)	
	Lateral	Obliquus abdominis externus (Subcutaneous) Superficialis = Supracostals Profundus = External intercostals Obliquus abdominis internus= Internal intercostals Transversus abdominis = Subcostals
	Ventral	Rectus abdominis

The epaxial and hypaxial muscles are innervated by the dorsal and ventral rami of spinal nerves, respectively.

21.4.2. Nomenclature of Muscles

Skeletal muscles may be nomenclatured on the basis of their location, attachments, number of parts, shape, size, direction of fibres, action and other considerations. Kent (1983) comments that homologies between muscle groups are more easily discernible than homologies between muscles. Embryonic origin and nerve supply are the most reliable criteria.

Like other amniote vertebrate muscles, the nomenclature in avian muscles has also undergone many changes. Thus an exhaustive list of synonymy of muscles has evolved primarily due to intensive studies in avian anatomy. Although naming of certain muscles is based on the skeletal elements to which they are attached (e.g., M. quadratomandibularis, M. occopitomandibularis, M. sphenopterygoquadratus, M. intermandibularis, M. branchiomandibularis, M. pterygoideus), and some on the location of the muscles with respect to an organ or a skeletal part (e.g., M. supracoracoideus, M. hyoglossus, Mm. syringeales, M. ceratoglossus); the large majority of the muscles are named on the basis of their actions (e.g., M. adductor mandibulae, M. depressor mandibulae, M. protractor pterygoidei et quadrati, M. tensor patagialis, M. constrictor glottidis, M. dilator glottiidis,

M. expansor secondariorum, M. pronator superficialis, M. extensor carpi radialis, M. flexor digitorum, Mm. levatores costarum).

However, whatever might have been the basis of nomenclature of the muscles, their actions are referred to as either of the following contrasting sets:

Adductor and **Abductor**
Levator and **Depressor**
Protractor and **Retractor**
Flexor and **Extensor**
Pronator and **Supinator**
Constrictor and **Dilator**,

and also **rotator**,**sphincter**, **tensor**, etc.

Muscular attachments to bones and cartilages are never direct, but they are always intervened by some sort of collagenous tissues, even in so-called 'fleshy attachment'.

A thin sheet of connective tissue, the **epimysium**, encloses a complete muscle, but, within the muscle, the bundles are separately surrounded by **endomysium** and **perimysium**. Usually a skeletal muscle spanning between two bones or bony parts is said to have an **origin** and an **insertion**. The origin-end of the muscle is the one that is less mobile, more stable (but not static), and the site from which the muscle is said to arise, whereas, the insertion-end usually causes movement of the skeletal part to which the muscle is attached through connective tissue fibres (**aponeuroses** and **tendons**). However, certain muscles cause movements of the skeletal parts at both the ends (e.g., M. pseudotemporalis profundus, M.adductor mandibulae posterior, M. pterygoideus). Aponeuroses and tendons may also develop strongly at the origin-end of many muscles (e.g., M. adductor mandibulae externus superficialis pars temporalis, M. pseudotemporalis profundus).

Among the connective tissue elements, **aponeuroses**, **tendons** and **ligaments** play important roles in the functioning of muscles in birds. The **aponeuroses** are fine, thread-like connective tissue fibres appearing as closely parallel, divergent or convergent threads or as thin flat sheet of connective tissue fibres which accomplish attachments of muscles fibres at the sites of their origin, insertion, or both. They are much finer than **tendons** which are usually thicker, cylindrical or subcylindrical assemblage of connective tissue threads that exert much stronger force or pull on the skeletal elements than that produced by the aponeuroses. Thus, through tendons, the contraction of the muscle is able to initiate sustained force of action on the skeleton whereas the aponeurotic or so-called 'fleshy' attachments cause weaker—but quicker—actions on the skeletal parts.

The arrangements of fibres along aponeuroses and tendons vary in diffeent muscles and in different birds. Such arrangements may be simple, parallel or pinnate (also termed as 'pennate'). Parallel-fibred muscle may be strap-like or band-like (e.g., M. mylohyoideus). The **unipinnate** muscle is one in which the fibres are arranged convergently either at the end or side of a single aponeurosis or tendon (e.g., M. pseudotemporlis superficialis; M. extensor digitorum longus). In **bipinnate** type, the fibres are either arranged along both sides of a flat tendon or aponeurosis or two muscle slips are attached to two separate aponeuroses that may unite to form a single, broad aponeurosis before insertion (e.g., M. flexor digitorum longus and flexor hallucis longus) and, in **multipinnate** type, bundles of muscle fibres are complexly arranged along several aponeuroses which may unite to form a single, broad aponeurosis or tendon before its attachment to a skeletal part (e.g., M. pterygoideus; M. deltoideus).

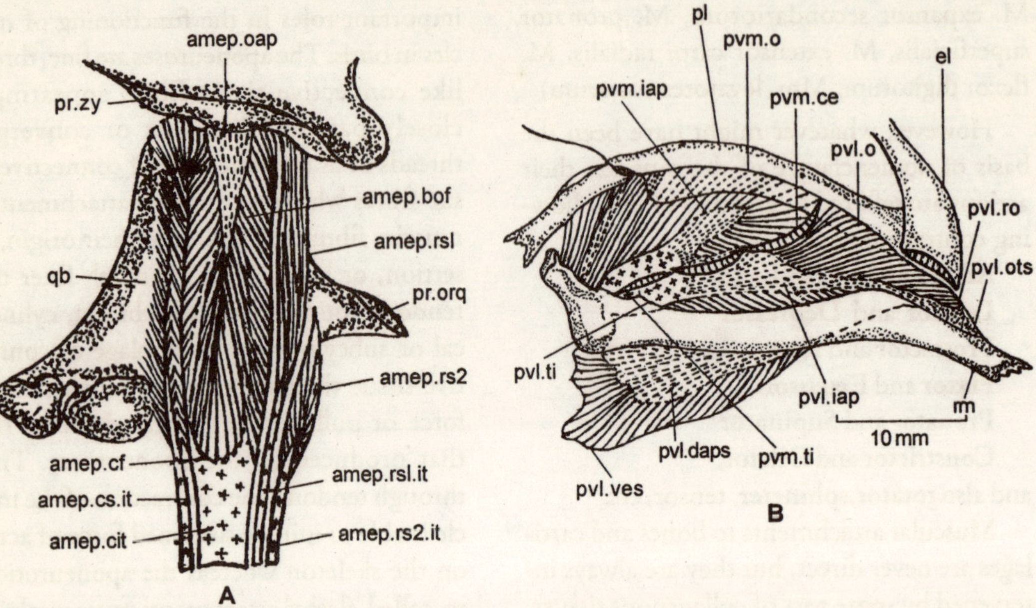

Fig. 21.9: Aponeuroses, tendons and fibre arrangements of muscles. A. Unipinnate slips of muscles (amep.rs1, amep.rs2) and bipinnate arrangement of fibres (amep.bof) along the origin-aponeurosis (amep.oap_; B. M. pterygoideus showing complex multipinnate arrangement of fibres between lateralis and medialis components of the muscle. Also, aponeuroses and tendons have been clearly shown in both the diagrams; amep.rs1 and 2, M. adductor mandibulae exeternus profundus, rostral slip 1 and 2; amep.bof, bipnnate origin-fibres of the same muscle; amep.oap, origin aponeurosis of the same muscle; pvl.iap and pvm.iap, insertion-aponeurosis of the lateralis and medialis parts of the pterygoideus muscle; pvm. ce, cut end of the mdialis part of the muscle where some fibres have joined those of the pvl (Bhattacharyya, 1998)

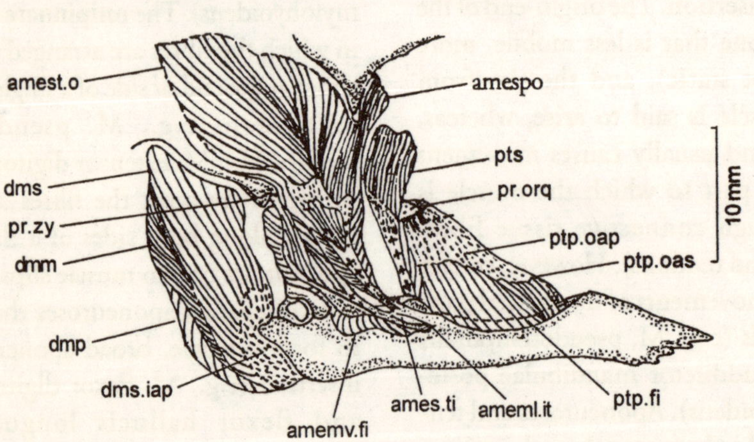

Fig. 21.10: Lateral view of the skull in Treron (Fruit Pigeon) showing a typically unipinnate muscle, M. pseudotemporalis superficialis. The spoon-shaped muscular part is attached to a thin, flat and strong insertion–aponeurosis. Also shown that an extensive aponeurotic sheath is formed at the origin site of the M. pseudotemporalis profundus; pts, M. pseudotemporalis superficialis; ptp.oas, M. pseudotemporalis profundus, origin aponeurotic sheath (Bhattacharyya, 1998)

According to recent terminology (NAA, 1993), the principal subdivisions of avian skeletal musculature are listed below with names of important muscles under each subdivision. The description of the muscles has not been included in this short text. However, it is relevant here to mention that studies on avian skeletal muscles can be fully understood in the context of osteology and arthrology of the skeletal system.

Subdivisions of Avian Muscles	Names of Muscles
1. MUSCLES OF THE SKIN AND SUBCUTANEOUS MUSCLES	M. adductor rectricium M. expansor secondariorum
2. MUSCLES OF THE EYEBALL	M. rectus dorsalis M. rectus ventralis M. rectus lateralis M. rectus medialis M. obliquus dorsalis M. obliquus ventralis Mm. membranae nictitantis
3. MANDIBULAR MUSCLES	M. adductor mandibulae externus Pars rostralis (temporalis) Pars ventralis (medialis) Pars profundus M. pseudotemporalis superficialis M. pseudotemporalis profundus M. adductor mandibulae caudalis M. pterygoideus M. ethmomandibularis M. protractor pterygoidei et quadrati M. depressor mandibulae
4. MUSCLES OF THE HYOID APPARATUS, LARYNX, TRACHEA AND SYRINX	M. intermandibularis ventralis (M. mylohyoideus) M. serpihyoideus M. stylohyoideus M. branchiomandibularis M. ceratoglossus M. hyoglossus M. genioglossus M. cricohyoideus M. constrictor glottidis M. tracheolateralis M. sternotrachealis

Subdivisions of Avian Muscles	Names of Muscles
	M. cleidohyoideus M. syringealis (dorsalis and ventralis) M. tracheobronchialis
5. VERTEBRAL MUSCLES	M. biventer cervicis M. complexus M. splenius capitis M. rectus capitis (dorsalis, lateralis and ventralis) Mm. cervicales dorsales, laterales and ventrales Mm. interspinales Mm. iliocostalis et longissimus dorsi
6. MUSCLES OF THE TRUNK AND CAUDALS REGION	M. scalenus (in *Gallus*) Mm. levatores costarum Mm. intercostales (externi and interni) M. costosternalis M. sternocoracoideus M. rectus abdominis M. obliquus (externus and internus abdominis) M. transversus abdominis
7. CAUDAL AND CLOACAL MUSCLES	M. bulbi rectricium M. adductor rectricium M. levator caudae M. depressor caudae M. caudofemoralis M. pubocaudalis (externus and internus) Mm. cloacales: M. sphincter cloacae M. transversus cloacae M. depressor cloacae M. levator cloacae M. dilator cloacae
8. MUSCLES OF THE WING AND THORACIC MEMBRANE	M. rhomboideus (superficialis and profundus) M. serratus (superficialis and profundus) M. scapulohumeralis (cranialis and caudalis)

Subdivisions of Avian Muscles	Names of Muscles
	M. pectoralis M. supracoracoideus M. latissimus dorsi M. deltoideus M. triceps brachii M. biceps brachii M. expansor secondariorum M. brachialis M. pronator (superficialis and profundus) M. flexor carpi ulnaris M. flexor digitorum (superficialis and profundus) M. extensor carpi radialis M. extensor carpi ulnaris M. supinator M. extensor (longus and brevis) alulae M. abductor alulae M. adductor alulae M. flexor alulae
9. MUSCLES OF THE PELVIS	Mm. iliotibiales M. iliofibularis M. ambiens M. iliotrochantericus M. iliofemoralis Mm. femorotibiales M. ischiofemoralis M. extensor digitorum longus M. fibularis (peroneus) longus and brevis M. gastrocnemius M. flexor digitorum longus M. flexor hallucis (longus and brevis)

22

Flight Mechanism and Types of Flight in Birds

The amazing history of the evolution of birds from their reptilian ancestors—more specifically from the dinosaurian reptiles—leaves the imprint that flight evolved from the Triassic archosaurian reptiles along two independent lines in the **pterodactyls** and birds. In the latter, true flight evolved only once. To conquer the air, the birds had gone through a series of remarkable changes in their bodily structures—both external and internal. The pterodactyls were not feathered bipeds like the birds, nor could they possibly walk on land or have any flapping flight. Thus the problem of flight in air has been solved differently by different groups of birds. But the principles and mechanics of flight in birds, and in the man made heavier-than-air flying machines, remain almost the same. And that is why Leonardo da Vinci (1452-1519) studied and sketched numerous bird-flights before designing, 500 years ago, his 'flying machines'.

The present discussion briefly explains the mechanism of flight in birds in general and that followed in a particular type of flight. An emphasis has been laid to explain the role of flight feathers, either of the wing or the tail, both in taking off or in landing.

Aymar (1935), Engels (1941), Brown (1953), Newman (1954), Fisher (1955), De Beer (1956), Bock (1964), Pennycuick (1968, 1972, 1975), Brodkorb (1971), Farner and King (1971-75), Ostrom (1979), Young (1981) and many others explained possible steps in the evolution of flight as well as aerodynamic, mechanical and psycho-biological factors associated with the flight mechanism of birds.

22.1. MODIFICATIONS OF THE EXTERNAL AND INTERNAL STRUCTURES OF THE BODY FOR FLIGHT

To achieve flight in air through a number of pro-avian grades, the bodily organisation in birds had to be perfectly modified in response to an unusual environment such as the aerial mode of life. The volant adaptations in birds are so remarkable that a carinate bird has been compared to a heavier-than-air flying machine which the nature has built up with every scientific precision.

The modifications encompass both external and internal structures:

22.1.1. External Structures

1. Bird's body is spindle-shaped and streamlined for smooth progression and sailing in air.

2. The forelimbs modified as wings are attached high up in the thorax for the suitability of their movements.
3. Feathers derived from epidermis are unique exoskeletal structures in birds. They regulate body temperature and the flight feathers act as the principal organs of flight. The feathers avoid friction caused by the air-current and take part in protection and sexual display.
4. The wing feathers (Remiges) act as oars and the tail feathers (Rectrices) act as the rudder and sail.
5. The wings and tail provide superb architectural design for supporting planes capable of adjusting to variations in the direction and strength of the wind flow (through slots, emerginations and cambering of feathers).
6. Both striated and non-striated muscles present at the bases of feathers control their position and thus regulate heat-loss, flight and other activities.
7. The larger **prepatagium** and the smaller **postpatagium** are important integumentary modifications which stretch as folds of skin between the upper arm and the trunk, respectively. Both perform the extension and shortering of wings.

22.1.2. Internal Structures

The internal structures, in both hard and soft parts, are modified such that the birds can skilfully and sustainably fly in air. The avian flying machine is basically built on the plan of a biped archosaurian reptile that gradually reached perfection to volant life. This switch-over through millions of years of evolution is comparable to the mechanical precision invented by man.

A. Skeletal System

1. In conformity with the spindle-shaped body, the internal structures within the body of a bird are very conveniently centralised so that, in flight, the wings are aided by such centralisation of organs.
2. The occurrence of struts and trabeculae and a number of air sacs and their extensions in the bones has rendered pneumaticity to the skeletal system.
3. Frequent fusion, increase in width and pneumaticity of bones provide lightness and rigidity to the entire skeletal system in birds. This is possible due to a distinct phenomenon that the entire skeletal system works on the principle of hollow girders of which there are several small girders of bones and two large plated girders–the **sternal girder** and the **synsacrum.** Most of the body weight is channelised through these large girders.
4. The broad and strong **sternum** with a vertical **keel**, the short and broad pillar-like coracoid, the fused **notarium** (in many birds) and the synsacrum, the V-shaped characteristic clavicles fused ventrally at the sternal end to form the **furcula** that acts as a spring during wing-strokes and the tail reduced to a fused, upturned **ploughshare bone**–all combine to form some of the distinctive features of the skeletal system in birds. Such modifications are all destined toward acquisition of flight in birds.

B. Digestive System

1. This system is highly developed because of the great expenditure of energy required in flight.
2. There being no mastication, the food is swallowed intact, passed down the long

distensible gullet either direct to stomach (or stored temporarily in the oesophageal crop in birds) which is divided into a glandular **proventriculus** secreting enough of peptic juices and a highly muscular, crescentic **ventriculus** or **gizzard** for churning the food. There being no gall bladder the bile from the liver and the pancreatic juice from the pancreas pour directly into the U-shaped loop of the duodenum by several ducts. The assimilation of the essence of food in the small intestine becomes much quicker.

C. Cardiovascular System

Cardiovascular system is very highly developed as in mammals; basal metabolic rate and temperature considerably higher than that in mammals (about 42°C/108°F); complete separation of respiratory and systemic circulation as in mammals, effecting higher pressure of blood in arterioles. This, in turn, causes rapid transfer of materials to the tissues.

D. Respiratory System

The presence of anastomosing and intercommunicating air capillaries in avian lungs, accompanied by the air sacs, has several advantages—such as one-way flow of air through the air capillaries, renewal of lung air twice at every breath, extension of air sacs in the pneumatic cavities of bones, high rate of respiration compatible with higher rate of O_2 demand than in other vertebrates.

E. Brain and Sensory Organs

1. Very large cerebral hemispheres (large relative to the body in comparison to other vertebrates, except in mammals) indicate highly organised voluntary movements or **conditioned reflexes, instinctive behaviour, intelligence, memory**, etc. The large forebrain in *Corvus* and in the parrots is especially significant in claiming a very high degree of intelligent behaviour in these birds.
2. The cerebellum in birds is large, folded and with **lateral flocculi** and a central body. The cerebellum in flying birds acts as a perfect organ of balance at all planes and spaces during flight.
3. The birds possess an outstanding visual sense. Large optic lobes in the brain and very large eyes (as in hawks and owls excel in size that of man) may be correlated with the telescopic vision in birds.

F. Muscular System

1. Muscles in the wings and legs of birds are highly specialised for flight, perching, walking, running, swimming, diving, climbing, etc.
2. There are nearly two dozen flight muscles, either directly or indirectly, taking part in the movements of the wings. Among these the pectoralis, coracobrachialis, scapulohumeralis, deltoideus, triceps, biceps, tensor patagialis, flexor and extensor groups of muscles are very important. These muscles, present in antagonistic sets, perform all sorts of movements of the wing. As in the wings, the tail feathers are also provided with **protractor** and **retractor muscles** attached at their bases.
3. The movement of the tail feathers is controlled by the **'tail bulbs'** present at the bases of the feathers and supplied by the internal and external oblique muscles and the iliocaudalis muscles.

22.2. MECHANISM OF FLIGHT IN BIRDS

Bird's flight has been compared to that by man-made heavier-than-air machine—the aeroplane. Indeed, the speed, skill, craftsmanship and precision achieved by birds in their variety of flights are no less remarkable than those invented by man in aeroplanes. In fact, the fastest-moving animal in the world is not the cheetah with its 70mph/110 kmph speed, but a bird—peregrine falcon—which swoops on its prey (at over thrice the speed of the cheetah) at 217mph/350 kmph!!

Aerofoil is the name given to an object with a plane surface moving through air in the direction inclined at an angle. The object moving so comes mainly under the action of two forces : (i) a '**lift force**' that tends to lift the object vertically and helps to propel the object forward. This lift force is equal to or greater than the weight of birds. (ii) a '**drag force**' or a backward force that tends to pull the forward moving object backward. The weight of a flying bird or of a man-made heavier-than-air machine is supported in air on this principle. But 'unlike the fixed wings of an aeroplane, the wings of a bird function both as an aerofoil (lifting surface) and as a propeller for forward motion. The avian wing is admirably suited for these functions, consisting of a light, flexible airfoil' (Pough, Heiser and McFarland, 1990). In birds with flapping flight, the primary feathers attached to hand mainly propel the body forward, whereas the secondary feathers attached to the forearm lift the body. Removal of a few primaries from the flight feathers of a bird of average size (e.g., pigeons, doves, etc.) incapacitates the ability of flight no doubt, but it has been observed that even after removal of more than 50 percent of secondary feathers from the body, the bird can still fly.

A fish progresses against resistance of water which is displaced through paddling. Flight in air is much more complicated due to differences in speed of air currents around the aerofoil. This is accomplished by the wings which by their complex morphological nature follow precisely the principles of aerodynamics. While a streamlined body like that of a bird cuts through the air, the air is equally driven upward and downward, thus reducing the air pressure equally on the dorsal and ventral surfaces of the body. This is a stable condition in flight in which no lift is possible. The air pressure on the upper surface and undersurface of the wings keeps the body afloat and, in such undisturbed condition of air currents, the bird smoothly steers ahead. However, the lift can be achieved either by changing the angle of the attack or by a **cambered wing** i.e., by a dorsoventrally convexo-concave wing. In either case, air pressure on the dorsal surface of the wing is much reduced compared to that under ventral surface of the wing, and this generates lift of the wing.

According to Lighthill (1975) and Pennycuick (1972, 1975), both the lift and drag forces are proportional to the square of the speed, and the requirement for sustained flight in still air is that the object shall have sufficient speed to generate a lift force equal to its weight.

The wing of a bird is not a simple aerofoil as it appears but its shape, size and morphological nature are much more complicated. The front edge of the wing is thicker and it gradually becomes thinner as it tapers behind. Its upper surface is convex and the lower concave. The air flows with much greater speed over the convex upper surface than on the lower. According to **Bernoulli's law** in physics (Bernoulli, a Swiss mathematician, 1738) **'in a fluid stream, the pressure is least where the velocity is greatest'**. As per this law, the speedy airflow over the upper surface of the wing causes decrease in pressure there, and this allows the lift force to keep the body of the bird aloft. If a bird tilts its wings more sharply, the air pressure on the concave underside of the wing increases and now the flow of air, instead of coursing along the upper surface of the wings, becomes **'turbulent'**, particularly at the

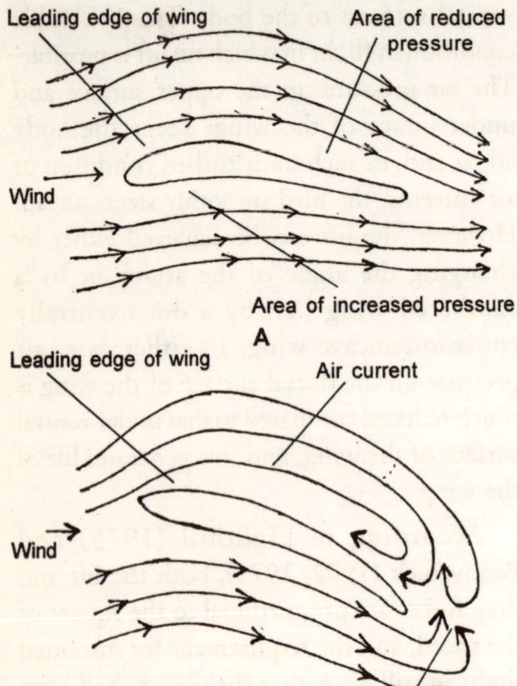

Fig. 22.1: Role of air current flowing over the wing. A. The wing being placed conveniently, the air current passes smoothly over and underneath the wing, thus helping in its lifting; B. The wing being sharply inclined turbulence sets in above and behind the wing, thus decreasing lift of the wing, (based on and modified from Young, 1981)

hinder edge of the wing and then the lift of the wing is destroyed. While in flight, the aerofoil is maintained above the critical speed.

If the aerofoil falls below the critical speed, the wings stall (i.e., drops suddenly without being supported from underneath), the bird's movement is slowed down and it falls rapidly towards the earth until the downward movement is restricted by several other means. Some birds may stall wilfully before diving or swooping on their preys. In landing, however, a bird slows down its speed without stalling to avoid crashing on land or on the tree-branch for perching. This mechanism of slowing down is achieved in several ways, one of which is by reducing turbulence. The eagles, vultures and swallows, while slowing down to perch, spread their primary feathers widely causing gaps or slots between the leading edges of these feathers. Air passes through these slots with much greater velocity, resulting a lift in the body at the slower landing speed. Alongside, some primary feathers may have their vanes reduced marginally or **'emarginated'** to help increase in the size of the slots.

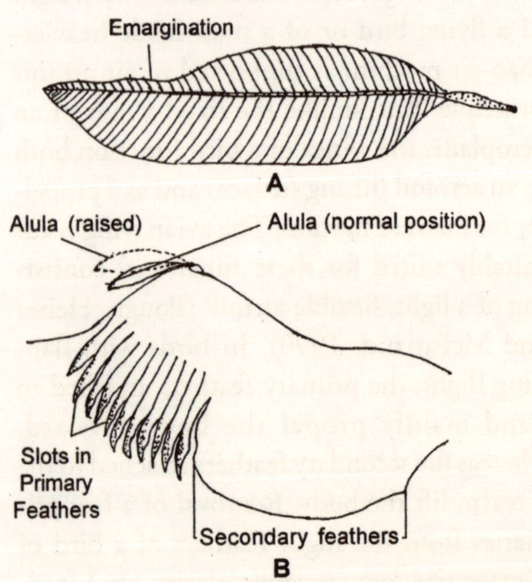

Fig. 22.2 : Role of emargination and slots in primary feathers. A. Emargination at the rostral edge of a flight feather in a passerine bird; B. the slot formed by raising the alula and the slots in primary feathers control turbulence caused by disturbed air pressure

The other method of producing a slot through **'venturi effect'** (i.e., by forming air tunnel) is possible by raising the **'bastard wing'** or **alula** which is present as a separate tuft of feathers attached to the pollex (thumb). In landing process, some birds may achieve an additional lift by fanning out and bending their rectrices.

The avian wing has several structural outfits that are very important factors in determining flight in birds. Such wing characteristics are briefly described:

1. **Shape of wing:** The shape of wing or aerofoil includes several gradients which play significant role in aerodynamic capacities and variations in the flight of birds. The morphology of the wing and its shape includes several features like the outline and taper of the wing, the aspect ratio, the wing area and loading, the wing tip, slots, camber, etc. which by their variable proportion make a difference in flight.

2. **Wing area and loading:** As the drag force on the area of the wing multiplied by square of the speed is least for high speeds, a small area of the wing is commonly an advantage for speedy flight (although there are several exceptions) as observed in the flapping flight of smaller birds. On the other hand, a large wing area results in slower flight, because the lift force acting on the wing multiplied by square of the speed is much less than that acting on smaller wing area. Here the drag force acting on the wing becomes much greater. This kind of slower flight is observed in hunting birds which slowly swoop on their preys (e.g., hawks, eagles, vultures) or in storks which, with their large wings, fly rather slowly or in medium speed to cover sometimes long distances. Also, the birds soaring on thermal currents fly at slower speed and depend largely on greater lift force.

Keeping pace with the variations in the wing area, the loading of the wing also varies considerably. However, the larger birds usually have larger wings. Loading is considered as weight of the bird divided by the wing area (g/cm^2). For example, a very small bird like the Ruby-throated humming bird, with a body mass of 3 g and a wing area of 12.4 cm^2, has a wing loading as 0.24, and a larger bird like a Golden eagle, having a body mass of 4,664 g and a wing area of 6,520 cm^2 has the wing loading as 0.71 (Pough et al, 1990). Thus it is evident that, in general, birds with lighter wing loading spend lesser power in sustained flight than birds with heavier wing loading. Yet a humming bird with least body mass (3 g) but with a heavier wing load is a more powerful flier than a barn swallow with much heavier body mass but with a lighter wing load and the latter quite often resorts to soaring flight. Hence, it is quite possible that wing loading is not only related to the body mass and wing area of bird but it is also related to the nature of flight of a bird (flapping, gliding, soaring, etc.).

3. **Aspect ratio:** The length/width ratio of a wing is referred to as its aspect ratio. Fast-flying smaller birds with small wing area have a proportionately reduced drag, but many fast-flying medium to large-sized birds have a large wing span which helps in gliding down slowly. Swifts and swallows—that are known as fast-fliers by flapping flight—possess long, narrow wings that have a high aspect ratio, and, consequently, a high lift/drag (L/D) ratio.

Among sea birds, the albatrosses are fast gliders and soarers having large wing span and a very high aspect ratio. However, wings span with high aspect ratio may stall at high speed in which case the stalling is rectified and the bird restores its normal position. Low aspect ratios generally occur in birds which fly slowly by soaring flight, either taking or without taking advantage of the upward thermal current of air. In either gliding or soaring, energy expenditure is much less, compared to that in flapping flight.

Young (1981) reports the aspect ratios of a few birds:

Albatross	:	25
Gull	:	11
Swift	:	11
Vulture	:	6
Sparrow	:	5

4. **Wing tips, slots and camber:** One of the important characteristics of the wing as an aerofoil is that the wing-tip is subject to opposing currents of air that create a vortex around it causing **turbulence**. The air pressure at the wing-tip is produced in two directions: outward flow of air from under the wing and inward flow from above. This has been referred to as '**induced drag**' which is maximum at the wing-tip. Such irregular vortices of air at the wing-tip and the consequent induced drag may be controlled either by lengthening the wing and thus increasing the wing area or by tapering the wing in such a way that the area at the wing tip region is highly reduced. The turbulence at the trailing edge of the wing is brought under control by producing slots or gaps between the primaries. The air can smoothly flow through these slots. Thus the slots act as auxiliary aerofoils on the leading edge of the wing and help to restore smooth flow of air, overcoming stalling due to turbulence. The alula or the 'bastard wing' also acts as a slotting device. A long, narrow and pointed wing as possessed by a falcon stalls first at its tip. Such a wing is suitable for fast fliers. A bird like a vulture—that makes skilful, planned excursions in air—has slower flight and possesses a short and broad wing with long arm feathers.

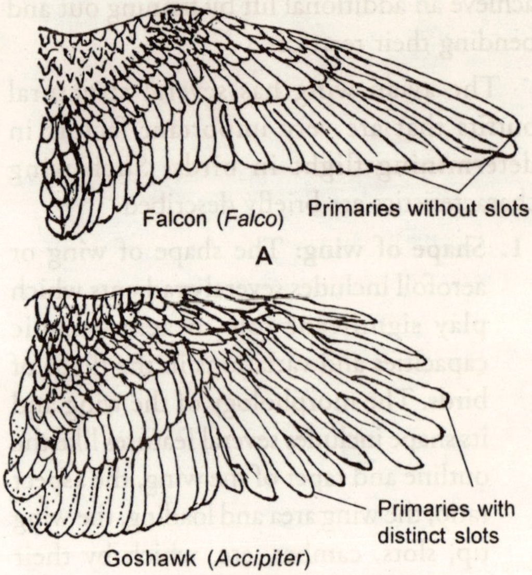

Fig. 22.3: Wing shape and role of slots in primary flight feathers. A. Wing longer and the primaries are practically without slots in the high-speed flight of a Falcon; B. Wing broader and the primaries form distinct slots in the manoeuverable flight of a Goshwak (*Accipiter*)

The camber of the wing is also an important factor in generating lift. Almost all the wings are cambered i.e., dorsoventrally convexo-concave. This results in an unequal pressure of air over the dorsal and ventral surfaces of the wing. The dorsal surface being convex, the air passes over it much faster than on the concave ventral surface. This results in a 'suction zone' of reduced air pressure on the dorsal surface, thus generating a lift of the wing. When the lift is equal to or it exceeds the bird's body weight, the bird becomes afloat in air.

The camber of the wing is variable in different types of flight, as well as along the length of the wing. Although, in general, the camber of the wing is the highest near the body and it decreases outwardly, variations occur in different species of birds. High camber, like the low aspect ratio, causes slower speed in flight. Pough et al (1990) point out that the proximal part of a moderately cambered wing generates greater lift than the distal part. 'Stalling angle' is the one that is formed by upward tilting of the leading edge of the wing. In that event, the lift of the wing is mainly caused due to decrease in pressure over the dorsal surface than the increase in pressure ventrally.

22.2.1. Types of Flight

Generally four different types of flight are recognised in birds. The flight types more or less (not strictly) correspond to the morphological characteristics of the wing, such as wing area and loading, aspect ratio, slots, camber and the nature of wing tips, etc. Further, it is to be noted that several variations may exist within each type of flight.

The types of flight are:

1. Flapping flight.
2. Gliding flight.
3. Soaring flight.
4. Hovering flight.

1. FLAPPING FLIGHT

Most of the terrestrial birds that we see around us have the commonest mode of flight, the flapping flight. The hatchlings, either nidicolous (altricial) or nidifugous (precocial) leave the nest for the first time in their life by flapping flight. And, because this inaugural flight is not learned from their parents or practised by experience, it is an outstanding phenomenon in the flight of birds. Even birds with other kinds of flight may often resort to flapping for a short duration.

'The flapping flight involves complex screw-like motion of the wing, downwards and forwards, and then upwards and backwards' (Young, 1981). However, there are so many variables within flapping flight that their working are not always very clear. Brown (1953), and more recently Pough et al (1990), have given a precise explanation of the mechanism of flapping flight at different phases. A bird can fly straight in air only when it produces a force of thrust brought about by the primaries at the wing tips. The lift is produced by the secondaries of the inner wing which is more proximal to the body and remains more or less stationary duiing flight.

Brown (1953) has given an account of the cycle of movements of the wing of a pigeon flying at 20mp/32km/hour:

1. At the beginning of the downstroke, the wings are held vertical and fully extended.
2. During middle of the **downstroke**, the wing tip feathers are twisted to initiate a forward drive.
3. Near the end phase of the downstroke, both velocity and angle of attack fall.
4. At the **upstroke** (also known as **'return stroke'** or **'recovery stroke'**), the wings are partially folded and their primary feathers are spread out in such a way that air may pass through them and this makes the lift possible with a thrust. This is immediately followed by a backward swing of the primary feathers. At the end of the upstroke, the wrist and elbow are extended and then slightly bent so that the upper surfaces of the twisted prima-

ries produce an upward thrust. The wrist is then pronated which causes the primaries to spread out for the next downstroke.

Pough et al (1990) have very clearly explained the mechanism and cycle of movements of the wings in flapping flight. According to these authors, the complex cycle of flapping flight may be summed up as follows. Although Brown (1953) laid down certain general principles of flapping flight in brief as mentioned above, the following summation by Pough et al (1990) is more elaborate, explanatory and with vivid illustrations:

1. Most of the lift, as also most of the drag, are due to forces acting on the inner wing and body of a flying bird.
2. The wing tips move forward with the bird, but at the same time also move downward relative to the bird.
3. From the positions of the inner and outer wings, during the upstroke and downstroke, it may be ascertained that vertical motion is applied mainly at the wingtip.
4. Most of the lift to counter the pull of gravity is generated by the inner wing and the body.
5. During tilting of the wingtips from level position in downstroke, the flow of air across the primaries is different from that across the secondaries and the body.
6. When a bird progresses in a constant speedy flight, the forces acting on the inner wing proximal to the body and that on the distal outer wing are summed up– and the resultant force produces a thrust that exceeds the total drag, and the lift becomes possible when the lift force is equal to or greater than the body weight.
7. At the downstroke, when the wings move downward and forward, the primaries are bent upward under air pressure, producing a thrust which exceeds the total drag and the bird accelerates. In the passive recovery stroke, which is upward and backward, the bird decelerates. In a large bird, where the wing action is slower, longer time is taken in the upstroke and deceleration of the wings.
8. During flapping flight the powerful downstroke of the wing is performed by the contraction of the M. pectoralis major and also the powered upstroke is caused by the contraction of the M. supracoracoideus (M. pectoralis minor).
9. Increase in speed of a flying bird is achieved by increasing amplitude of its wing beats, but the frequency of wing beats remains the same.
10. In many birds the frequency of respiration bears a relation to the wing beat frequency.

2. GLIDING FLIGHT

After flapping the wings in quick succession, all birds are able to stretch their wings and glide for a considerable time and distance. In stable air current, the speed in gliding remains constant. Some species of birds take the advantage of gliding and soaring with the rising air currents. In gliding, the glide path shows a curve from a higher to a lower altitude at an angle that is determined by the ratio of lift to drag (L/D ratio). Pough et al (1990) have given a simplified formula in gliding flight to show that the drag multiplied by the distance travelled is equal to the loss of potential energy of the body which is equivalent to the body weight multiplied by the loss in height.

This may be represented as:

$$\frac{\text{body weight}}{\text{drag}} = \frac{\text{distance travelled}}{\text{loss in height}}$$

Hence, different species of birds have different gliding patterns and strategies, depending on the lift/drag ratio which, in turn, considerably depends on the aspect ratio of the wing and on the size and weight of the birds. A low or a high **sinking speed** in flight means the ability of a bird to fly a long distance horizontally in relation to the distance dropped in such a flight. The **glide polar** has been defined as a curve showing relation between sinking speed and forward speed through air. In general, the soarers are perfect gliders. The gliding pattern observed in albatrosses, vultures and a few other falconid birds is far superior to that by flapping birds like pigeons, gulls, etc.

While summing up gliding in birds, Young (1981) states that it depends on several factors : (a) velocity acquired by previous wing beats, (b) descending from a higher altitude to a lower one, and (c) taking advantage of air currents. Birds may take to gliding for a reasonable length of time, but usually for a short period. For example, many shorebirds glide at the time of landing; ducks, gulls, herons and a few other water birds and coastal birds glide over water; swallows and swifts glide in air and a falcon glides to swoop on its prey.

The **minimum sinking speed** is one at which the loss of height is the minimum in forward movement of a bird. This speed is slightly higher than the stalling speed, and the best glide ratio is still higher.

3. SOARING FLIGHT

Soaring or sailing is the most specialised form of flight practised by both land birds and sea birds—albatrosses, gulls, shearwaters, petrels, vultures, falcons, storks, eagles, etc. Quite often, soaring is accompanied by gliding. Many soaring birds travel long distances in great circles without beating wings. Young (1981) distinguishes two main types of soaring birds : terrestrial birds using **thermal up-currents of air**, and sea-birds using **wind variations above the sea-level.** Thermal up-currents of air result from differential heating of the earth's surface in the sun. Soaring land birds (e.g., vultures) rise upward above the ground taking in advantage of such up-currents of air in a series of small circles up to the height of about 1,000ft/300 metres or more, and then glide slowly. Storks, known for long distance migration, economise energy in this way by soaring in their migratory flights.

Pough et al (1990) have clearly distinguished three types of soaring: **dynamic soaring, slope soaring and static soaring.** However, albatrosses and a few large sea birds may use both dynamic and slope soaring.

Albatrosses and shearwaters usually ascend through dynamic soaring which is possible only where there is a distinct vertical wind gradient of about 50 ft/15 metres or so above the sea surface. Such birds have large and long, pointed wings without slots in their outer primaries and with high aspect ratio. Sea birds, particularly albatrosses,taking advantage of dynamic soaring, are mostly found in the regions where the wind action is very strong and persistent. On reaching the top of the wind gradient, the albatross glides downward with great speed and then, from near the sea surface, it turns and glides upward, again making use of the vertical air current. The upward soaring bird does not stall until it has reached the top of the wind gradient where the velocity of air is stable.

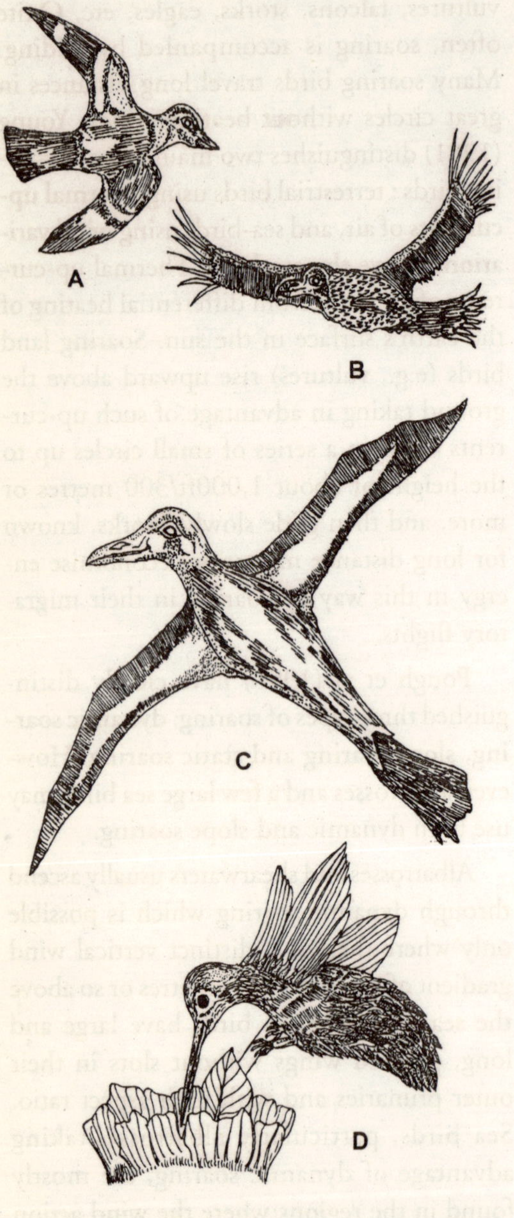

Fig. 22.4: Principal types of flight. A. Flapping flight of a pigeon; B. Gliding flight of a Vulture; C, Soaring flight of an Albatross; D. Hovering flight of a Humming bird—the Humming bird sips nectar from a flower

The **slope soaring**, also used by albatrosses and petrels (order Procellariiformes) is a rather easier method of soaring flight in which the birds take advantage of slope-lift of the wind caused by sea waves which deflect the wind upward.

Birds like vultures, eagles and storks, which are very good fliers, sometimes covering long distances, very often make use of **static soaring flight**. They possess high-lift wing having slots in the primaries and an aspect ratio intermediate between the elliptical wings with low aspect ratio (e.g., pheasants) and the long, narrow and flat wings without slots in the primaries as in the albatrosses, shearwater and petrels with high aspect ratio. The wings adapted for static soaring are provided with very good camber and slots. The birds which take the advantage of static soaring do so by gliding when the speed of the up-current of air exceeds that of the sinking speed of the bird. Through static soaring, a bird spends least energy compared to flapping flight. Soaring and gliding are very effective modes of flight for larger birds like vultures and storks, which cover long distances in their annual migrations.

4. HOVERING FLIGHT

It is an extremely specialised mode of flapping flight in which the body is kept vertical and the wings beat in very quick succession, perpendicular to the body. This characteristic hovering flight has been achieved by the humming birds (Trochilidae) and also by the old world nectar-feeders (Nectariniidae) for probing flower-nectar through their long beaks and tongue. The wing skeleton in humming birds is so constructed that it can be moved freely at the shoulder joint. During hovering at one place near the flower for quite some time, the wings beat in a horizontal plane backwards and forwards at an extraordinary speed, sometimes reaching even 150-200 beats per second (usually in 20-50 secs.)!

In hovering flight, no lift is gained by the body in absence of any forward movement through the air; hence, energy consumption is the maximum in this type of flight. Recent investigations into the flight physiology of humming birds reveal that the mechanical power expenditure during hovering increases with greater body mass and decreased air density. Due to higher rate of O_2 influx through the cardiovascular and respiratory systems, the humming birds (like helicopters) exhibit the highest rate of aerobic metabolism in relation to their body mass compared to all other vertebrates. They have also achieved high enzymatic capacity for glucose and fatty acid oxidation and high mitochondrial volume and cristae surface area in the flight muscles (even exceeding the range of those in mammals). The humming birds like helicopters possess the unique capacity to generate vertical forces and thrust during forward flight, as well as extremely rapid rotational movements.

22.2.2. Basic Types of Wings in Relation to Flight

The remarkable architecture in avian wings is reflected in four basic types of structural and functional designs that bear close relationships with the types of flight. These wing-types are :

1. **Elliptical wings** possessed by pheasants and grouses have low aspect ratio, high camber and well-developed slots in the primaries. This kind of wing is usually associated with slow flight, as these forest birds have to apply skills and manoeuevrs to evade obstacles in forests.
2. **High-speed wings** possessed by swifts, swallows, etc. have reasonably high aspect ratio, with very little camber and mostly without slots in the outer primaries.
3. **Soaring wings (both dynamic and slope soaring)**: These two types of soaring are practised by the sea birds like albatrosses, shearwaters and petrels which possess large wing span with the characteristics of a long, narrow and flat wing devoid of slots in the outer primaries. Wings of such marine birds have a very high aspect ratio. The albatrosses—one of the largest among flying birds—have their pectoral muscles comparatively smaller and weaker for which they can take off only by running or paddling or by flapping in strong winds.
4. **High-lift wings and static soaring:** High-lift wings possessed by large land birds like vultures, hawks, storks and a few

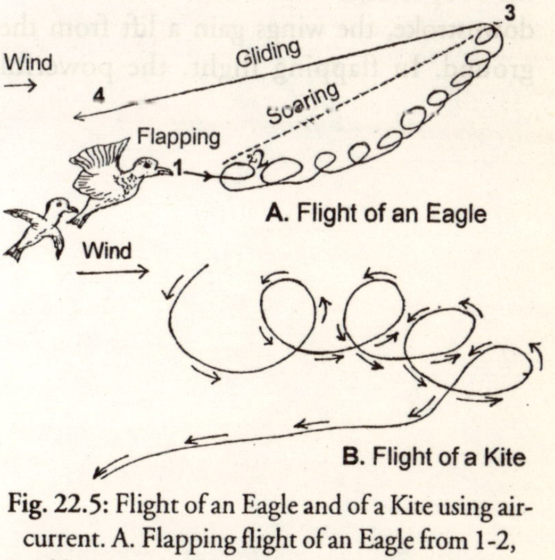

Fig. 22.5: Flight of an Eagle and of a Kite using air-current. A. Flapping flight of an Eagle from 1-2, steadily soaring with up-current in small circles from 2-3, and then gliding downward from 3-4; B.Soaring of a kite using upward and downward air current and then gliding and flapping steadily (Based on and modified from Young, 1981)

others are also of static soaring type. Such wings are characterised by an intermediate range of aspect ratio, a deep camber and distinct slotting in the primaries.

22.2.3. Mechanism of Take-off and Landing

Different birds apply different methods in taking off, but, in a generalised way, a bird has to gather sufficient forward momentum that actually causes the lift while taking off. Some larger and heavier birds (e.g., Great Bustard, Eagle, Stork) can only take off after speedily running on the ground. Albatross, being a very large and heavy bird, cannot directly take off from a calm sea surface; it can take off with the support of wind flow against waves. Larger marine birds follow more or less the same principle.

The take off sequences in the pigeons have been clearly demonstrated. While taking off, the body is held vertical and, with a forceful downstroke, the wings gain a lift from the ground. In flapping flight, the powerful downstroke results from the contraction of the M. pectoralis major. The M. supracoracoideus (M. pectoralis minor) by its contraction pulls the thick tendon of the muscle at the humeral end. This results in adduction, folding and flexion of the wings and these, together with the súpination of the wrist, bring about a quick rotation of the humerus upward and forward; the humerus is also pronated causing extension of the wings. The main function of extension of the wings, however, is performed by the M. triceps and other extensor muscles. The sum total action is translated in the form of a jerk to forward movement. In taking off from the flat ground, the bird jumps ahead to acquire the required velocity.

In **landing process**, a bird slows down its movement while coming low by fanning out its rectrices which act as both lift and brake before the bird perches on a tree-branch or comes down on the ground. Before perching or alighting on the ground, the legs are lowered and the bird comes to rest.

23

Volant Adaptation

In the discussion of the chapter on *Archaeopteryx* and the origin of birds, it has already been stated that, according to most authors, the neognath birds are monophyletic in origin, assuming that feathers and true flight developed only once and established perfection through time and space and through several post-*Archaeopteryx*, pro-avian stages.

Alongside, two other groups of vertebrates also achieved success in flight–the pterosaurs of early Jurassic (Order Pterosauria; Class Reptilia) represented by the pterodactyls and the bats (Order Chiroptera; Class Mammalia) which appeared in early Eocene, and the bats are still surviving. All these three groups of vertebrates are said to have volant adaptation by virtue of their acquisition of true flight. But mention must be made here in this connection that the pterosaurs with their extraordinarily large wing span (The *Pteronodon* of 70 million years ago had a wingspan of 23ft/ 7 metres. A larger pterodactyl, whose remains were found in Texas, USA, in 1970s, had a wingspan of 49ft/15 meters) were rather skilful gliders and soarers. Their ability to flap the wings was practically nil and, hence, they lacked in sustained flight. The bats, on the other hand, can flap their wings and have sustained flight for quite a long distance, but unlike birds, cannot cross the expanse of 20-30 miles (32-48 km) at a stretch in air. And though the petrodactyls and the birds possibly took to the air quite independently, a strange parallelism in structure and function existed between the fully evolved fliers of the two groups. Hence the birds have survived as the unchallenged conqueror of air through a series of remarkable structural and functional modifications in their bodily parts.

23.1. ESSENTIAL PRE-REQUISITES FOR FLIGHT ADAPTATIONS

In volant adaptations, certain essential prerequistes are to be fulfilled by animals having capacity for sustained flight in air. Such adaptational pre-requisites are :

1. The flying animals must overcome the downward pull of gravity which has been, and is, the problem of first importance. A lift force has to be generated by the flying animals in opposite direction, exceeding the force of downward pull.
2. During flight, the drag force pulls the body both downward and backward. The wings suitably manipulate to overcome the drag force and keep the lift/drag ratio stable.

3. The capacity to increase or decrease the flight speed, even within a small area, if so required.
4. The capacity to maintain stability in flight, even by changing flight technique, if so required.
5. True flight also demands perfect take off and landing.
6. In flying vertebrates, the lightness and rigidity of the body (and the skeletal framework) are the two key factors which develop in relation to the strength of the muscles that move the wings.
7. In flying vertebrates, the wings or the like (e.g., Patagia) are essential for flight; the wings in birds develop by transformation of the forelimbs.
8. Flying involves the senses too, like the vision of birds or the sonar system of bats. Flight presupposes a delicate sense of balance and nervous control correlated with high specialisation of the cerebellum and of the entire nervous system.
9. Flying requires a high rate of body metabolism associated with the increased efficiency of digestive, circulatory and respiratory systems.

23.2. VOLANT ADAPTATIONS IN BIRDS

Birds, being endowed with the remarkable gift of feathers, reached perfection in flight through millions of years. In flying birds, along with feathers, appeared some such congenial innovations in the entire structural and functional organisations that all these modifications may be described as volant adaptations in birds.

The volant adaptations in birds may be dealt with under the following heads:

1. Structural and functional adaptations of the wings.
2. Adaptations for weight reduction: lightness and rigidity of the body.
3. Adaptations for prolonged power-producing system and for conservation of heat: fuel and propeller system.
4. Adaptations in skeleton and muscles.
5. Adaptations for gaseous exchange: heart and lungs.
6. Adaptations for balancing: Brain and sense perception.
7. Adaptations for shape, speed in flight and streamlining of the body.

23.2.1. Structural and Functional Adaptations of the Wings

In birds the forelimbs have been modified into wings which are provided with large flight feathers or remiges and several rows of smaller wing coverts. Feathers serve not only as the chief organs of flight but they also provide a non-conducting insulation to the body for conservation of heat and a waterproof covering for protection. The structural novelty of the wing in relation to different types of flight has been discussed earlier in some details, Besides feathers, the wing comprises skeletons, muscles and nerves.

Feathers are perhaps the greatest gift of nature to birds. These are derived from epidermis, replacing scales of reptiles. The relics of scales still persist in certain localised parts of birds (e.g., scales in tarsometatarsal region). Of different types of feathers performing different functions, the flight feathers have a definite orderly arrangement on the wings and around the tail of birds. The primary remiges attached to the hand skeleton (carpals and metacarpals) are mainly designed for propulsion in flight, while the secondaries attached to the ulna of the forearm are used for producing a lift force. The tertiaries attached to the humerus also assist in the process. The characteristic shape and contour of the wing in different species of birds offer a *prima facie* indication of the type of flight.

In most birds, the wings provide an outer convex and an inner concave surface with a thicker and harder leading edge in front and a thinner trailing edge behind. The convexo-concave pattern (camber) of the wing is significant, as the low, moderate or high camber of the wing acts as a determinant for the type and speed of flight. Together with this, the wing-tips with or without slots in the primaries, aspect ratio of the wing, the presence of alula, etc. are also important characteristics of the wing influencing flight in birds. In level flight, the birds experience reduced air pressure over the convex dorsal surface and moderately increased pressure on the concave ventral surface of the wing. This results in a smooth flight, free from turbulence.

The movement of the inner segment of the wing nearer the body is controlled by its junction at the shoulder at the base of the neck. This segment holds the wing rigidly at an angle, thus lifting it upward. Outwardly, the wing becomes somewhat arched or elliptical or it may be gradually narrower and longer. The alula present at the front end of the carpometacarpus also serves as a small aerofoil assisting in flight and landing of birds. The primary remiges maintain the required angle of attack in flight and function as the main propeller in flight. In the construction of a feather, the caudal segment of the vane nearer the quill is much thicker and broader compared to the thin, gradually narrowing rostral segment. Any increase in air pressure is confronted and conveniently adjusted by the broad caudal segment of the feather. At the distal end of the wing, the wing-tip feathers may be provided with or without slots and emarginations, characteristic to the specific type of flight. The slots act as additional aerofoils which prevent stalling due to turbulence. Thus the flight feathers are–structurally and functionally–so designed that they largely contribute to the changing shape, size, angle and speed of the wing during flight.

23.2.2. Adaptations for Weight Reduction: Lightness and Rigidity

Volant adaptation with sustained flight in air demands reduction in body weight, both in hard and soft parts. The entire skeletal framework in the body of a bird has been engineered so skilfully that the bones are much lighter in relation to the size of the body, and in contrast to bones in the body of a mammal of the same size. This has been possible due to numerous air spaces and extensive network of struts and trabeculae within the bones. Yet, none the less, significant is the compactness of the bones due to fusion of several vertebral segments (e.g., notarium, synsacrum, pygostyle) and the girdle bones.

The skull too has been rendered light and compact in the carinate birds due to absence of sutures. The lightness of the skull can further be attributed to the absence of teeth and presence of comparatively lighter jaws and jaw muscles than those in many terrestrial reptiles. In fact, the skull weighs only 0.2% of the body-weight of a bird.

Besides compactness of bones, the rigidity in the skeletal system has been contributed by the alignment of several bones forming T and **L-beams** (the keeled sternum is an example of **T-beam**). Thus, in flying birds, the hollow girder system of bones with enough strength is capable of withstand variations in air pressure and rigours of flight. The humerus and scapula are broad and pillar-like and their broader heads form suitable sites for the attachment of important flight muscles. The V-shaped furcula is characteristic, as it provides additional bracing support and strength to the adjoining wings.

The scapula, coracoid, clavicle (furcula) and the humerus are all structurally modified in the manner that they can suitably negotiate the changing movements of the wings due to variable air pressure during flight. The ribs are long, flat, light and are provided with uncinate processes for additional articulations and strength. The ribs with sternum are capable of slight movement rostro-caudally during wing movements in flight.

That is why 'the bones of modern birds are so completely adapted to the need to conserve weight in flight that a bird's feathers usually weigh more than its entire skeleton....' The resulting design is so light that a golden eagle with a wingspan of 7.5 ft weighs less than 8.8 lbs/4 kg !!

23.2.3. Adaptations for Prolonged Power Producing System and For Conservation of Heat: Fuel and Propeller System

The energy consumption of a flying bird is much more than of the equal size of a mammal. The chemical energy is stored as fuel in the form of fat and carbohydrate in the large muscles of the body, particularly M. pectoralis major, which alone accounts for about 2/5th of the total body weight. Thus the huge stored chemical energy is continually converted into mechanical energy that produces excess power used by the wings in sustained flight. The wings act as the propeller in forward progression. The chain of actions in the power producing system also help to conserve more than normal rate of heat in the body compatible with the high rate of metabolism. Of all the systems working together for the purpose of perfect volant adaptation, the skeletomuscular, digestive, circulatory and respiratory systems have undergone profound modifications leading to a very quick conversion of the potential energy into kinetic form which ensures a stable production and supply of power.

The birds maintain body temperature varying between 39°C (102°F) and 42°C (107.6°F) with some exceptions below the minimal or above the maximal ranges (e.g., 44°C /111°F in some passerine birds). With the increasing heat in the body, the combustion of food becomes easier and quicker. In the process of conservation of heat, the vasomotor system and the non-conducting feathered covering of the body are of great assistance.

To augment efficiency of the source and supply of energy, the undermentioned systems and organs have been greatly modified:

i. Digestive system: Here the varieties of raw food materials engulfed act as fuel that is transformed through chemical cooking into essential constituents which, ultimately, produce energy. The alimentary organs are modified such that they are capable of churning, digesting and absorbing the essence of food in the quickest possible time. Primarily granivorous birds like pigeons and doves have an oesophageal pouch, the crop that serves for temporary storage of grains and cereals peeked from the ground and a highly muscular gizzard for thorough churning of food. The absence of a gall bladder accomplishes direct supply of bile to the duodenum for a quicker digestion. In fact, the secretions of various digestive glands, enzymes and the hormones activate thorough and complete digestion and absorption of food with minimum egestible materials.

ii) **Respiratory system and air sacs:** The respiratory system in birds comprising the compact lungs, the thin-walled air sacs and a complex system of air-tubes and channels is also unique among the living vertebrates. The narrow extensions of the air-sacs filled with hot air penetrate into the air-spaces of most of the bones, thus rendering lightness and buoyancy to the body. Unlike in the mammalian lungs through which air flows in an out (tidal air flow), **air flow through avian lungs is in one direction only.** The air-sacs, besides their several important functions, also supply oxygenated air to all the parts of the body, but gaseous exchange does not take place in the air-sacs as these are practically devoid of blood capillaries. Gaseous exchange profusely takes place within the parabronchial lung. Further, the rise in temperature in the body due to enormous muscular activity and high rate of metabolism is largely dissipated by the flow of air through the air sacs.

[See Chapter 30, Fig. 30.25.]

iii) **Circulatory system:** Avian heart is much larger in size compared to that of the similar body size of other vertebrates. The heart and large blood vessels have the capacity to carry and supply much greater quantity of oxygenated blood and return venous blood than in other vertebrates. The percentage of haemoglobin in blood is almost equal to—or a little less than—that of mammals. But the concentration of carbohydrate in avian blood is more than double compared to that of mammalian blood (Welty, 1982).

Rostrally, in the head region, the jugular veins from both sides are joined by a transverse connecting vessel called '**Interjugular anastomosis**' where the two rostral cephalic veins also meet (Baumel et al, 1993). This arrangement can minimise the pressure changes in blood due to the turning of the head through 180° degrees. The internal carotid arteries, too, for some distance in the neck region, run through the hypophysial canal (Canalis caroticus cervicalis) in the middle segment of the cervical vertebrae. In this region, the two internal carotids converge and course through this canal where the left one lies just ventral to the right and the two arteries then emerge from this canal at the rostral segment of the cervical vertebrae.

[See Chapter 30, Figs. 30.29, 30.30.]

23.2.4. Adaptations in Skeleton and Muscles

Along with the transformation of the forelimbs into wings, the functions of the forelimbs were largely taken over by the beaks formed by rostral prolongations of the skull bones. Thus the upper jaw formed mainly by the premaxilla, and the lower jaw formed by the mandible were given characteristic shape of the upper and lower beaks in different birds differently through keratinised thickenings called **rhamphotheca**. The beaks, claws, scales and feathers are all exoskeletal structures derived from the epidermis, and the first three are derived from the ancestral reptiles. **The feathers are outstanding and unique acquisition in birds.** All these structures have been strongly modified in different birds suited to their habits and habitats.

Among the endoskeletal structures, remarkable modifications have taken place in the skull and in the entire skeletal framework. Some of these changes have been acquired by birds through specialisation of several reptilian ancestral characteristics and most in response to their flight adaptations.

The bones in the skull and in a number of endoskeletal parts of nearly all the birds are hollow and pneumatic with numerous air spaces inside. Larger birds have greater pneumatisation of bones. In fact, pneumaticity of bones varies in different degrees in different species of birds, extending

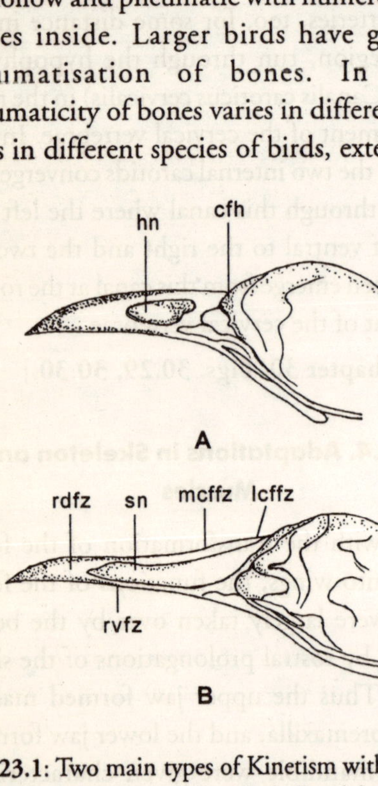

Fig. 23.1: Two main types of Kinetism with flexon zones in avian skull. A. Prokinetic skull with holorhinal nostril (e.g. crow) where at the craniofacial hinge (cfh) a distinct hiatus is formed; B. Rhynchokinetic skull with schizorhinal nostril (e.g. Plover) where the upper jaw bends at several flexon zones unlike a single cfh as in A. cfh, Craniofacial hinge; hn, holorhinal nostril; lcffz, lateral craniofacial flexion zone; mcffz, medial craniofacial flexion zone; rdfz, rostrodorsal flexion zone; rvfz, rostroventral flexion zone (Modified from Bühler, 1981; NAA, 1993)

not only throughout the axial but also the appendicular skeletons, including even the phalanges. In general, the diving birds (e.g., diving ducks) have their bones less pneumatic and, among the flightless birds, the Kiwi's skull is apneumatic. The pneumaticity and lack of sutures have rendered the skull bones light and compact so that, in modern birds, a **prokinetic** skull has evolved where the upper jaw as a whole can be moved up and down usually at the nasal-frontal hinge (e.g., in most passerine birds as *Corvus*) or it can be moved in part at definite elastic flexon zone(s) in a **rhynchokinetic** upper jaw (e.g., most charadriiform and many other birds). In either case, the upper jaw kinesis is brought about by the contraction of appropriate jaw muscles that move the chain of kinetic bones in the skull (e.g., quadrate-pterygoid-palatine and maxillopalatine complex).

The lightness and rigidity of the skeletal framework may be mainly attributed to the pneumaticity and fusion of bones, respectively. Except the flexibility of vertebrae in the neck region and at the joints between the caudal thorax and synsacrum and between the synsacrum and the free caudal vertebrae, the entire vertebral column acts as a rigid beam which is supported by the thoracic cage, the broad flat bones of the pectoral and pelvic girdles and by the wings for taking over the load of the viscera. The upturned pygostyle is also significant as it functions as a rudder by controlling movements of the rectrices. The large and flat sternum with the vertical keel and with articulation for the massive coracoid provides a superb example of a T-beam, accommodating extensive attachments for the large pectoralis muscles.

The endoskeletal modifications are in tune with the muscular modifications, either in the skull, trunk or in the legs, and all these changes have appeared as part of flight adaptations in birds. The changes in the physical and chemical nature of the fleshy parts of the muscles as well as in their connective tissue elements (tendons, aponeuroses, ligaments, etc.) for attachment to bones are also varied in different species of birds performing different locomotory functions such as flying, perching, diving, swimming, running.

The birds—according to their modes of flight and other means of locomotion—may possess different percentages of body and leg muscles correlated to their body size. Furthermore, it has been observed that a distinction can be made between the flying birds and running birds on the presence of myoglobin in muscle fibres in the former and its absence in the latter—based on aerobic and anaerobic metabolism, respectively. Sustained fliers have their breast muscles with a high rate of aerobic metabolism. Birds that are quick peckers of seeds and grains from the ground, or those that swoop insects in air, have a well-developed upper jaw kinesis and their jaw muscles are comparatively simpler with more of fleshy fibres for quicker movement of the jaws. But in flesh-eaters, fruit-eaters, nutcrackers, etc., the tendons, aponeuroses and ligaments in the jaw muscles are better developed for better force production in opening and closing the beaks. Such variations have also been reflected in the flight muscles, trunk muscles and in the leg muscles of different species of birds.

23.2.5. Adaptations for Gaseous Exchange: Heart and Lungs

Earlier in this chapter it has been mentioned that, in birds (particularly in flying birds), both the circulatory and respiratory systems have undergone several changes which may be correlated to their higher body temperature and thermoregulation, as well as their adaptation to aerial mode of life. The heart is quite large (particularly the two atria) compared to the body size and are truly four-chambered with complete separation of the right side of the heart from the left—so a perfect double-circuit blood circulation is established here with no possibility of any mixing of blood between the pulmonary and the systemic circuits. The increased capacity of the heart and of the blood vessels carrying blood to and from the heart is a great advantage in volant adaptation as it helps in maintaining the high rate of metabolism in birds. The thick and broad right arch of the aorta, the larger axillary, subclavian, pectoralis and other arteries supplying the wing muscles are distinctive features. The coronary, portal and the lymphatic systems are also well-developed—almost comparable to those in mammals.

As regards gaseous exchange, it has been mentioned that it takes place only in the parabronchial lung. In birds, both inspiratory and expiratory air flows through the parabronchial lung in the same direction. However, during inspiration, with the increase in volume of the thoracic chamber, following contraction of the external intercostal muscles, air is drawn in by the posterior thoracic and abdominal air sacs and by the parabronchial lung. Therefrom the air enters into the anterior thoracic and clavicular air sacs. Gaseous exchange takes place through an extensive interwoven network of air capillaries and blood capillaries which flow in opposite direction. A cross-current system of exchange is established between the air flow and the blood flow. This arrangement increases the efficiency of gaseous exchange in several species of high altitude fliers.

(See also Chapter 30; Fig. 30.)

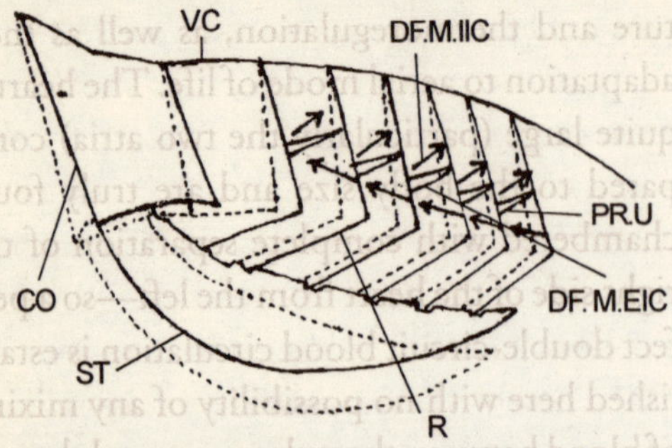

Fig. 23.2: Schematic diagram showing inspiratory movements in the thoracic cage of a pigeon, mainly initiated by the external intercostal muscles. The dotted lines depict forward movements of the ribs and downward movement of the sternum, thus increasing space in the thoracic cavity. CO, coracoid; DF.M.EIC, Direction of force of the external intercostal muscles; DF. M.IIC, Direction of force of the internal intercostal muscles; PR. U, Uncinate process; R, Rib; ST, Sternum; VC Vertebral column.

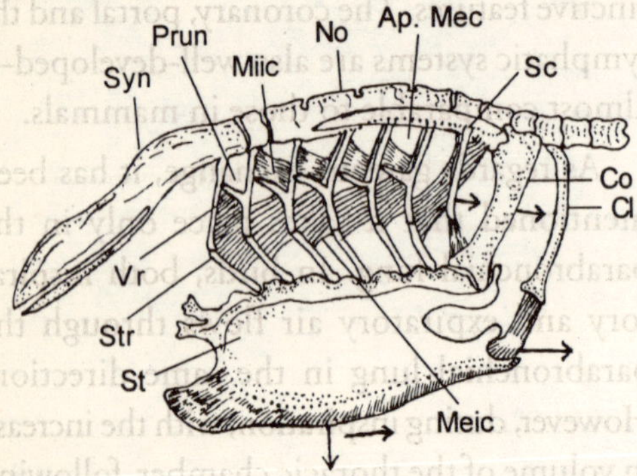

Fig. 23.3: Lateral view of the thoraco-abdominal region of Gallus showing external and internal intercostal muscles. Ap. Mec, Aponeurosis of the external intercostal muscle. Cl, Clavicle; Co, Coracoid; Meic External intercostal muscle; Miic, Internal intercostal muscle; No, Notarium; Pr. un., Uncinate process; St, Sternum; Str, sternal rib; Vr, Vertebral rib. The arrowheads indicate direction of forces due to muscular action
(simplified and modified from Yasuda, 2002)

23.2.6. Adaptations for Balancing: Brain and Sense Perception

The brain and the sense organs coordinate movements of birds–either in flight or at rest—like instruments of high precision. The heavier viscera being placed nearer and the wings being positioned above and in front of the centre of gravity, the tail in reptiles—that was so essential for the maintenance of balance—is nearly lost in birds.

The wings, hindlegs, the pectoral and pelvic girdles and the sternum are remarkably modified in birds. The elongated, fused and rigid pelvic girdle, together with tarsometatarsus of the hindlegs, are admirably suited for safe and accurate landing—either on the perch or on land. The muscles of the hindlegs are located proximally but their long tendons, extended distally up to the toes, control movements of the feet and toes. The air sacs, being located in bilateral symmetry around the lungs, are important in controlling and adjusting balance of the body like a ballast by shifting air from one side to the other. The large gizzard and the symmetrically placed air-sacs push the livers lower and more towards the right side of the body, and a balance has been maintained by placing the ovary and the oviduct on the left side only. This helps not only in the maintenance of balance of the body but also in egg-laying—through widening of the body in this region that keeps the feet apart.

The **uropygium** with its rectrices acts as a rudder which functions as a great balancing organ by moving the rectrices vertically as well as laterally. By lifting the rectrices, the body is raised and broadened obliquely and, by lowering the rectrices, the body may assume a downward curve. The bird maintains its balance in landing process by fanning out or spreading the rectrices. But in altitudinal flight, the fanning out of rectrices in turbulent air flow may be the reason for toppling of the body upside down. The birds, under such circumstance, skilfully employ the devices (e.g., slots, alula, emargination of feathers at wingtips) for restoration of their normal position.

In maintenance of equilibrium of the body, the brain and sense perception in birds play a significant role in the evolution of reflexes for the purpose.

i). **Brain:** In birds, the larger size of the cerebral hemispheres, optic lobes and the cerebellum in the brain (than those in the brain of ancestral reptiles) account for a very high degree of nervous coordination. The large cerebellum—subdivided into a central **vermis**, a pair of **lateral lobes** and a pair of **floculli**—is very well-developed in birds, as their volant adaptation demands a better sense of equilibrium. The nerve centres controlling functions of different muscles that maintain balance of the body are located here.

ii). **Vision:** The birds are well-known for their acuity (sharpness) of vision. The large eyes are lodged in larger orbits which are separated by a thin interorbital septum. The large **optic lobes, optic tracts** and **optic thalami** are indicative of a very effective coordination of both near and distant vision in birds. However, comparatively reduced eye muscles responsible for moving and turning the eyeball has been adequately compensated by the extraordinarily quicker movement of the neck through 180° degrees which covers a very wide area of vision.

The hard sclerotic plates of the eye can withstand pressure changes and these—together with the striated ciliary muscles (**Crampton's muscle** and **Brücke's muscle**)—effectively take part in the prompt accommodation of the lens. The lens in avian eye differs from that of mammalian eye by the possession of an **annular pad (Pulvinus anularis lentis; NAA, 1993)**, one on either side, around the central core of the lens. This pad has some hydrostatic function in transmitting pressure changes from the ciliary muscles to the central core for accommodation of the lens.

Another structure of importance in avian eye is **the pecten (Pecten oculi)**, a comb-like pleated vascular structure that projects into the vitreous body from near the entrance of the optic nerve. The pecten either acts as a nutritive organ for the retinal wall or has something to do with the intraocular pH regulation.

[See Chapter 30, Fig. 30.45.]

iii). **Organ of hearing and maintenance of balance:** The other sense, next in importance to sight, is the organ of hearing which is also linked up with a highly developed sense of equilibrium in birds.

The ampullary cells of the semicircular canals of the inner ear are very sensitive to the changes in locations of the body and these cells serve to restore normal position of the bird. When a bird is tilted to one side, the **endolymph** of the semicircular canal of that side flows towards the **macula lagena** and the **otoliths** present in the endlymph excite the cells present in the macula wherefrom stimulus is conducted to the brain which directs the bird to recover from its tilted position.

Bologna (1995) reports that the nocturnal birds of prey have an asymmetrical arrangement in their ears. Barn owls can detect prey in dark through hearing very feeble noises produced by the movements of the prey. Some oilbirds, nightjars and Old World swifts have developed 'Sonar' device—like bats—to detect obstacles in their environment with their sense of hearing.

23.2.7. Adaptations for Shape, Speed in Flight and Streamlining of the Body

The spindle-shaped, streamlined body of a bird is a very successful adaptation for flight in air. The light head, long neck and the short body are devoid of any projecting organ from the body surface (e.g., pinna-like external organ) that may offer resistance to air-flow.

The speed in flight, however, varies in different species of birds. While most smaller birds fly at an average speed of 25-30 miles (40-50 km) per hour, many passerine birds and hawks fly between 30 and 40 miles (50-65 km) per hour, or even faster. Most shorebirds fly between 40 and 50 miles (65-80 km) per hour and the migratory ducks and geese, the woodcock and the plovers fly between 50 and 60 miles (80-100 km) per hour. A few species of falcons, which are usually speedy fliers, can suddenly increase their speed two or three times–reaching up to even 100 miles (160 km) per hour or more–when they swoop on their prey. In fact, the fastest-moving animal in the world is the peregrine falcon which swoops on its prey at speeds up to 217 mph (350 km/hr)–three times the speed of the fastest land-animal—the cheetah—with its speed of 70 mph (110 kmph)!

The air-sacs of birds, having been placed between the muscles and bones, prevent friction between the muscles and thus increase efficiency of the flight muscles. While in flight, the feet of birds are folded and kept symmetrically aligned with the streamlined body. The surface of the body has been rendered smooth and free from resistance to air flow by the caudally directed contour feathers which maintain perfect streamlining of the body.

23. 3. CONCLUDING REMARKS

Thus in birds, having perfect volant adaptations, almost all the systems in the body have undergone some changes, great or small, and these changes have occurred through time and space as also due to their specific habits and habitats.

The adaptational changes or modifications in the bodily organisation of birds are remarkable—the feathered wings, pneumatic bones, lightness and rigidity of the skeletal framework, warm bloodedness, larger heart and blood vessels with higher capacity of blood flow than in other vertebrates (except mammals), a very effective system of respiratory organs and air sacs that ensure complete and quicker supply of oxygen to cells and tissues compatible with high rate of metabolism and thermal regulation in the body, and a system of very powerful and massive pectoralis muscles which provide unlimited source of energy for the propulsion of wings in sustained flight.

24

The Riddle of Migration

Professor W. Rowan's 'The Riddle of Migration' published as early as 1931, is relevant even today when much of this riddle has been explored. But not all. What Roger Tory Paterson (1957) stated about five decades ago may be repeated even today: "...We are coming closer to an understanding of the bird navigation but are still far from its ultimate secret ...The 5-inch warbler singing among the new leaves of the big oak may have just arrived from Venezuela or Brazil ... it will probably travel another thousand miles before it puts down at its own private airport somewhere in Canada".

For birds, the desire to migrate is as irresistible as their hereditary impulse to breed in the spring. But all birds are not migratory in the sense we use the term 'migration' here; only about 15-20 percent of birds of the world indulge in this kind of migration. However, considering all kinds of migration—local, seasonal, altitudinal, cyclic, etc.—it is **about one-third of the total species of birds that migrate** and the numbers of individuals are about tens of thousands of millions. It has been estimated that about 40 percent of the bird species in the Palaearctic region are migratory and about 5 billion birds migrate from this region every year.

24.1. WHAT IS BIRD MIGRATION

Bird migration usually means a two-way journey, an yearly round trip. It is a north-and-south movement, as most birds breed in the north and winter in the south. Landsborough Thomson (1936, 1949), an eminent ornithologist, defines migration as **"changes of habitat periodically recurring and alternating in direction which tend to secure optimum environmental conditions at all times"**. This **back and forth** movement is distinctive, because it distinguishes migration of birds from that of the swarms of locusts which neryer return to their original starting point of migration.

24.2. ORIGIN OF MIGRATION

According to Jean Dorst (1970), the origin of bird migration can be traced back to the Tertiary period during which seasonal alterations were possibly the prime cause for migratory movements, very similar to those found today. In Quartenary period, such rhythmical movements became clearly established, particularly during glaciation period when, due to extreme cold, the species in the north had to move out to the warmer south for food and shelter and then after a few months returned to their northern home when the climate was bearable. Thus it appears that initially the impetus to migration in birds might have originated due to a combind effect of the climate, the environment and the physiological conditions.

Recently, scientists are probing through various advanced techniques as to how natural selection favours the evolution of migration from non-migration, and vice versa. These new techniques are applied now to explore the origin of avian migration and evolutionary change in migration patterns. This requires an in-depth knowledge of historical biogeography, phylogeny, and ecology of the migrants and non-migrants.

Changes in migration patterns in birds have also been recorded through ring recoveries. Changes in migratory behaviour have been observed in different groups of birds with regard to the timing, direction and distance of migration, as well as percentage of migrants in partially migratory populations.

Fancisco Pulido (2002) of Max Planck Research Centre for Ornithology, Germany—working on adaptive evolution of migratory behaviour in birds–states that migratory behaviour has always been subject to evolutionary changes. In recent times, such changes in avian migration patterns have been more pronounced due to the global warming. He also asserts that the tempo and the mode of adaptive evolution are determined both by the genetic architecture of migratory traits and the ecological conditions determining their fitness functions.

24.3. HISTORICAL

The riddle of migration fascinated mankind far back from the days of Aristotle (384-320 B. C.) who first acknowledged that some birds migrate. Bird migration has been quoted in the Bible. Frederick II, the Roman emperor, included 11 chapters in his monumental work 'The Art of Falconry' (1245). Down the Middle Ages, many fantastic theories of bird migration were published. Even as late as in the 18th century, though Linnaeus (1707-1778) and Gilbert White (1720-1793) of Selborne accepted bird migration, they claimed that the house martin and swallows hibernated during winter and came out in the spring.

24.4. TYPES AND PATTERNS OF MIGRATION

Although birds are not much affected by the extremes of heat or cold, the scarcity of food at their nesting or breeding ground in the north during winter season compel them to migrate southward where plenty of food is available. Although, after nesting, most birds leave their breeding areas and migrate far or near, yet some species of birds live close to their nesting areas—these are called '**nonmigratory**' or '**sedentary**'. Some species have erratic movements without following any particular direction; this is called **nomadism**. Migration proper, as already defined, is characterised by a seasonal swing between the northern nesting ground and a southern feeding ground and a return flight vice versa.

Taking all these considerations in view, bird migrations may be classified into the following types and patterns:

24.4.1.1. Cyclic Migrations

This has been referred to as secular emigrations by Woodbury (1941) in which periodic overpopulation in a herbivorous form and consequent devegetation of its feeding grounds cause a suicidal exodus out of the normal range. Such migrations may be sporadic, i.e., irregular or rhythmic—correlated with a cycle of definite length. For example small Arctic rodents, after every 3-5 years, reach population peaks and undertake death migration.

24.4.1.2. Seasonal Migrations

Seasonal migrations may be either irregular or follow a regular rhythym with change of season.

The seasonal migrations may be chiefly of the following types:

1. **Vagrant migration:** These are also known as nomadism, as mentioned above. These include irregular seasonal movements or random wanderings in search of food. A few species of birds, either scattered or in flocks, living in boreal forest or Southern deserts, show this kind of unpredictable movements in search of food or breeding areas.
2. **Moisture rhythm migration:** Woodbury (1941) suggests that these are regular migrations of the seasonal type in tropical or sub-tropical countries where breeding season coincides with the rainy season.
3. **Latitudinal migration:** Most regular seasonal migrations are north and south or latitudinal. It involves a swing from a breeding or nesting place, the bird's home to a feeding or resting place, its winter-quarters. Thus, in the Northern Hemisphere, the breeding grounds of birds lie nearer the Arctic or temperate zone and their winter quarters nearer the Equator. A lesser and opposite movement occurs in the Southern Hemisphere where the seasons are reversed. Several species of North American and Eurasian birds cross the Equator to spend the winter in warmer parts of South America and Africa (e.g., white stork, Pectoral sandpiper, Golden plover). The American Golden Plover (*Pluvialis* sp.) passes 9 months of Arctic winter 8,000 miles (12,800 km) south in the pampas of Argentina, thus enjoying two summers each year (Arctic summer: May-July; Arctic winter: August-April). Many shorebirds of the North also do the same. Some birds of Siberia are winter visitors to the plains of the Himalayas in India (e.g., Siberian White Crane, Common Teal, Mallard Duck, Tufted Pochard, a few species of Sandpipers).

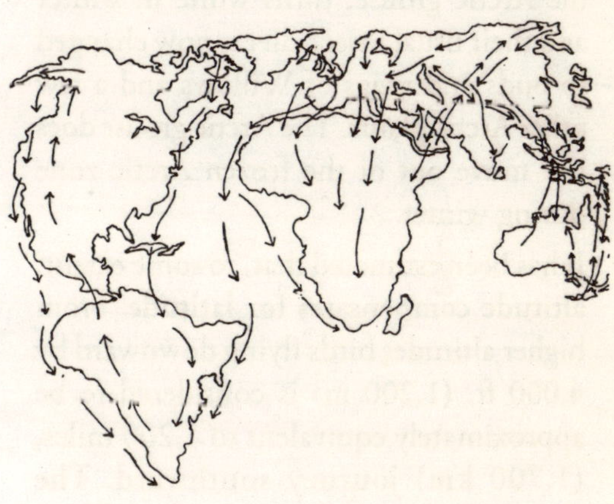

Fig. 24.1: The conventional routes used by most migratory birds in their north-south and south-north latitudinal migrations

4. **Longitudinal migration:** These are East-West migrations, or vice versa. The Redhead Ducks which nest in the Bear river marshes in Utah winter on the Atlantic Coast. The Starlings move from their breeding place in Eastern Europe or Western Asia towards the Atlantic coast to avoid the continental winter. The Californian gulls are also East-West migrants. Several species of Pochards are

winter visitors to Northern India and North Africa from the Eastern Palaeartic region.

5. **Altitudinal migration:** Some birds exhibit altitudinal or vertical migrations—flying up into the mountain regions (from a few hundered to several thousand feet/metres) during the summer and returning to the lowlands or the valleys in the winter. Such migration is quite common to the Grebes (water bird; Podicipediformes) and Coots (Duck-like water bird; Gruiformes) of Andes in Argentina, violet-green Swallows of U. K. and the Willow Ptarmigan (a kind of grouse; Galliformes) of Siberia. The brown plumage of Ptarmigan, like that of the Arctic grouse, turns white in winter and their usual insect-diet is now changed to buds and twigs of Willows and a few other Arctic plants. The Arctic grouse does not move out of the frozen Arctic zone during winter.

   It has been estimated that, to some extent, **altitude compensates for latitude.** From higher altitudes birds flying downward by 4,000 ft. (1,200 m) is considered to be approximately equivalent to 1,200 miles, (1,700 km) journey southward. The Tawny Eagle (*Aquilla* sp.) and the Kestrel (*Falco* sp.) are the winter visitors from the Himalayas in the plains of India.

6. **Partial migration:** Most species of birds are resident and locally migratory, but in a number of species of birds, some individuals migrate (sometimes across long distances) and some do not. Hence, during the period of winter migration (in the Northern Hemisphere), the number of resident birds in a particular geographic area in the south may considerably increase due to arrival of migratory birds of the same species for a few months. However, some migratory individuals feeding and resting in the south may not return to their northern breeding ground at all, provided sufficient food is available at their southern wintering quarters.

   The Barn Owls, Eastern Bluebirds, Eurasian and Siberian Jays, the Blue Jays and a number of other bird species are partially migrants to the south.

24.4.2. Patterns of Migration

The seasonal migration in birds is more pronounced in the Northern Hemisphere, as much greater areas of land masses are annually covered with winter snow, causing extreme scarcity of food. Thus, in the freezing winter, a considerable number of Eurasian and north American birds migrate southward around the Equator or cross the equator to spend the winter in South America or southern parts of Africa. Another pattern of migration is also quite common between western and northern Eurasia and South, as well as, Southeast Asia.

Much lesser number of birds from the southern continents migrate to pass their non-breeding season in the Northern Hemisphere. However, some southern sea-birds cross the Equator to reach the northern oceans and may travel as far north as the banks of Newfoundland or Greenland. The Australian 'Slender-billed Shearwaters' (*Procellaria tenuirostris*; Procellariiformes) travel in a great loop through the Pacific from Southeast Australia, pass Japan to the Bering Sea and then via the west coast of North America back to their starting place across the Pacific in a loop. The most remarkable feature of these long-distance migrants is their accurate periodicity of return.

Avian migration patterns have been subject to changes through hundreds of thousands of years due to variations in the seasonal cycle in the climate which has changed a number of times during the past two million years. Moreau (1972) states that the current avian migration patterns are of comparatively recent origin and are probably not more than 15,000 years old!

24.5. MECHANICS OF MIGRATION

The migration route follows coastlines as a rule, and such locations as islands, capes, prominent landmarks and other resting places. Results from marking individual birds with numbered legbands have disclosed that very often the same birds stop at the same places several years in succession.

Most species have established routes for migration and travel more or less on schedule, appearing or disappearing according to the calendar. This is especially true of smaller land birds, whereas ducks, waterfowls, and many shorebirds are mostly influenced by storms in autumn and open water in spring.

Different species of birds choose different periods—such as day or night—for their migration, which may be summarised:

1. Small birds like warblers, thrushes and sparrows usually migrate at night and use daytime for their stop-over to feed and rest.
2. Crows, blackbirds, hawks and geese travel by day, but stop to forage in suitable places.
3. Swallows and swifts also travel by day, but they capture their insect-food in air, as they travel.
4. Ducks, gulls, shorebirds and many other birds migrate either at night or by day with suitable stopover sites–old or new.
5. The vast majority of birds are mainly nocturnal—the small passerine birds which, by flying at night, escape their possible enemies.
6. The diurnal migrants travel in flocks which may be well-organised (e.g., ducks, geese and swans) or loose (swallows).
7. Regarding species composition of migratory birds: The nighthawks, swifts and kingfishers travel separately, while the swallows, turkeys, bluebirds and may others travel in mixed companies of several species.

Comments: The Swiss Ornithological Institute has recently conducted experiments to explore migration mechnisms and behavioural adaptations in passerine bird migration through **computer simulation.** It has now been possible to investigate how the first-time migrants cross the vast geographical and ecological barriers to reach their winter quarters. In such an experiment with the Western Palaearctic-African migration route, the endogenous direction has been found to be most crucial.

24.6. ALTITUDE IN MIGRATION FLIGHT

While some birds migrate quite close to the surface of the earth or the sea, somewhere between 500 feet (150 m) and 1,000 feet (300 m), most migratory birds in the flight maintain an altitude between 3,000 feet (900 m) and 5,000 feet (1500 m). However with the help of modern devices like radar, speed indicators, altimeters, etc., it has been possible to record the flight altitudes of certain migratory birds as mentioned below:

1. Ducks and Geese from the eastern Palaearctic region travel southward directly across the Himalayan mountain barrier often at an altitude of 9,500 feet (2,900 m) to 16,500 feet (5,000 m), and sometimes even up to 19,000 feet (5,800 m).

2. The Himalayan Bearded Vulture (Lammergeier; *Gypaetus barbatus aureus*) and Himalayan Griffon Vulture (*Gyps himalayensis*) fly between 18,800 ft (5,700 m) and 22,000 feet (6,700 m).
3. Choughs i.e., the Red-legged Crow (*Pyrrhocorax pyrrhocorax)* have been found to follow the climbers even up to the highly reduced atmosphere at an altitude up to 25,666 feet (7,800 m). In the Everest expedition of 1953, Sir Edmund Hillary observed a chough following him at 26,600 feet (8,100 m), which is an example of superb physiological adaptation of a bird to sustain in extremely rarified air.

   At lower altitude, the author has observed these birds above 10,000 feet (above 3,000 m) in the Garhwal Himalayas.

24.7. LONG-DISTANCE FLIERS AND THEIR ROUTES OF MIGRATION

In a summary of the chapter on bird migration, Pough et al (1990) comment that 'migration is the most dramatic manifestation of the mobility of birds, and some species travel tens of thousands of kilometers in a year'.

In recent times, bird ringers have ringed millions of birds so that their flight could be followed and destinations recorded. The routes and flight distances of a few well-known migratory birds are described below:

1. **Arctic Tern** (*Sterna paradisaea*): It is the champion among the long-distance fliers. It leaves its northern colonies far inside the Arctic Circle in late summer and travels south right across the world to spend the Antarctic summer (winter months of the Northern Hemisphere) and back again, a distance of about 11,000 miles (17,600 km) each way! Some have their route through the Pacific, while others take a diversion to the west coast of Europe and Africa and may even stray to the Indian Ocean. Some of them cover more than 14,000 miles (22,400 km) each way from the Arctic coast of Siberia to Australia through a circuitous route!!

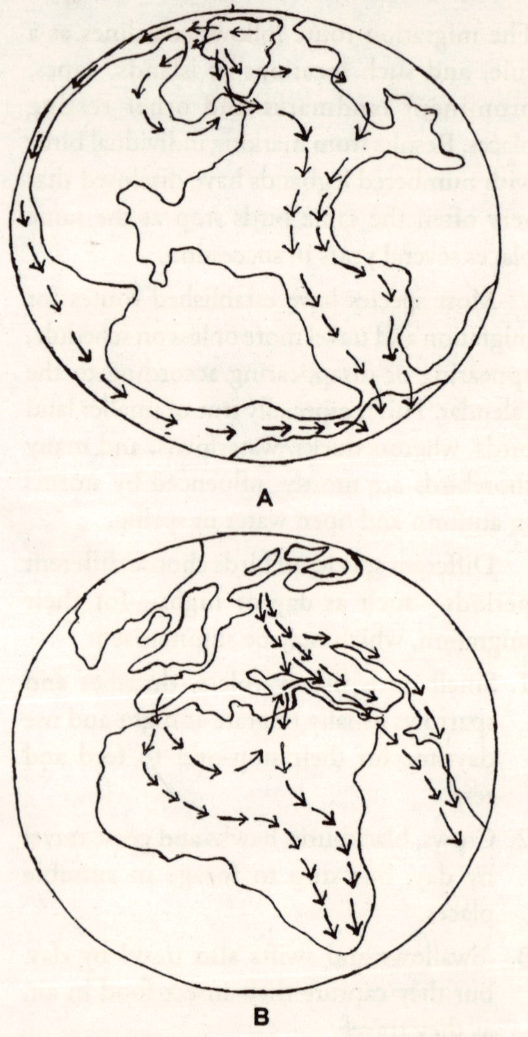

Fig. 24.2: Migratory routes followed by two long distance fliers. A. Arctic Tern (*Sterna paradisaea*); B. White Stork (*Ciconia ciconia*)

2. **White Stork** (*Ciconia ciconia*): It is the most spectacular of all migrants in Europe. It breeds in central and eastern Europe,

Spain, Turkey, central and eastern Asia, and northwest Africa. The bird winters in the south of South Africa. This bird too covers a long distance of about 8,000 miles (12,800 km) each way from Netherland, Germany or Spain to South Africa across the Mediterranean. The flight posture of this bird is spectacular because its neck remains fully outstretched in flight.

3. **Pectoral Sandpiper** (*Calidris melanotos*): It is a shorebird (order Charadriiformes) that breeds in Arctic Tundra (central and eastern Siberia) and in the north of Northern America and migrates to its winter home in South America. This bird usually follows the coastline and does not lose sight of the land as it migrates.

   The **sharp-tailed Sandpiper** (*Calidris acuminata*) breeds in northeast Siberia and winters in the south in Malaya Archipelago, New Guinea and Australia. Both the species mentioned above cover a distance of about 6,000-9,000 miles (9,600-14,400 km) each way.

4. **Bobolink** (*Dolichonyx oryzivorus*; Order Passeriformes): It is a thrush-like singing bird and is said to be the leading migrant among land birds. It breeds in Canada and migrates to the grassland of Argentina to spend the winter, covering more than 7,000 miles (11,200 km).

Some **Barn swallows** (*Hirundo rustica*) breeding north to Alaska or in Europe and north Africa spend the winter in southern South America, South Africa or in north Australia, thus travelling up to even 9,000 miles (14,400 km), or more. If this bird survives long enough, it may undertake the migratory journey twice a year.

5. **Greater Golden Plover** (Eurasian Golden Plover) (*Pluvialis apricaria*): This migratory bird has an outstanding record of non-stop flight at a reasonably high speed between 60 and 70 miles (96 and 112 km) per hour. The Eurasian or the Greater Golden Plover breeds in northern Eurasia and winters in the south Palaearctic region. The Lesser or the American Golden Plover (*Pluvialis dominica*) breeds in northern North America (Hudson Bay, Alaska) and winters in South America—either in Hawaiian Islands, covering a distance of 2,175 miles or a distance of about 2,400 miles (3,800 km), (along short axis of an elliptical route) in about 32-35 hours non-stop flight. The migration route along long axis of the ellipse is however about 8,000 miles (12,800 km).

   The **Pacific Golden Plover** (*Pluvialis fulva*) breeds in northeast Eurasia and winters in India, Myanmar, Thailand, Malayasia, Australia and New Zeland.

6. **Water Rail** (*Rallus aquaticus*, Rallidae, Gruiformes): This bird is also a long-distance east-west migrant. It is discontinuously distributed from Iceland to Japan and the bird winters around the Mediterranean as well as in Southern Asia.

7. **Red-backed shrike** (*Lanius collurio*): This medium-sized passerine bird breeds in southern England, Europe and temperate Asia and winters in tropical Africa and southern Asia.

24.8. SPEED IN MIGRATION FLIGHTS

The speed of most small migratory birds seldom exceeds 30 miles (50 km) per hour, although swallows and starlings fly faster. The flight-speeds of a few migratory birds are:

1. **Hawks**............30-40 mph (50-65 km/h).

2. **Peregrine Falcon** (while diving or swooping on its prey)........... 100-275 mph (160-440 km/h).
3. **Most shorebirds**............40-50 mph (65-80 km/h).
4. **Ducks and Geese** ....50-60 mph (80-100 km/h).
5. **Wood cock**.........50 mph (80 km/h)
6. **Plover** ....50-60 mph (80-100 km/h)
7. **Stork** .... 26 mph (40 km/h).
8. **Swifts** (2 species) ....171-200 mph (270-320 km/h)

Few other species having non-stop flight 65-70 mph (105-112 km/h).

The migrating birds usually fly 6-11 hours a day, resting in between for food or drink, so that the 'migration front' progresses rather slowly.

24.9. ENERGETICS, PHYSIOLOGY AND BIOCHEMISTRY IN THE MIGRATORY FLIGHT OF BIRDS

Many new investigations in the fields of energetics, physiology and biochemistry of bird flight, particularly in migratory flights, reveal the following facts:

1. Migration consists of two phases, **flight** and **stopover**, which the migratory birds alternately undergo a number of times to complete a full migration cycle.
2. It has been possible now through special designing of wind tunnel to study the physiology of birds flying through long distances and a long period of time with much greater accuracy than before.
3. Direct measurements of energy expenditure, body mass and changes in the composition of breast muscles in relation to long-distance flights have been worked out in many birds.
4. Besides environmental factors, habitats, feeding grounds along routes of migration, etc., the long-distance migrants need to be equipped with higher fat reserves in their body (compared to that in short-distance migrants), particularly while crossing high altitudinous mountains or the vast expanse of a desert. Other migrants with low and medium fat reserves avoid such hazards. They adjust their flight and stopovers for feeding accordingly.
5. The energetics and physiology of migratory birds of different species or the population of the same species are largely determined by different flight capabilities, fat reserves and direction of migration.
6. The day migrants and the night migrants have different strategies for flight, feeding and resting which are partly influenced by the ecophysiology.

24.10. HOW DO MIGRATORY BIRDS FIND THEIR WAY: NAVIGATION IN BIRDS

One of the most intriguing questions in the migratory behavior of birds is: 'How do migrating birds find way to reach their destination? This is known as navigation or orientation in birds which means their direction-finding ability. Along with this extraordinary capacity they also possess a remarkable time-schedule. The migrating birds use a variety of cues for navigation, such as the position of the sun, polarised light, magnetic field of the earth, ultrasound. Since the 1960s experiments on orientation research have brought the scientists closer to the problem, yet much of the mystery still remains to be solved. Of several hypotheses explaining the probable navigational cues for the migratoty birds, some are described below. However, it is logical to assume that different species of birds may be guided by one or the other cues as mentioned below, but these features are not absolute in direction finding; rather, it is quite possible that the underlying genetic factor inherited by a particular species may account for the direction-finding in migration, particularly by the young birds that did never explore the migration routes earlier.

1. **Visual landmarks:** Distinct topographical features like mountains, coastlines, ridges, large rivers, chain of oceanic islands, etc. act as visual landmarks in direction-finding for the diurnal migrating birds. Such visual contours have been called **migrating 'routes'** or **'flyways'**. Quite a few other species of migrating birds fly over vast stretches of the sea, but the night-migrants with broad 'fronts' appear to fly with advancing night-time not with the guidance of visual landmarks but by changing positions of the stars in the sky!

2. **General airflow:** Telescopic and radar experiments have shown that most nocturnal movements—either southward or northward—proceed on a broad front with the general air-flow. The night-migrants do not form massive flocks, nor do they follow narrow paths.

3. **Thermal up-currents of air:** Some scientists speculate that the soaring birds migrate with the help of the thermal up-currents of air. But unless this air-current is very strong and unidirectional, migration flights for a considerable distance and time cannot be sustained.

4. **Learning from experience:** Some naturalists in the past held another view that the migrating birds acquired recognition of their routes or flyways from past experience of their flight. This assumption seems to be untenable due to several reasons such as: (i) neither all nor the same migrating birds are seen in the migrating flock every year, (ii) very young birds having no previous experience where their parents unmistakably guide the older birds to their winter-home destination, and (iii) the night-migrants for whom the experience is not based on the visiblity of landmarks, etc. during night.

5. **Ancestral home theory:** Yet another view was held by some authors that the migrating birds returned to their northern ancestral home intinctively, as in the case of certain insects and the carrier pigeons. But homing experiments with the carrier pigeons have proved that their return to home-site is mainly guided by their vision in detecting their familiar landmarks, and adjusting to direction and position of flight accordingly. Experiments with piegons have also shown that on sunny days, the pigeons use the sun as a compass.

6. **Geomagnetism:** The earth's magnetic field and its relation with the orientation in birds' flight is a potential field of research since long past. Some workers in the past held the view that birds navigate by responding to the earth's magnetic field and their inner ear is supersensitive to the change in intensity of the earth's magnetic field. This is known as **'Coriolis effect'** produced by the rotation of the earth whereby freely-moving airmasses are deflected to the right in the Northern Hemisphere and to the left in the Southern Hemisphere.

Sengupta (1979, 1986) reported an interesting phenomenon: a 'suicidal flight' by a large number of birds from several species in a basin-like small village in Jatinga Hills which is an extension of the Jaintia Hills, north-eastern Assam, India. Sengupta analyses the possibility of a change in the intensity of the geomagnetic field in this area in a particular season of the year. He points out that the variation in the electric field possibly affects geomagnetic field of Jatinga, and this variation is sensed by birds in the neighbouring areas and they leave their roost. He quotes Darov and Kholodov who support Sengupta's hypothesis and maintain that there is a link between the geomagnetic field and the orientation in birds that may be affected to the extent of about 1/1000 of the intensity of the normal earth field.

Sengupta also quotes a recent work by Luigi (2006), a French Geologist who suggests that seismic fault area ruptures during earthquake, thus creating fissure through which gases like carbon dioxide, hydrogen sulphate and methane may be emitted. Besides these, ethylene that partially paralyses the central nervous system also evolves. The inhalation of this gas in large dose makes the birds immobile, they lose their reflexes, become senseless and finally die. It appears Luigi's view supports contention of Sengupta.

Although there is no direct evidence of the impact of a change in the geomagnetic field on the flight of birds, it is quite logical to assume on the foregoing analyses that the birds shift their orientation in parallel with the altered magnetic field.

7. **Celestial bodies like sun, stars and constellations:** In modern orientation science, much greater emphasis is being laid on the increasing evidences gathered from the navigation of birds with the help of celestial bodies like sun, stars and constellations. 50 years ago, Gustav Kramer (1957) experimentally showed that the day-migrants use the sun as a compass and they also possess an internal 'time clock' to orient themselves to the changing position of the sun at different times of the day. According to this author, the migrating birds flying south or southwest in Sep-Oct have the sun on the left in the morning, fly directly towards the sun at noon and have the sun on the right in the afternoon. Von Frisch (1967) also observed similar behaviour in the movements of the bees.

   Just as the sun is the 'orientator' for the day-migrants, the night-migrants navigate with the help of the stars. As the night progresses, the birds are able to adjust their positions with the changing positions of the constellations. However, such observations are not confirmatory for all birds, as there are plenty of exceptions. For example, the young birds fail to use stars in their first migration at night.

24.11. RECENT ADVANCES IN ORIENTATION RESEARCH

Orientation researches at the Latvian Academy of Sciences by Katz et al (1988), and by the scientists at the Max Planck Institute, Germany, and other institutes in U.S.A. reveal many interesting facts:

1. The sun is important for orientation even by some nocturnal migrants such as European Robins; an experiment with Great Reed Warblers suggests that both stellar rotation and a directional magnetic field are needed for normal ontogenic development of orientation. It is suggested that time compensation mechanisms are linked with stellar navigation systems (Katz, Y.; V. Liepa and J. Viksne, 1988).

2. The results of experiments with both European Robin (*Erithacus rubecula*) and White Stork (*Ciconia ciconia*) indicate that either the rising or the setting sun is important for both nocturnal and day migrants. When the sky was fully overcast, the orientation was random (Moore, 1978; Bingman and Able, 1979).

3. According to the **menotactic hypothesis** (i.e., recording events according to calendar months), at the moment of sunrise or sunset there is a direct connection between the **'azimuth'** of the sun and the preferred direction. Azimuth is the arc of the horizon that a vertical plane passing through the heavenly body makes with the meridian of the place of observation. At other times of the day, the bird's 'internal clock' could participate in selecting the direction. However, the bird's clock indicates the passage of time from the sunrise, not absolute time (Katz et al, 1988).

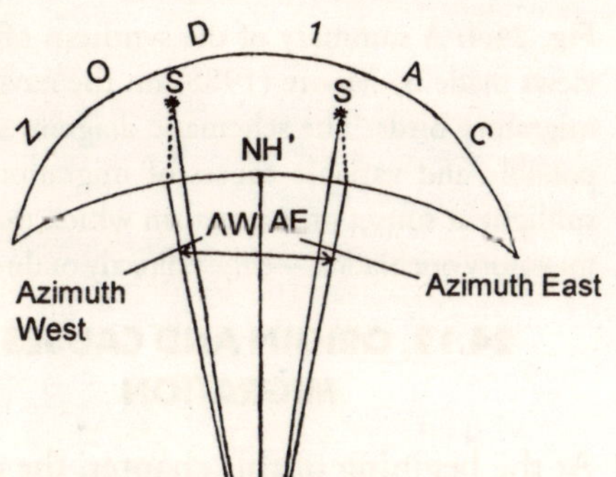

Fig. 24.3: Azimuth: From the observation point (PO) if one looks straight to the North horizon (NH) and to the sun or any stellar body (S), then the arc formed by the imaginary vertical line (broken line) from the sun touching the North Horizon and the NH from the point of observtion is the Azimuth, which may be either on the West (AW) or on the East (AE) (Moore, 1979; Bingman and Able, 1979; Katz, 1988)

4. Although there is evidence that birds perform time-compensated stellar orientation, time-independent mechanism of celestial orientation has also been pointed out by some authors (Matthews, 1963; Emlen, 1967, 1970, 1975; Wallraff, 1977) in which case, according to Katz et al (1988), the birds need to memorise the spatial relationships of stars with regard to a primary reference system (possibly rotational and also magnetic reference systems).

5. Beck and Wiltschko (1988), experimenting with young flycatchers, support the earlier studies by Gwinner and Wiltschko (1978) that the magnetic field serves as an external reference system for the genetically encoded direction–in Garden Warblers, the endogenous time programme alone could induce a change in direction.

6. A discovery in the features of birds' vision is the capacity to distinguish the plane of light polarisation and the sensitivity to ultraviolet light may be related to perception of celestial rotation, assuming some kind of polarisation filter in the bird's eye (Delius and Emmerton, 1978; Kreithen, 1978).

In bird migration, the orientation research has been advanced by many notable workers of whom Keeton (1969, 1971), Sauer and Emlen (1971), Papi et al (1972), Leipa (1978, 1982), Gauthereaux (1980), Katz (1981, 1985), Wiltschko (1982, 1983), Bingman (1983), Berthold (1984, 2003), Able (1986), Gwinner (1986), Beibach et al (1986), Beck and Wiltschko (1988), Terril (1988), Moore

(1988), Able and Able (1990) and Cochran et al (2005) are to be specially mentioned.

7. Beibach et al (1986) and Beibach (1988) studied **ecophysiology** of the resting warblers in Sahara Desert and concluded that these long-distance migrants, while flying from Europe to Africa, cross the Sahara Desert to reach their winter home. The birds found in the desert are not stragglers, but they take it as strategy to cross the desert by flying during three nights and resting in the desert during the intervening two days. It has been confirmed from the estimation of their fat reserves that their energy and water budgets allow them to do so.

   Different bird migration across the Alps at night with low, medium and high fat reserves are significant in ecophysiological studies (Bruderer and Jenni, 1988).

8. Terril (1988) discusses the role of **photoperiod** and the importance of ecological factors in bird migration. He quotes Gauthreaux (1980) and states that seasonal changes in the environment are major, predictable and cyclic. Migrants have developed an internal clock that organises their annual cycle (i.e., breeding, migration, moult, etc.) to match seasonal changes in their environment.

   Meddle and Wingfield (2002) reported that avian hypothalamus not only coordinates daily patterns of activity, endocrine functions and other physiological activities, but it also integrates seasonal patterns of reproduction, moult and migration. For most birds breeding in temperate zones, seasonal breeding is mainly determined by changes in day length.

9. Moore (1988), in a diagrammatic scheme(Fig. 24.4), synthesised the opinions of different authors to explain the causal relationships between the sunlight, magnetism, stars and migration.

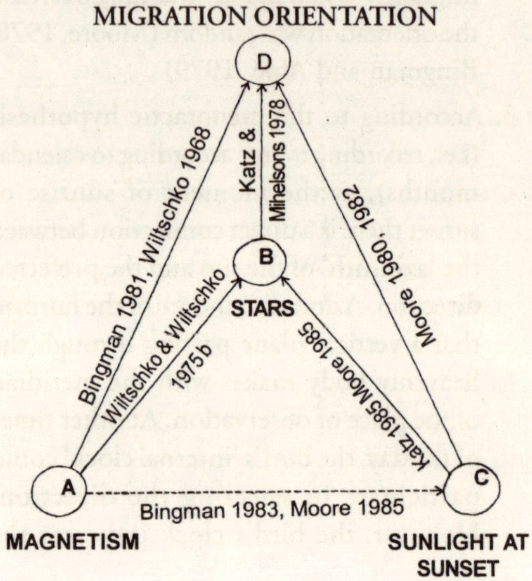

Fig. 24.4: A summary of the synthesis of different views made by Moore (1988) on the navigation of migratory birds. The schematic diagram shows the possible and variable causes of migration such as sunlight at sunset or magnetism which may lead to migratory orientation—either directly or through stars

24.12. ORIGIN AND CAUSES OF MIGRATION

At the begining of this chapter, the origin of migration has been precisely attributed to climatic, environmental and physiological conditions. This means that the urge for seasonal migration in birds is evoked by both external and internal stimuli. 56 years back Lowery (1951) stated that 'the root of the problem is that bird migration, whether it occurs by day or by night, is an unseen movement'. Things have changed a lot since then as, by experiments with modern equipments, it has been possible to study bird migration with greater precision.

The following theories, both old and new, and recent observations to explain the origin and causes of bird migration may be considered relevant:

1. **Northern Ancestral Home theory:** Considering that most bird migrations in the Northern Hemisphere are from the North to the South, the northern Arctic and temperate zones for such migrating birds are believed to be their ancestral home for breeding and nesting. It was suggested that the ice-sheets during Pleistocene period might have been originally responsible for migration. The birds moved southward with advancing glaciers and northward when the glaciers receded.

   This theory does not explain migration in those parts of the world where there is no glaciation. Further, as mentioned earlier, the origin of bird migration may be traced back to Tertiary period and this phenomenon might have been established from the Quarternary period. Hence, bird migration is older by at least 60 millions of years than the Pleistocene period.

2. **Southern Ancestral Home theory:** According to this theory, the birds had their ancestral home in the tropics which became so congested that the birds travelled northward for abundant food and isolation during their short breeding season. They returned to the tropics where they stayed longer at home or, alternatively, the birds had their breeding and nesting in the south and migrated to north for food.

   In the light of modern researches on avian migration, both the ancestral home theories seem to be untenable, because **migration is a seasonal phenomenon, influenced by several factors of which climate, environment, physiology and behaviour are the main.**

3. **Photoperiodism:** Long ago, Rowan (1931) and Bissonnette (1937) confirmed gonadal responses with light treatment in some birds, not in all. Wolfson (1952) showed that it is the summation of day-lengths—rather than gradually increasing day-lengths—that stimulates gonadal development. According to this view, circannual stimulus due to endocrine changes is not the cause of migration in all species.

   As objections to this theory of photoperiodism it has been mentioned that: (i) it is not applicable to birds that breed in the tropics where day-length is relatively uniform, and (ii) certain birds do not conform to annual light cycles and breed irregularly. Barn owls breed in every month of the year, in both northern and southern countries, while other owls are all spring-breeders.

4. **Woodbury's Periodic Response theory (1941):** This theory suggests that migrations are in tune with various environmental periodicities, and not necessarily day-length alone. Such rhythmic behaviour patterns, according to this view, have arisen through mutation, preserved by Natural Selection, and have been so fixed in heredity that they may occur independently of environmental stimuli.

   In this context, Meddle and Wingfield's view (2002) has already been mentioned that for most birds breeding in temperate zones, seasonal breeding is determined by changes in day-length, although supplementary cues—such as temperature, food and mate availability—can accelerate or perturb photoperiodic response. These authors maintain that the discovery of **'Immediate Early Genes'** (IEG) weakly expressed in neural tissues may influence the brain circuits involved in seasonal breeding and associated behaviour.

Noskov et al (2002), while discussing principles of photoperiodic control of Annual Cycle of events in birds, distinguish the annual cycle in two parts of photoperiodic control—summer-autumnal and winter-spring–each differing from the other in the duration of day-lengths. These authors mention that the positive effect of photoperiodic reaction is observed within certain limits of the values of photoperiods called 'threshold values.'

Experiments on photoperiodic responsiveness, migratory restlessness and melatonin hormone secretion in nocturnal migrants reveal some interasting facts:

i. Some birds induced with various photoperiods and light intensities show fattening and gain in body as well as testicular mass.

ii. Photosensitivity decreases as short day exposure increases.

iii. It has been observed that, in captivity, the nocturnal migrants show restlessness in migratory period. The hormone melatonin, which regulates circadian rhythm in birds, is reduced at this period.

iv. There is a functional relationship between **melatonin concentration** and migratory restlessness in a nocturnal migrant.

5. While discussing the development of migratory orientation, Able and Bingman (1988) stated that the complex mechanisms in birds' migration are probably associated with the relevant stimuli and the underlying genetic programme. Terrill (1988) mentioned that the migration of birds is a response to changes in the environment. Natural selection has produced a complex system which enables the birds to avoid harmful and exploit beneficial environment, and to time their migration accordingly.

Gwinner (1986), Berthhold (1984) and Gauthreaux (1980) are of the opinion that migrants have developed an internal clock that organises their annual cycle (i.e., breeding, migration and moult) to temporarily match seasonal changes in their environment. These authors stated that heritable circannual programmes provide the foundation for not only the timing of migration, but also the direction, distance, duration and rate of migration.

24.13. ADVANTAGES OF MIGRATION

Several advantages are enjoyed by the migrating birds. Such advantages may be enumerated:

1. The seasonal migratory birds are able to avoid extreme cold climate, snow-storm and ice-sheeting of the ground in northern temperate and Arctic latitudes during winter.
2. By migrating from the north before the winter, the birds benefit by avoiding short day-lengths, while the source of food is also considerably depleted due to freezing of water, snowfall and loss of vegetation.
3. In both spring and autumn migrations, the energy expenditure is very high, and this is compensated by birds moving to suitable foraging habitats with abundance of food. Thereby the birds gain in body mass and fat storage. Thus the migratory birds can capitalise longer day-lengths of both the hemispheres for search of food.
4. Cox (1968) pointed out that competition for food resources play an important role in migratory birds. He observed that, during the breeding season, when the food requirment is very high, the competition

is more intense between species with similar bill shape than with bills of different shapes. According to Cox, the variation in bill form was greatest in the resident tropical birds, and the migrating birds derived much greater advantage by possessing similar bill shape in the group, so that they are not required to be ecologically separated by bill morphology while foraging during migration (thereby avoiding competition for food).

24.14. BIRD MIGRATION IN INDIAN SUB-CONTINENT AND BIRD RINGING

The birds arriving in the Indian sub-continent as winter visitors—mainly from the eastern and north-eastern Palaearctic region—have different status in different parts of our country. For example, birds coming to central India as winter-visitors may not be true winter-visitors—which are supposed to stay for 4-5 months for their nesting and feeding. Instead, these birds may stay there for a few days and leave the place for their destination to a far-off land. Thus these birds are called **'passage migrants'** which join the resident birds of the locality and cause temporary increase in bird population for a few days only. Again, in other parts of the country, the visitors that arrive may be truly winter-visitors in combination with the resident birds.

Compared to the enormous traffic of migratory birds coming from Eastern Europe, Central, Northern and North-eastern Asia, either crossing or bypassing the Himalayan, Hindukush and Karakoram mountains every year to reach our country, all the factual data about their migration have not yet been available. Salim Ali and Ripley (1987) reported that the main streams of such migration take place at both ends of the Himalayan chain through the valleys of the Indus and Brahmaputra. These winter-visitors—having settled in considerable numbers in western, central and eastern India—converge down the two sides of the Indian Peninsula to settle in the southern states and as far south as Sri Lanka in smaller numbers.

Out of nearly 2,500 odd species and subspecies of bird fauna of the Indian subcontinent and Sri Lanka, approximately 400 species are extralimital seasonal immigrants i.e., they breed outside our territory, mostly in the palaearctic region beyond the Himalayas. Such regular winter-visitors to our land are represented by a few to several species of birds belonging to a number of orders of the class as given next. The regular winter-visitors have their breeding grounds mostly in the Palaearctic region:

BIRD-GROUPS—ORDERS

EXAMPLES

	ORDERS
1. Ducks and Geese Mallard duck; Ruddy Shelduck, Barheaded Goose, Gadwall, Wigeon, Tufted Pochard, Redcrested Pochard	**ANSERIFORMES**
2. Wading or Shorebirds Golden Plover, Blacktailed Godwit, Common Sandpiper, Brownheaded Gull, Caspain Tern	**CHARADRIIFORMES**

3. Cranes Damoiselle Crane, Common Crane, Coot	**GRUIFORMES**
4. Passerine birds Swallow: Redrumped swallow; Common swallow, Crag-Martin. Flycatchers, Warblers, Thrushes, Pipits, Wagtails, Finches and Buntings. Paradise Flycatcher, Red-breasted Flycatcher, Bluethroat, Lesser Whitethroat, Rosy Pastor, Indian Pipit, Yellow Wagtail, Eurasian Gold finch, Blackheaded Bunting	**PASSERIFORMES**

Many other species—whose palaearctic breeding range encroaches boundaries on the Afghanistan—West Pakistan border—behave like true migrants, arriving in our land in autumn and leaving for their northern breeding grounds in March/April.

24.15. BIRD RINGING

While the observational study of bird migration extends over several centuries in the past, a more systematic and scientific study of bird migration is of comparatively recent origin.

One of the most commonly used and effective method of studying bird migrations is by **'ringing'** or **'banding'** birds. This method had been somewhat casually practised by various ornithologists in the past, but scientifically practised only since 1899. In that year, the Danish ornithologist H. C. C. Mortensen ringed 164 Common starlings (*Sturnus vulgaris*) with zinc rings that showed the year and place of ringing.

Since then, the method spread through the scientific world, using light aluminium-based rings. Each ring attached to the instep or trasal region of a bird, caught in a trap or net, bears a serial number and abbreviated name of the institution providing the ring. A special register is maintained wherein details of each serial number are noted, such as developmental stage of the bird (whether nestling, immature or adult), state of health, moulting, size, weight, etc. and the bird is then released. These data provide interesting information about the age, routes of migration and various other particulars at the recovery ends. Thus ringing is a very important scientific approach to study bird migration. Exhaustive records of ringing collected through several years reveal the routes of migration followed by different species of birds.

Our knowledge about the migration of Indian birds is largely based on fragmentary records. To overcome this shortcoming, Bombay Natural History Society (BNHS) initiated bird ringing as early as 1926, but very little bird ringing could be done prior to 1958, and that too mainly concerning migratory ducks. The early ring recoveries confirmed that most of our migratory ducks are derived from Siberia and from central and north-eastern Asia over distances of 1,875-3,125 miles (3,000-5,000 km).

Salim Ali and Ripley (1979, 1987) and Grubh and Hussain (1986) of BNHS report that large-scale ringing in organised field projects started from 1959 with not only ducks, geese and wildfowls but also with a variety of birds, including the small passserines. This continued up to 1973 and again resumed from 1980 with funds from U. S. Fish and Wildlife Service and from the Goverment of India.

The internationally recognised institutions from all over the world have till now banded several millions migratory birds. Salim Ali and Ripley (1987) mention that, besides determining the geographical origins of birds and their different routes of migration, bird ringing also provides information about spreading of virus diseases of man and animals through ticks and other blood-sucking parasites. Out of about half a million birds ringed in India, about 15 percent or so are water-fowls whose breeding grounds have been found to be Central Asia and Eastern Siberia.

More recently, the use of plastic rings of different colours has an advantage by which the banded bird can be recognised with a few particulars from a distance, without capturing the bird.

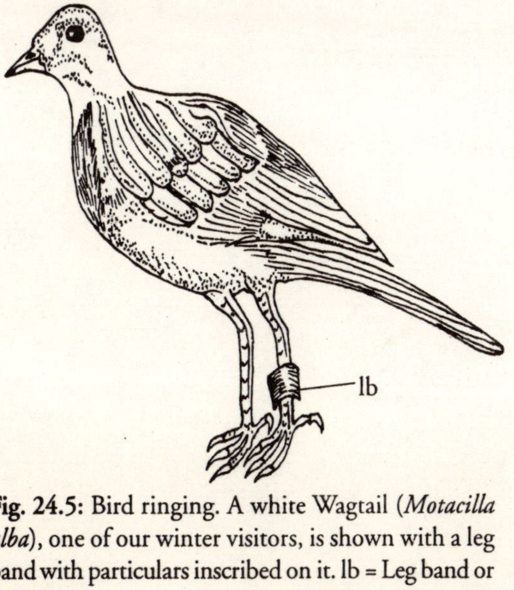

Fig. 24.5: Bird ringing. A white Wagtail (*Motacilla alba*), one of our winter visitors, is shown with a leg band with particulars inscribed on it. lb = Leg band or tarsal band

24.16. MIGRATION AND CONSERVATION OF CRANES AND STORKS

Higuchi and Berthold (2002) reported that the large migratory cranes and storks are threatened in many parts of the world, and, as such, migration studies are very essential for the conservation of these birds. According to these authors, more detailed information about migration and conservation is available now by using satellite tracking of migration, picturing habitat characteristics of satellite images, and by application of migrational researches to conservation.

Through satellite tracking of **Oriental White Storks** (*ciconia boyciana*) and **Cranes**, it has been possible to work out the detailed migration routes, migration patterns through time, migration distances, etc. The Oriental White Storks and the Red-crowned Cranes leave their breeding areas in the Russian Far East, independently, follow similar migration routes, and winter in Chinese wetland habitats. While these two species do not meet any threat to their existence in Eastern Siberia, they have been declared endangered species in China (their wintering ground) presumably due to considerable decrease in wetlands, hunting and use of pesticides. Migrational research on White Stork (*Ciconia ciconia*) in aviaries, using MRI (Magnetic Resonance Imaging) and Spectroscopy has been much helpful to study fat deposition and body composition in migratory flight.

24.17. DISCUSSION AND CONCLUDING REMARKS

Many scientists have highlighted the intricate phenomenon of migration as extremely variable among species, populations, age groups, and sexes. They emphasise that, in long distance migration, one of the crucial parts is stopover and refuelling which, to a large extent, act as deciding factors for the speed and overall success in migration. At stopover sites which are unknown food habitats, the migrant birds have to compete with the resident birds in foraging, besides being vulnerable to attack by predators, as well as hunters. Thus competition, as pointed out earlier, plays an important role in migration.

Hence, in spite of the fact that various optimality models have been devised and predictions made in the migration of a species or a population of birds, various intrinsic and extrinsic factors and their interactions influencing migration as a part in the complete life-history of a species still remain unexplored. This requires a comparative as well as integrated approach, combining advanced technologies in all aspects of migration in birds—such as aerodynamics, feeding, ecology, physiology, behaviour and conservation.

Although much of the riddle of migration has been explored in recent years through exhaustive banding and ultra-technological experiments, migration, as a whole, as Wallace (1953) predicted, is a phenomenon of multiple origin—no single factor can account for the complex seasonal movements of birds.

Recent studies, however, indicate that the annual stimulus for migration is genetically programmed, favoured by the environment and stabilised by the selection force. The details of this assertion, like most of nature's secrets, still remain to be fully explored.

25

The Birds that Lost Flight

The flightless birds, by nomenclature known as the **Palaeognathae** or the **Ratitae**, form a very interesting group that diverged as an early offshoot of the Neornithes or the modern birds. The evolution of these birds can be traced back to the late Cretaceous or the Palaeocene periods. At present, the living representatives of the flightless birds are restricted only to the Gondwana land such as, Africa, South America, Australia and New Zealand. Prof. T. H. Huxley classified (1867) birds, based on the structure of the palate and the jaws and what he referred to as the **dromaeognathous** type of palate being the most ancient (and, hence, primitive) type —in fact equivalalent to W. P. Pycraft's (1900, 1901) palaeognathous type–the term commonly used now. **'Ratitae'**, meaning extremely reduced or lost breast-bone, is included within the Palaeognathae except tinamou–which possesses a palaeognathous palate but has a keeled sternum like the carinatae; it is a reasonably good flier.

Thus the modern palaeognaths are formed of a somewhat unrelated assemblage of mostly flightless birds such as ostriches from Africa (the largest among birds), the rheas from South America, the cassowaries (the smallest of the ratites) and the tinamous from South and Central America.

ORIGIN, DISTRIBUTION AND EVOLUTIONARY POSITION

M. J. Benton (Vertebrate Palaeontology, 1997) quotes Olson (1985) who comments that the palaeognathous palate shows primitive theropod and avian characteristics. But there are several **synapomorphies** (i.e, the shared derived characteristics) such as the extensive vomer-pterygoid joint, as also some characteristics in the other parts of the skull as confirmed by Cracraft (1988).

The origin of the palaeognathous birds is rather shrouded in mystery which led to conflicting opinions. From the end of the nineteenth century up to middle of the twentieth, some authors (Furbringer, 1888; Parker, 1892; Pycraft, 1901; Stressemann, 1927-34 and Gregory, 1935) believed in the degenerate nature of the ratites which might have evolved from flying ancestors, while others (Lindsay, 1885; Lowe, 1935 and Friant, 1945-46) claimed the ratites to be primitive and that they branched off from the main stem of bird evolution before the birds acquired flight. Lowe (1935) claimed that from the primitive palaeognathous palate, the neognathous palate of the flying birds has been derived and that the ancestors of ratites were also flightless. But, after excellent analysis by de Beer (1956) on 'The evolution of ratites' and conclusions drawn by recent investigators like Romer and Parson (1986), Gauthier (1986), Cracraft (1986), Sanz et al (1995), Chiappe (1995), Benton (1997), Zweers et al (1997) and Kardong (1998), Lowe's contention appears to be untenable. Recent evidences from all quarters confirm Owen's (1866) classical statement that the 'Ratites are those birds in which power of flight is abrogated.' Then de Beer went further to say that the ratites have degenerated from a carinate condition and, according to him, 'the palate of the ratites is not primitive but neotenous and represents an early stage through which the palate of many carinates passes during development.'

Thus, considering *Archaeopteryx* as an evolutionary stage in the direction of true birds (confirmed by *Simpson*, 1930 and Steiner, 1950), the aviform line first diverged into **enantiornithine** and **ornithurine** offshoots, the latter represented by the toothed birds, *Hesperornis,* which is aquatic, diving and swimming, and *Ichthyornis,* which is a flying bird with keeled sternum. Benton (1997) classified *Ichthyornis* as a carinate. However, both these forms have been grouped into a superorder Odontognathae (= toothed birds).

Next appeared the flightless birds, possibly representing several different lines of evolution, possessing the so-called 'primitive' type of palate, a reduced sternum without keel, a reduced wing and with an unfinished plumage. This group includes tinamou—which is an exception in that it has a keeled sternum (though the palate and other features are similar to other palaeognathous birds) and is capable of flight.

All these birds having diverged early with the so-called 'primitive' nature of palate have been grouped under the superorder 'Palaeognathae' by most of the recent authors (Romer and Parson, 1986; Cractaft, 1986; Chiappe, 1995; Benton, 1997; Zweers and Vanden Berge, 1997/98 and Kardong, 1998), although a few authors (Young, 1981; King and McLelland, 1984) have omitted the super order Palaeognathae and have included tinamous and all the flightless birds under the superorder 'Neognathae', considering that all these birds have arisen from flying ancestors.

According to Benton (In: Vertebrate Palaeontology, 2nd ed., 1997), most of the living palaeognaths have short fossil records, extending back to the Miocene or Pliocene periods only, and, at present, the living forms have a rather discontinous distribution, restricted to Africa, South and Central America, Australia and New Zealand. The oldest fossils from the Palaeocene and Eocene, however, suggest that palaeognaths originated in the Northern Hemisphere, either in Europe or in North America.

A few examples of the above are cited here with their geological periods and places of occurrence:

FOSSIL FORM	GEOLOGICAL PERIOD	PLACE OF OCCURRENCE
1. *Lithornis*[1]	Palaeocene, Eocene	Wyoming, USA
2. *Remiornis*	Palaeocene	France
3. *Eleutherornis*	Eocene	Switzerland
4. *Palaeotis*[2]	Eocene	Germany

Notes: 1. Retained power of flight.
2. Crane-sized flightless ostrich

About tinamous of South America, Romer and Parson (1986) comment that 'they have the power of flight (although of a very poor standard) and yet have the 'old-fashioned' type of palate; they may represent an ancestral group from which the ratites have developed. It is quite probable that the various groups of ratites are not closely related; probably the palaeognaths are not a natural group.' According to de Beer (1956), 'the palaeognathous palate is found not only in the ratites, but also in tinamou which is a carinate with good power of flight. This fact in itself is sufficient to indicate that the ratites subsequently lost the power of flight; otherwise, one is led to believe that the tinamou has acquired flight from the flightless ratites. To explain the loss of flight, it has been conjectured that the ancestors of flightless birds moved to southern continents, particularly South America, Australia and New Zealand where there were very few ground-living predators. Thus once the flying ancestors may have lost flight due to absence of such predators, the descendants evolved into native ratites. Romer and Parson (1986) also share this opinion.

CLASSIFICATION

Earlier in the text, in Chapter 16 'A Brief Classification of Birds', several orders of flightless birds (5 living and 2 extinct orders, as listed by King and McLelland, 1984) have been described with a few distinctive characteristics and examples of each. Here only the names of the orders have been mentioned with examples and geographical distribution of each. A short discussion on the sequence of the orders used by different authors has also been appended.

Classification of flightless birds (used in this text):

CLASS AVES

Subclass Archaeornithes (ancient birds)

Subclass Neornithes (modern birds)

Superorder Odontognathae (toothed birds; Cretaceous)

Superorder Palaeognathae (Tinamous and Ratite birds; Palaeocene—Recent)

Order 1. Struthioniformes e.g., *Struthio*; Africa.

Order 2. Rheiformes e.g., *Rhea*; South and Central America.

Order 3. Casuariiformes e.g., *Casuarius* (Cassowary); *Dromaius* (Emu); Australia and New Guinea.

Order 4. Dinornithiformes e.g., *Dinornis* (Moa); New Zealand (declined and became extinct before 1775).

Order 5. Aepyornithiformes e.g., *Aepyornis* (Elephant bird); Africa, Madagascar (Malagasy); extinct since several centuries.

Order 6. Apterygiformes e.g., *Apteryx* (Kiwi); New Zealand.

Order 7. Tinamiformes e.g., *Tinamus*; *Crypturellus*; South and Central America.

SUPERORDER NEOGNATHAE

All modern carinate or flying birds.

Monroe and Sibley (1993) in 'World Cheeklist of Birds' have proposed a very different systematic arrangement of avian taxa, based on Molecular Systematics. They have recognised only two orders of living palaeognathous birds:

ORDER STRUTHIONIFORMES

Family: Struthionidae (Ostriches)

Family: Rheidae (Rheas)

Family: Casuariidae (Cassowaries and Emu)

Family: Apterygidae (Kiwis)

ORDER TINAMIFORMES

Family: Tinamidae (Tinamous)

It may be mentioned here that the sequence of orders of the flightless birds (both living and extinct) has been used differently by different authors. Although Young (1981), King and McLelland and a few others, on the basis of neognathous ancestry of the flightless birds, have recognised only two superorders of the Subclass Neornithes: superorder Odontognathae and Superorder Neognathae, yet most of the recent authors have retained the Superorder Palaeognathae as discussed earlier in this chapter. As regards orders of flightless birds, most authors have included 5 living orders and 2 extinct orders with minor variations in their taxonomic sequences.

The 'World Checklist of Birds' (published by Monroe and Sibley (1993), also known as 'Sibley-Ahlquist-Monroe (SAM) classification' is an extended and revised edition of the original text–'Phylogeny and Classification of Birds' by Sibley and Ahlquist (1990). This classification is based on the differences in **DNA sequences** and **DNA hybridization** techniques followed in birds.

SOME SALIENT FEATURES OF PALAEOGNATHOUS BIRDS

1. The ostriches are the largest living birds of Africa and West Asia; they are above 7 feet (2.1 m) tall, and may reach up to 10 feet (3 m) and weighing more than 330 lbs (150 kg). Among other living birds, Rhea is about 5.5 feet (1.7 m) tall, Cassowary about 3 feet (0.9 m), Emu about 6 feet (1.8 m); Kiwi and Tinamou are comparatively smaller birds.

   Among extinct orders, Moas were about 8 feet (2.4 m) tall and Elephant bird about 10 feet (3 m), having the largest egg 9.5 inches (24 cm)/ 13 inches (33 cm).
2. These are flightless, terrestrial birds (except Tinamous) adapted to running in moderate to high speed.
3. The wings are small or rudimentary; head, neck and body may be almost featherless, partly feathered or densely feathered ; aftershaft may be present or absent. The feathers are primarily of downy type.
4. The sternum is without keel (except Tinamous), and with usually 3-4 toes on each foot (2 in Ostrich).
5. The pygostyle is rudimentary or absent.
6. The condition of the palate is called palaeognathous (W.P. Pycraft, 1900, 1901) in which the pterygoids extend forward and make contact with the hinder ends of the prevomers, while the palatines lie farther to the sides.
7. The upper jaw kinesis is absent, there being no mobile joint at the base of the upper jaw (except in Cassowary where very slight mobility is observed far forward in the upper jaw at the base of the nostrils).

PROFILE OF A FEW LIVING PALAEOGNATHOUS BIRDS

A short profile of a few ratite birds, as well as of tinamous, is given below to depict individual peculiarities of these birds. These birds are characterised by discontinuous distribution and variable habits and habitats.

1. Ostrich (*Struthio camelus*)

This is the largest among living flightless birds, measuring about 3 meters (10 feet) in hight or even more, and weighing up to 380 lbs (150 kg).

Taxonomic status: In the 'World Checklist of Birds' by Sibley and Monroe (1993), only one species (*Struthio camelus)* has been recognised, having its distribution in Africa, mainly northern and southern parts. Others have recognised 2 species, *Struthio camelus* and *S. molybdophanes*, known as Sonali Ostrich of north-east Africa.

Distribution: Africa, mainly northern and southern parts. The distribution of the species also extended to Arabia where the bird has vanished due to indiscriminate hunting.

Habitat: Savanna or the tropical grassland with scattered drought-resistant trees, vast stretches of open scrubby land and dry, sandy regions with sparsely distributed vegetation constitute the natural habitat of the bird.

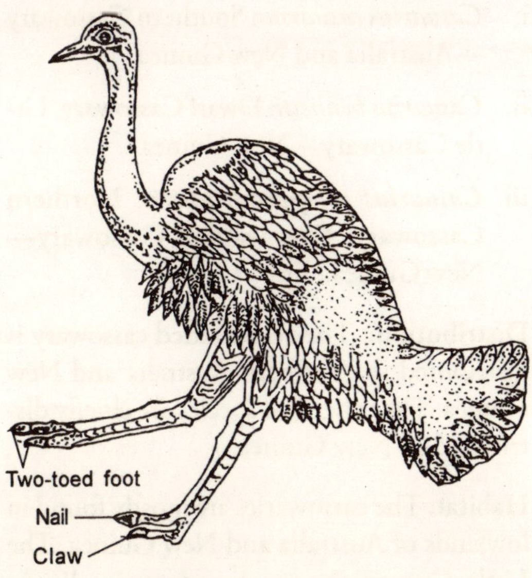

Fig. 25.1A: Palaeognathous birds. Ostrich

Characteristic features and habits: Male and female ostriches differ in their plumage colour. While the male has a distinct black and white plumage, the female's plumage is brownish. Although pterylosis is continuous in the adult bird, the head, neck and the legs are very thinly feathered. The feathers are plume-shaped and without aftershaft. The ostrich has its legs long and strongly built, which may be correlated with the high running speed of the bird (about 50 miles 80 km/hr). The wings are rudimentary.

The foot is provided with only **two toes, each terminating in a flat nail.** The ostrich is one of the few birds possessing a retractile penis. A unique character among birds is that **the urine in ostrich is concentrated in the cloaca and is excreted separately** (not with the excrement). The ostrich is **polygamous,** the male mating several females.

The food mainly consists of vegetable matters and invertebrates, and occasionally also a few small vertebrates. The ostriches are gregarious, living in groups, and quite often in association with the foraging mammals. It digs a hollow nest on the ground within which it lays up to 20 eggs, each weighing about 2.2-2.5lbs (1-1.5 kg). Both the parents incubate the eggs in turn. The youngs hatch after about 42 days and the male bird rears them.

2. Rhea (*Rhea americana;* Greater Rhea)

This tall, elegant and the so-called 'South American Ostrich' has a height up to 6 ft 7 in. (1.7 m) and a weight up to 55 lb (25 kg). The bird has an almost uniform coating of soft plumage in its head, neck, trunk and thighs. Each foot is provided with three clawed toes directed forward.

Taxonomic status: Sibley and Monroe (1993) have recognized only two species of *Rhea*, *Rhea americana* and *R. pennata* or the **Lesser Rhea.** Bologna (1997) has mentioned the generic name of the Lesser Rhea as *Pterocnemia* and others have recognised one more species of Rhea as *R. tarapacenis* (Puna Rhea).

Distribution: Rhea is distributed in the south, south-central, south-western and in the west of Andes of South America.

Habitat: The rheas live in almost the same ecological conditions in South America as do the ostriches in Africa. The rheas prefer to live in open grassy steppes and pampus or scrubby land in the foothills.

Characteristic features and habits: The rheas too, like the ostriches, have long legs, the tarso-metatarsus varying from 12"-15" (30-38 cm) in length. In rheas, there are about 22 transversely arranged horny plates. The Lesser Rhea is smaller in size than that of Greater Rhea. The rheas, like ostriches, are also good runners.

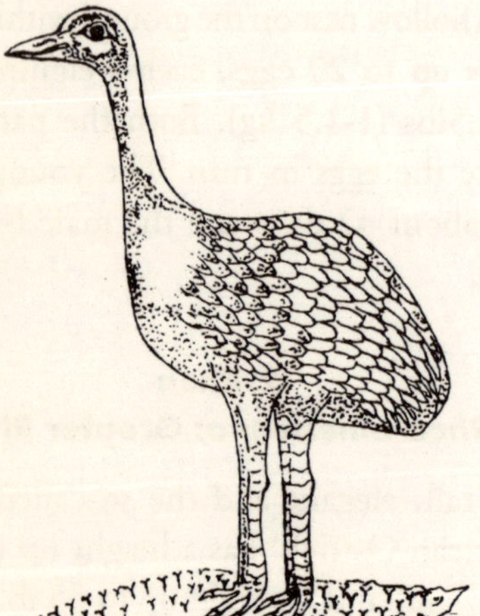

Fig. 25.1B: Rhea

The wings are reduced but slightly better developed than in the ostrich. While running in speed, the bird turns one of the wings upward, using it as a sail. The Rheas lack distinct tail feathers unlike in ostrich.

The Greater Rhea feeds on grass, herbs, shrubs and on small invertebrates. Usually the male digs a hole-nest in the ground in which the female lays about 15-20 eggs at an interval of a few days. The incubation is done by the male for about 40 days and the youngs are also reared by the males. The adult size is reached in about six months.

3. Cassowary (*Casuarius casuarius*)

The Cassowaries of Australia are flightless, terrestrial birds, slightly smaller in size than the Emus (the emus being the second largest living flightless bird, next to ostrich). The two-wattled cassowary—also known as the Southern Cassowary—measures about 1 meter (40") in height and may reach up to 187 lbs (85 kg) in weight.

Taxonomic status: 3 species of Cassowaries are recognised:

i. *Casuarius casuarius*: Southern Cassowary —Australia and New Guinea.

ii. *Casuarius bennetti*: Dwarf Cassowary; Little Cassowary—New Guinea.

iii. *Casuarius unappendiculatus*: Northern Cassowary; One-wattled Cassowary—New Guinea.

Distribution: The two-wattled cassowary is distributed in Northern Australia and New Guinea. The other two species are chiefly distributed in New Guinea.

Habitat: The cassowaries are mostly found in lowlands of Australia and New Guinea. The Little Cassowaries are also adapted to live in mountain regions up to an altitude of about 10,000 feet (3,050 m).

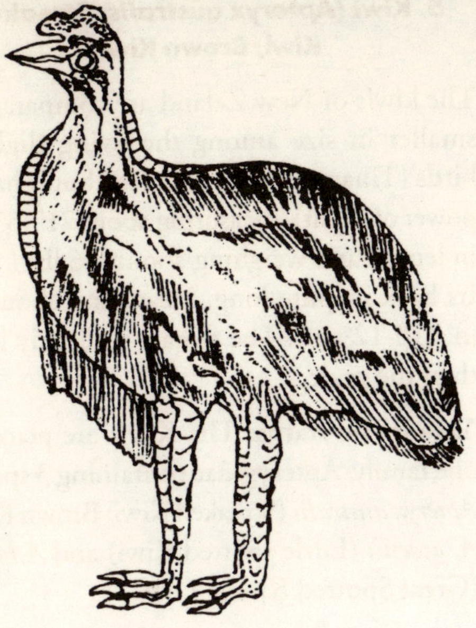

Fig. 25.1.C : Cassowary

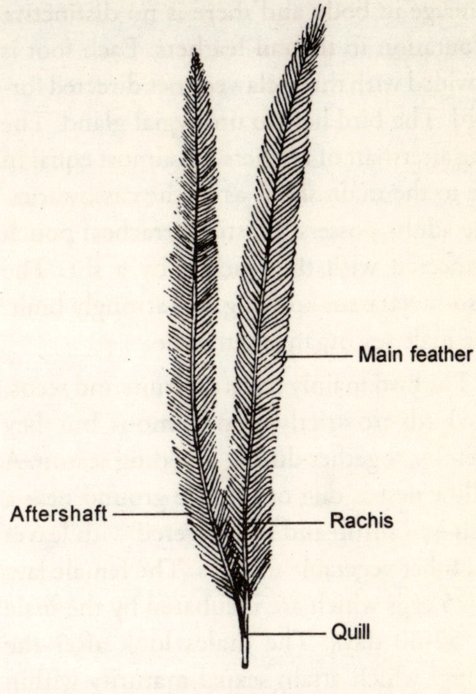

Fig. 25.1.C(a): Aftershaft of Cassowary

Characteristic features and habits: The cassowaries have small wings and their neck and body are densely feathered, the feathers being provided with exceedingly large aftershaft which is almost equal in size to the main shaft. Legs are very strong and each foot is provided with three toes of which the inner toe bears a straight claw that may reach up to 4 in (10 cm) in length. The two-wattled cassowaries bear two smooth caruncles on the front side of the neck; one-wattled cassowaries bear one such caruncle.

The cassowaries feed mainly on fruits and partly on smaller invertebrates and vertebrates. They have a very short intestine. They are usually monogamous, but they assemble together during mating season.

The nest, like other flightless birds, is a depression in the ground, covered with vegetable matters. The female lays 3-8 eggs which are incubated by the male for about 49-56 days. The youngs are also reared by the males.

4. Emu (*Dromaius novaehollandiae*)

Emus are the second largest living ratite bird, measuring about 6 ft (1.8 m) in height and weighing up to 120 lbs (55 kg). The trunk is longer from the base of the vertical neck than that of cassowary.

Taxonomic status: Emu of Australia is now represented by only one species. The Emu of Kangaroo Island, *Dromaius minor*, is now extinct.

Distribution: The bird is distributed throughout Australia.

Habitat: The birds are mostly found in the scrubby steppes in Australia. The habitat niche of the bird in general are the same as in Cassowary.

Characteristics features and habits: Both the males and females have a brown-coloured plumage in body and there is no distinctive colouration in the tail feathers. Each foot is provided with three-clawed toes directed forward. The bird has no uropygial gland. The long aftershaft of feathers are almost equal in size to the main shaft, as in the cassowaries. The adults possess a distinct tracheal pouch connected with the trachea by a slit. The tarso-metatarsus are long and strongly built. The male sex organ is retractile.

The bird mainly feeds on fruits and seeds. The birds are strictly monogamous, but they assemble together during breeding season. A hollow nest is dug out in the ground near a bush or a shrub and it is covered with leaves and other vegetable matters. The female lays 15-25 eggs which are incubated by the male for 52-60 days. The males look after the youngs which attain sexual maturity within 2-3 years. The nestlings have a distinctive stripes of brown and white plumage which turns completely brown in the adult.

The Emus have a running speed of about 30-35 miles (48-56 km) per hour. Both the skin and flesh have economic importance for which the birds are gradually decreasing in number, falling easy prey for hunting.

Fig. 25.1D: Emu

5. Kiwi (*Apteryx australis;* Streaked Kiwi; Brown Kiwi)

The kiwis of New Zeland are comparatively smaller in size among the living flightless birds (Tinamou is the smallest, but it has the power of flight), measuring about 21"(53 cm) in length and weighing about 6.6 lbs (3 kg). Its beak is quite long, measuring about 4-5 in. (10-12 cm). The female is slightly larger than the male.

Taxonomic status: The Kiwis are placed in one family, Apterygidae containing 3 species: *Apteryx australis* (Streaked Kiwi; Brown Kiwi); *A. owenii* (Little spotted Kiwi) and *A.haastii* (Great Spotted Kiwi).

Distribution: All the 3 species are distributed in New Zealand, the last two species inhabiting southern part of the island.

Habitat: The bird lives in open grassland, shrubs and woody mountainous regions.

Characteristic features and habits: The long bill of the bird is curved downward and is provided with long bristles at the base. A peculiar feature is that the nostrils are situated near the apex of the bill. The palate shows large **basipterygoid process**. The legs are strongly built, somewhat backwardly placed and with sharp claws. There are 13 remiges in each of the wings which are rudimentary; distinct tail feathers are lacking. The Kiwis possess a very well-developed sense of smell which is a unique characteristic. Both the males and females have a reddish-brown or greyish-brown colour of soft plumage in both the young and adult stages.

The Kiwis are nocturnal and burrowing in habit, digging holes in the ground and resting there during daytime. Their food mainly consists of invertebrates like worms, slugs, insects as well as fruits of the nature of berries. A hole-nest is built by digging in the ground underneath the roots of trees or in the vicinity of bushes and rocks. The female lays one or two eggs, each weighing about 450 gms and covers them with grass and vegetable matters. The male incubates the eggs for about 75 days. The hatchlings covered with soft downy feathers are reared by the male in the nest for about 5 days after which the male guides them in short flights. The Kiwis take about 5-6 years to reach sexual maturity.

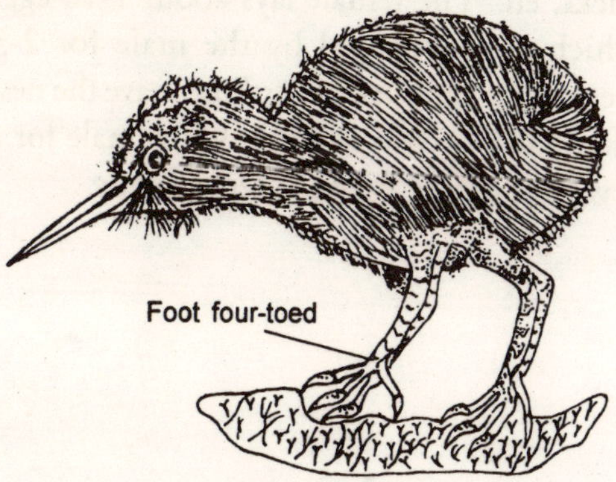

Fig. 25.1.E: Kiwi

Extinct flightless birds: Among the extinct flightless birds, the moas (*Dinornis*, order Dinornithiformes) of New Zealand and the huge elephant bird (*Aepyornis*, order Aepyornithiformes) of Africa and Malagasy were remarkable. Both have been extinct only several centuries ago. *Aepyornis* finally became extinct from Malagasy. Some elephant birds were even taller than ostrich, i.e., more than 10 feet (3 m) in height and they produced eggs, each weighing more than 22 lbs (10 kg) which happened to be largest single cell ever existed.

Fig. 25.1.F: Moa

6. **Tinamou** (Tinamous; e.g., *Eudromia elegans*; **Elegant Crested Tinamou**)

The Tinamous of South America are also comparatively smaller birds which are somewhat pheasant-like in plumage and guinea-fowl-like in form. These birds possess keeled sternum and are able to fly short distances. As such, they are palaeognathous birds, but not within the ratite group. They are about 8"-21" (20-53 cm) long and their weight varies between 1.1-6.25 lbs (0.5-2.5 kg).

Taxonomic status: The Tinamous are placed in the Order Tinamiformes and family Tinamidae. It is quite a large family containing 9 genera and 47 species and 5 Subspecies (Sibley and Monroe, 1993) or 52 species (as considered by others). The Genus *Eudromia* has only 2 species; *Eudromia elegans* (Elegant Crested Tinamoe; South-west South America) and *E. formosa* (Quebracho Crested Tinamou; South-central South America).

Distribution: All the species are distributed in different parts of South America, from southern Mexico to Patagonia and from the tropical rain forests up to 14,000 feet (4,270 m) altitude in the Andes.

Habitat: *Eudromia* prefers to live on scrubby land, lowland and steppe.

Fig. 25.1.G: Elegant Crested Tinamou

Characteristic features and habits: The Elegant-Crested-Tinamou has a few long feathers on the head, forming a crest. The wings are short, as in all other birds of this family, and the bird lacks hind toe. They are not good fliers, although the flight muscles attached to the keeled sternum are reasonably well-developed. The body is compact with a short tail and the legs are strong.

The bird, as well as other tinamids, feed mainly on vegetable matters—fruits, seeds, leaves, flowers and roots of plants. They also feed on insects and molluscs collected from the ground. The males of these birds are polygamous, the females polyandrous. The nest is a depression on the ground, lined by leaves, sticks, etc. The female lays about 1-16 eggs which are incubated by the male for 2-3 weeks. The nidifugous nestlings leave the nest after a day or two and follow the male for a few days before flying independently.

26

A Profile of Some Common Indian Birds

1. Dabchick or Little Grebe
2. Grey Pelican
3. Little Cormorant
4. Cattle Egret
5. Openbilled Stork
6. Flamingo.
7. Barheaded Goose.
8. Common Pariah Kite.
9. White-rumped or Bengal Vulture.
10. Indian Peafowl (Common Peafowl).
11. Common Sandpiper.
12. Blue Rock Pigeon.
13. Spotted Dove.
14. Rose-ringed Parakeet.
15. Koel
16. Barn Owl,
17. Little Swift.
18. Hoopoe.
19. Little Green Bea-eater.
20. White-throated King Fisher.
21. Golden-backed Woodpecker.
22. Red-Vented Bulbul.
23. Magpie-Robin.
24. Tailor Bird.
25. Purple Sunbird.
26. House Sparrow.
27. Baya Weaver Bird.
28. Common Myna.
29. Pied Myna.
30. Black-headed Oriole.
31. Black Drongo.
32. House Crow.

The Indian subcontinent has a very wide range of biodiversity and a rich characteristic avifauna. Considering India as a sub-region, Blanford (1901), over 100 years ago, divided it into five primary subdivisions:

1. **The Indo-Gangetic Plain**—extending across the entire northern India from the Arabian Sea to the Bay of Bengal.
2. **Peninsular India**—extending from south of the southern boundary of the Indo-Gangetic Plain to the southern limit of India across the whole of south Indian states.
3. **Ceylone (Sri Lanka)**.
4. **The Himalayas**—the mountianous range from the west to the east along the entire foot-hills.
5. **Assam and Burma (Myanmar)**.

Salim Ali (1987) comments that the Punjab, Sind and Rajasthan have an avifauna differing from that of the other parts of India and resembling birds of West Asia and North Africa. Birds of the upper Himalayas and upper Indus Valley are nearer to those of Central Asia, i.e., parts of Palaearctic region.

India is rich in avifauna with characteristic biodiversity and seasonal changes in different parts of the country. Of about 22-23 orders of carinate birds, about 18-19 orders are represented in India possessing nearly 3,000 species and subspecies—of these more than 450 species of birds are extra-limital winter visitors.

A short profile of a few common Indian birds with their salient features, distribution, habits and habitats is given. The Bengali name (B. name) of the bird has also been given within brackets.

1. Dabchick or Little Grebe (*Podiceps ruficollis*) (Bengali and Hindi names: Pandubi)

Classification: Order Podicipediformes; Fam. Podicipedidae.

Morphological Characteristics: The bird is of the size of a piegon or slightly smaller than that of a domestic duck (9-25 inches / 22.5-62.5 cm). The upper part of the trunk plumage is dark grey or black, the underparts white or brownish-white. In the breeding season, the head and neck plumage is dark brown. The head is usually provided with lateral tufts of feathers. The bill is sharply pointed, longer in larger species and with a prominent gape of mouth. The wings are short, tail absent, or rudimentary. The toes are lobed and the claws extremely flattened.

Distribution and Habitat: This water bird is found throughout India and in its neighbouring countries, as well as in Europe and North America. They inhabit lakes and ponds with abundant emergent vegetation. During winter the migratory species from north use coastal waters.

Food: The bird feeds on aquatic insects and their larvae, fish, crustacea, molluscs, frogs, tadpoles and aquatic vegetation.

Habits and Nesting: The bird is a graceful swimmer on surface water and a very fast diver-swimmer underwater while chasing its food prey. The nest is built either in shallow water on a rough pad of weeds and decaying vegetation or on a heap of floating vegetation appearing like a raft or mound. This species lays 2 clutches of 4-6 eggs which are incubated by both the parents. The nidifugous nestlings are reared by both the parents and the youngs fly in about 44-48 days.

Fig. 26.1: Little Grebe

2. Spot-billed or Grey Pelican (*Pelecanus philippensis*)

Classification: Order Pelecaniformes; Fam. Pelecanidae. Of 8 species of pelicans, *P. onocrotalus, P. crispus* and *P. philippensis* occur in south and south-east Asia.

Morphological Characteristics: The grey Pelican has the size of a vulture (50-72 inches/ 125-180 cm) and appears rather greyish-white with a brown nuchal crest. Bill very stout, long, straight and hooked rostrally. A large gular pouch underhangs the entire bill like a large sac of purplish skin. The neck is long but the legs are short and very stout; toes large, long and fully webbed. The wings are large and blackish, but the tail is short and greyish-brown.

Fig. 26.2: Grey Pelican

Distribution and Habitat: This bird occurs throughout India and its neighbouring countries like Bangladesh, Pakistan, Sri Lanka, Burma (Myanmar). Their habitat consists of large lakes, jheels, as well as coastal waters.

Food: The food of the bird comprises fish, molluscs, crustaceans, jelly fish, small turtles and even young birds.

Habits and Nesting: These birds are gregarious and either swim in flocks or rest on banks. In flocks they drive the fishes in a semicircle towards shallow region of the waterbody by vigorous beating of their wings and then collect their food individually by scooping.

The nest is built far away from water as a stick platfrom on tall trees and there may be a number of nests in one tree. Usually 3 eggs are laid with the variation of number in other species from 1 to 4. The eggs incubated and cared by both the parents. The hatchlings are nidicolous.

3. Little Cormorant (*Phalacrocorax niger*) (B. name: Pankouri)

Classification: Order Pelecaniformes; Fam Phalacrocoracidae; 36 species, of which the **Great Cormorant** *(P. carbo)* and the **Indian Cormorant** (Indian Shag), *P. fuscicollis* are often found in association with the Little Cormorant.

Morphological Characteristics: A shiny black crow or duck-like water bird that has a long, stiff tail and a moderately long, laterally compressed bill which is hooked rostrally. This species has few feathers at the back of the head, appearing as a crest. The bird has its neck and body long, wings short and feet large and fully webbed. A small white patch on throat is characteristic.

Fig. 26.3: Little Cormorant

Distribution and Habitat: The bird is found throughout India and adjacent countries. Other species are nearly worldwide, found in coastal waters, large lakes and rivers.

The Little Cormorant is found in all inland waters, in ponds, lakes, jheels or in lagoons of brackish water.

Food: Food comprises fish, crustacea and smaller amphibia.

Habits and Nesting: Usually gregarious, this bird is an excellent diver and underwater swimmer, chasing fish and other food-preys.

The nest is a twig platform on trees or of seaweeds, sticks, etc. on rocks in other species. Eggs 4 or 5 (usually 2-4), incubated by both parents; youngs are nidicolous.

4. Cattle Egret *(Bubulcus ibis)* (B. name: Gobak)

Classification: Order Ciconiiformes; Fam. Ardeidae. This family contains about 12 species of egrets, 42 of herons and over 12 species of bitterns.

Morphological Characteristics: This bird is slightly larger than a hen and has a yellow bill and white plumage (non-breeding) which is rendered orange-buff coloured during breeding season. Bill moderately long and spear-shaped. Legs long and narrow with long clawed toes.

Distribution and Habitat: This species has a worldwide distribution and found in great numbers throughout India and in adjacent countries. As mostly found in association with cattle, the usual habitat of the birds comprises grazing pastures, marshy land, etc. where they move along the cattle and catch insects either from their body or from the disturbed surroundings.

Food: Largely grasshoppers, flies and other insects, also crustacea, fish, molluscs and small amphibians and lizards.

Fig. 26.4: Cattle Egret

Habits and Nesting: This species is usually gregarious, swallowing the prey whole. Nesting season varies according to monsoon. The nest is an irregular and clumsy twig platform similar to that built by crows. Salim Ali (1987) reports that, in mixed colonies with cormorants, paddy birds, etc., nest may be built in large leafy trees away from the feeding ground. Eggs 3-5, incubated by both the parents; youngs nidicolous.

5. Openbilled Stork (*Anastomus oscitans*) (B. name: Shamukkhole)

Classification: Order Ciconiiformes; Fam. Ciconiidae.

This species is also known as Asian Openbill.

Morphological Characteristics: This bird is much smaller among the storks with a height of only 2.5 feet (76 cm). It is a white or greyish white bird with black in the tail and along posterior border of the wings. The diagnostic feature is the peculiar reddish-black bill in which the upper and lower beaks form an arch

Fig. 26.5: Openbilled Stork

in opposite directions, leaving a gap between them (hence the name openbilled).

Distribution and Habitat: It is found throughout India and in the adjacent countries. Their common habitats are the jheels and marshes. They are resident as well as locally migratory.

Food: The characteristic beaks are very useful for opening shells of bivalve molluscs or large *Ampullaria* snails and extracting their muscles and viscera which form major part of their food. However, the bird also feeds on small frogs, crabs, insects. etc. to a small extent.

Habits and Nesting: The bird has its general habits like those of other storks. The bird breeds in mixed colonies of other species of storks, cormorants, egrets, etc. The nest is a twig platform of round shape having a central depression and the margin is lined with leaves. Eggs 2-4, white.

6. Flamingo (Phoenicopterus roseus)

Classification: Order Phoenicopteriformes; Fam. Phoenicopteridae; 6 species, including *P. roseus* as a distinct species, but according to Monore and Sibley (1993), it is a subspecies of *P. ruber*. However, both of them are known as **Greater Flamingo.**

Morphological Characteristics: A large bird of a length of about 3-4 feet (90-120 cm) and a standing height of about 4 feet (1.2 m). The bird is unmistakably indentifiable by its long legs (longer and more slender than those of storks), long neck that is curved in the middle and a stout pink-coloured bill that is turned down at a sharp angle from the middle. The neck and body of the bird is rosy white, while that of the legs pink. The wings are large, pink and bordered black.

Fig. 26.6: Flamingo

Distribution and Habitat: This bird is found through Indian subcontinent except Burma (Myanmar). It is a resident bird, though some species are migratory. Their usual habitats are marshes, jheels, large lakes and ponds, estuaries, tidal mudflats, etc.

Food: Fish, molluscs, crustacea, insects, marshy plants, algae and diatoms, etc.

Habits and Nesting: These birds are gregarious, forming colonies during breeding season. Flamingos fly with their neck outstretched. While feeding in shallow water, the head is immersed and the bill inverted, so that the ridge of the bill (culmen) is able to scrape the mud bottom. The muddy liquid thus collected in the hollow of the mandible contains food particles and organisms which are then strained by the bill lamellae and fleshy tongue and taken as food.

The nest is a truncated cone of mud [about 6 to 12 inches (15-30 cm) in height] built on an extensive area of mudflat by the colony. Hundreds of such nests built close to one another look like a 'Flamingo City'.

Eggs 1 or 2, white and incubated by both the parents; hatchlings nidifugous.

7. Barheaded Goose (*Anser indicus*) (B. name: Bari haans)

Classification: Order Anseriformes; Fam. Anatidae.

This is a large family containing several subfamilies and tribes. The genus *Anser* has more than a dozen species.

Morphological Characteristics: Slightly smaller than the Greylag Goose *(Anser anser)*, this bird has its head, sides of the neck and trunk white, the rest of the body brown and the rear borders of remiges and rectrices black. Two distinct broad and black bars are diagnostic.

Distribution and Habitat: Like many other ducks and geese, this birds is also a winter visitor to almost whole of India and its adjacent countries from south and central Eurasia and north Africa. The common habitats include jheels, large lakes, rivers and the cultivated fields.

Food: The young shoots of winter vegetation, soft creepers and winter crops form major bulk of food.

Habits and Nesting: This species—like most other ducks and geese—migrate in flocks. In West Bengal, they are found in greater numbers in the Gangetic delta. During daytime, they rest on sandbanks in large rivers and at night visit the cultivation fields and feed on winter crop.

They build the nest in Ladakh and Tibet with the help of grass, herbage, vegetables and feathers around a depression on the ground. Eggs 3 to 4, light creamy-white.

Fig. 26.7: Barheaded Goose

8. Common Pariah Kite (*Milvus migrans;* also known as Black Kite) (B. name: Chil)

Classification: Order Falconiformes; Fam. Accipitridae.

This is a very large family containing more than 200 species, including kites, vultures, harriers, goshawks, hawks, eagles, etc.

Morphological Characteristics: This common scavenger bird is about 2 feet (60 cm) long and distinguished from similar birds by its long, forked tail. Its colour is deep brown, bill hooked and toes have sharp, curved claws.

Distribution and Habitat: This bird is found all throughout India and its adjacent countries and up to 8,000 feet (2,440 m) in the Himalayas. It is a resident bird, living close to human habitation and quite common in villages and towns alike.

Food: It feeds on anything eatable from the garbage as well as on rats, mice, small birds, lizards and some insects.

Fig. 26.8: Common Pariah Kite

Habits and Nesting: A very energetic bird, well-known for its flight manoeuvers with great skill and speed for swooping on the prey and anchoring it with the help of its curved clawed toes.

The nest is a twig platform built with wire, rags, rubbish and twigs on high trees, roof tops, corners and crevices of buildings. Eggs 2-4, pinkish-white or reddish-brown; cared by both the parents; youngs nidicolous.

Another very well known species of Indian kite is the **Brahminy kite** *(Haliastur indus)* having almost the same distribution as of the Common kite, but the bird has a graceful look and a rounded tail (not forked). Its entire head, neck and the anterior abdomen are white, while the rest of the body is reddish-brown.

9. White-rumped or Bengal Vulture (*Gyps bengalensis*) (B. name: Shakun)

Classification: Order Falconiformes; Fam. Accipitridae.

Morphological Characteristics: It is a large and heavy blackish-brown vulture (length about 2 feet (60 cm)) with a characteristic hefty and thick downwardly curved bill and with a featherless head and rump. A distinct white patch on back is a diagnostic character. The rest of the body has brown and dull-grey feathers. Legs stout and strong, toes with sharp talons.

Distribution and Habitat: It is the commonest species in India and also occurs in adjacent countries. The bird is found everywhere in the countryside and on the outskirts of towns and villages.

Food: This common scavenger bird is a flesh-eater and it feeds on carcasses of animals, tearing off flesh from the dead bodies with great speed, thus leaving behind only the skeleton in no time. They attack the carcass in a group, making much noise.

Habits and Nesting: These birds are swift fliers but very often they are found gliding and soaring in air with their acute vision fixed on any possible source of food. While feeding, their bill and feet function concertedly in tearing and grasping. A single white egg is laid and cared by both the parents. The youngs

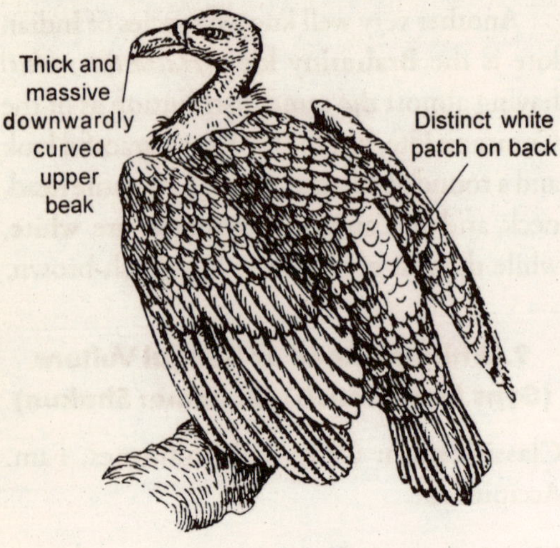

Fig. 26.9: Bengal Vulture

are nidicolous (staying longer than average in the nest).

10. Indian Peafowl; Common Peafowl (*Pavo cristatus*) (Hindi name: 'Mor' or 'Mayur'; Bengali name: Mayur)

Classification: Order Galliformes; Fam. Phasianidae.

This is the largest family in the order, containing 44 genera and 176 species.

Morphological Characteristics: The peacock (male peafowl) is one of the most spectacular birds of India with the beautiful and gorgeous colour patterns of its fanned-out tail feathers. It is the National Bird of India. The male's head bears a beautifully arranged V-shaped crest of feathers. The peahen too bears a small crest. The male is slightly larger in size than the female. The peacock has its head, neck ventral trunk and upper side of the wing coloured blue, the back of the bird and lower edge of the wing has a brown colour. The long (2-4 feet /60-120 cm long), ceremonial 'train' of the tail has the characteristic bluish-green colour with a symmetrical design of oval brown spots (also known as **eye-spots**).

Distribution and Habitat: The bird is found all throughout India, particularly in drier parts and up to 5,000 feet (1,500 m) in the Himalayas. It also occurs in Sri Lanka and Bangladesh, but a separate species, *P. muticus*. occurs in Myanmar. It has a pointed crest.

The usual habitats are the dense shrubby plains or thicker forests at the foothills or in deciduous forests.

Food: This bird has a wide range of food-choice. It feeds on worms, insects, small amphibians, lizards, snakes, vegetable shoots as well as grains and seeds.

Habits and Nesting: It is shy but quite alert. When disturbed, it flies away with admirable promptness through hedges and jungles. These birds are polygamous and are usually found in a group of 1 peacock and 3-4 hens. The peacock is famous for its dance before the hens during breeding season. At other times, during cloudy or rainy days, the peacock fans out its tail feathers to the full extent, covering a wide space and then moves on its legs in its typical dancing mood.

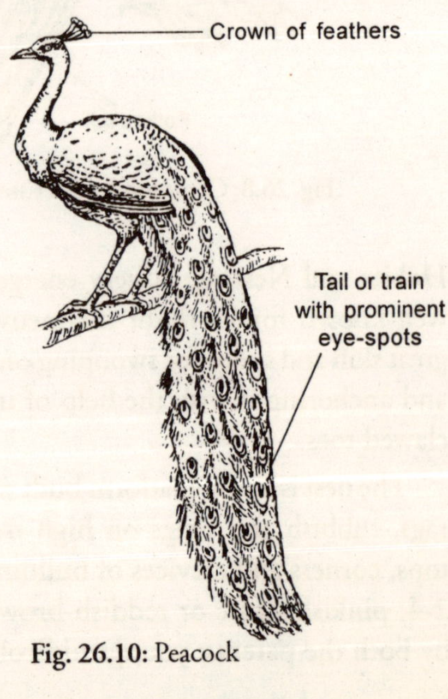

Fig. 26.10: Peacock

The nest is a shallow depression on the ground scraped out by its feet. The nest-bed as well as the edges of the nest are neatly prepared by sticks, leaves, vegetable matter, etc. Eggs 3-5, light cream and cared by both the parents or by the male, youngs nidifugous, downy.

11. Common Sandpiper (*Tringa hypoleucos*)

Classification: Order Charadriiformes; Fam. Scolopacidae; 15 species.

Morphological Characteristics: It has the size of a quail, measuring about 8" (20 cm) long. The head, neck and upper part of the body and tail have a characteristic greyish-brown plumage, the lower part is white. The plumage colour is more uniform in the winter. A distinct wing-bar may be present in the hinder region.

Distribution and Habitat: This species is found in Europe, Africa and Asia. It is found throughout India and its adjacent countries as a winter visitor. The usual habitats are large ponds, lakes, rivers, hill streams, marshes, rocky sea shore, etc.

Fig. 26.11: Common Sandpiper

Food: The bird feeds on insects, worms, molluscs and fishes.

Habits and Nesting: The bird walks in a special gait, keeping the head low and, in flight, the wing-bar is shown as a white streak. Salim Ali (1987) reports that this bird arrives in our country earliest ie., in August, and leaves late i.e., in May. Some are resident in our country all the year.

The nest is a depression in the ground vegetation, lined by leaves and rubbish. Usually one clutch of 4 eggs are laid, incubated by both the parents; the young are nidifugous.

12. Blue Rock Pigeon (*Columba livia*) (B. name: Paira)

Classification: Order Columbiformes; Fam. Columbidae. The genus contains 54 species; the Family contains about 310 species.

Morphological Characteristics: It is a very common, docile and graceful bird measuring about 13" (33 cm) long. The plumage is uniformly slaty-grey with shiny green and purple colour around the neck. Two black, transverse bands are found across posterior third of the wing and there is a black band at the end of the tail. The rump is white, while the legs are purple-red. Both adults and young have similar plumage.

The species has given rise to the Domestic Pigeon from which, through selective breeding, various other varieties of pigeon have developed.

Distribution and Habitat: This species is found throughout India, Bangladesh, Pakistan, Sri Lanka and Burma (Myanmar); this bird with other species are now cosmopolitan in distribution. In India, it occurs up to

13,000 feet (3,960 m) in the Himalayas. It is a resident bird, but also locally migratory.

The usual habitats include coastal rocks, cliffs (particularly for wild species), human habitations in cities, towns and villages, corners and cracks in buildings, farm houses, etc.

Food: Usually granivorous birds, pecking grains of cereals, pulses and seeds from the ground. The bird also feeds on berry-type fruits.

Habits and Nesting: The feral or semi-feral pigeons that we find in the vicinity of human habitations have been well-adapted to live not only in deserted buildings, but also in noisy factories or residential buildings.

They are gregarious by nature and their nests are built in cracks and crevices of cliffs (wild species) or in ceilings of buildings and farmhouses, storehouses of grains, railway stations, etc. The nest is an untidy platform of sticks and vegetable matter. Eggs 2; incubated by both the parents and the nidicolous hatchlings are fed by both with pigeon's milk.

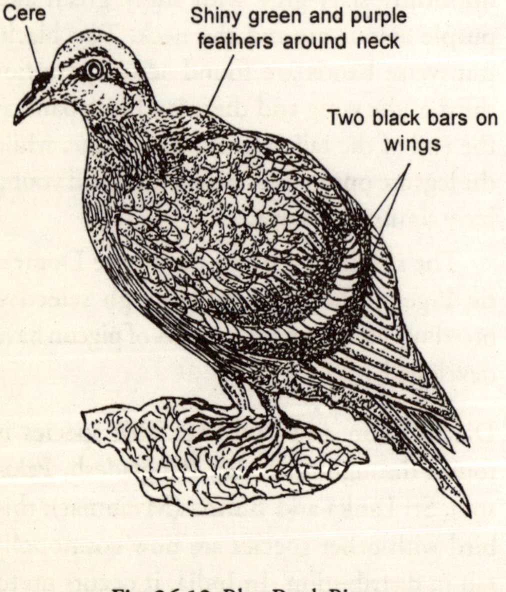

Fig. 26.12: Blue Rock Pigeon

13. Spotted Dove (*Streptopelia chinensis*) (B. name: Chhitey Ghughu)

Classification: Order Columbiformes; Fam. Columbidae. This genus contains 15 species.

Morphological Characteristics: This bird is slightly smaller than the Rock Pigeon and is readily distinguished by its black and white spotted design on the hinder part of the neck and white-spotted brownish body on the back and sides and greyish ventrally. Both the sexes have similar colour pattern.

Distribution and Habitat: It is a resident bird of India and occurs also in Bangladesh, Sri Lanka and Burma (Mayanmar). It is absent in Pakistan and in the drier north-western part of India. It is widely distributed in West Bengal.

They usually move in a group or in pairs in open fields in the vicinity of human habitation and in parks and woody tracts.

Food: The food habit is similar to pigeons.

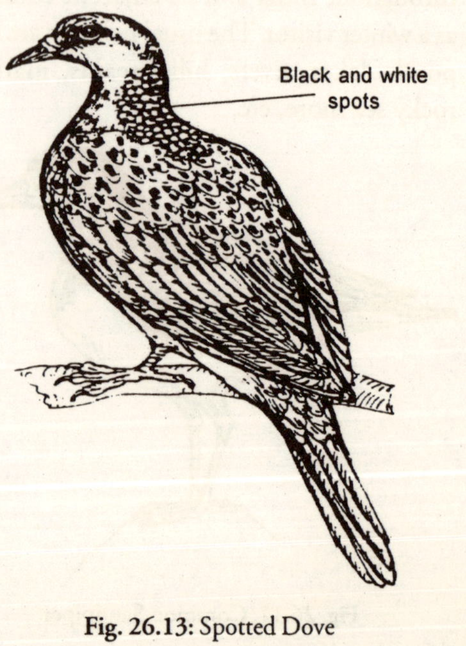

Fig. 26.13: Spotted Dove

Habits and Nesting: Their nesting season remains throughout the year. The nest is built with sticks, twigs and other flimsy materials in low bushes, lower branches of trees, corners of beams and corridors and cracks of buildings.

Eggs 2, incubation and care of nidicolous hatchlings shared by both the parents.

14. Rose-ringed Parakeet (*Psittacula krameri*) (B. name: Tia)

Classification: Order Psittaciformes; Fam. Psittacidae.

It is a large family containing about 80 genera and 360 species; the genus *Psittacula* contains 14 or 16 species.

Morphological Characteristics: It is a common Indian bird of the size of a Myna and smaller in size than the large Indian Parakeet *(P. eupatria)*. The male bird has a characteristic black and rose-ringed collar around the neck; the female lacking this collar. The entire body is green, the bill bright-red and sharply hooked and curved downward. The long, narrow tail is also characteristic of the parrots.

Fig. 26.14: Rose-ringed Parakeet

Distribution and Habitat: The bird is found throughout India, from the foothills of the Himalayas up to hills and plains of the south; also occurs in Bangladesh, Pakistan, Sri Lanka and Burma (Myanmar).

The usual habitats of the bird include cultivation and orchards in countryside, as well as in trees around parks in cities and towns. In large groups, they destroy agricultural crops and fruits.

Food: Groundnuts, grams, maize, green chillies and some fruits are favoured by them as food.

Habits and Nesting: These birds fly very fast and make lot of chirping, either at rest or in flight. It is one of the common resident birds in India. The parrots and parakeets are quite popular as cage birds. The nest is built in tree trunks or they use the hollow of the trunk as nest; also build nest in cracks of walls or deserted buildings. Eggs 4 to 6; young nidicolous and looked after by both the parents.

15. Koel, Asian Koel (*Eudynamys scolopacea*) (B. name : Kokil)

Classification: Order Cuculiformes; Fam. cuculidae.

The family contains about 17 genera and 79 species.

Morphological Characteristics: This bird is famous for its well-pitched and modulated calls at the advent of spring. The bird is slightly smaller, lighter and more slender than the crow. Sexually dimorphic, the male is shiny black with slightly curved yellowish-green bill and reddish eyes; the female, from head to tail has numerous brown spots mixed with white which is more distinct ventrally. The tail is long in both the sexes.

Distribution and Habitat: It is a resident and a locally migratory bird all throughout India and it also occurs in Bangladesh, Pakistan, Sri Lanka and Burma (Myanmar). This bird has a greater arboreal habit than many others. Large gardens with plenty of high trees, open countryside with large bordering trees, forests, etc. form their usual habitats.

Food: This bird is mainly frugivorous, mostly liking berries or soft and succulent fruits. They also feed on insects and their caterpillars.

Habits and Nesting: The bird's call in a nicely modulated and gradually increasing pitch in both urban and village life gives a signal for the advent of spring. Salim Ali (1987) reports that the call of the male bird begins with a low 'kuoo' which rises in scale with each successive 'kuoo' until it reaches the fever pitch at the seventh or eighth and then breaks off abruptly.

The bird is brood-parasitic, laying several eggs in crow's nest. The crows incubate the eggs and rear the nidicolous hatchlings, mistaking the eggs as their own.

Fig. 26.15: Koel

16. Barn Owl (*Tyto alba*) (B. name: Lakshmi Pacha)

Classification: Order Strigiformes; Fam. Tytonidae.

This is a comparatively smaller family containing 2 genera and 17 species; the genus *Tyto* contains 15 species.

Morphological Characteristics: The bird is of the size of a crow or pigeon (about 14"/35 cm) and is the common owl with distinctive features. The plumage on the back and on the wings are pale golden-brown with patchy greyish tinge and on the lower ventral side whitish. The characteristic heart-shaped face and neck are pale-white with brownish patches. Bill short but powerful and the feet strong with strong curved claws in toes.

Distribution and Habitat: This bird has a worldwide distribution. It occurs all throughout India, Bangladesh, Pakistan, Sri Lanka and Burma (Myanmar). Its habitat comprises of dilapidated buildings, corners and cracks of habitable houses, hollows in old tree-trunks, etc.

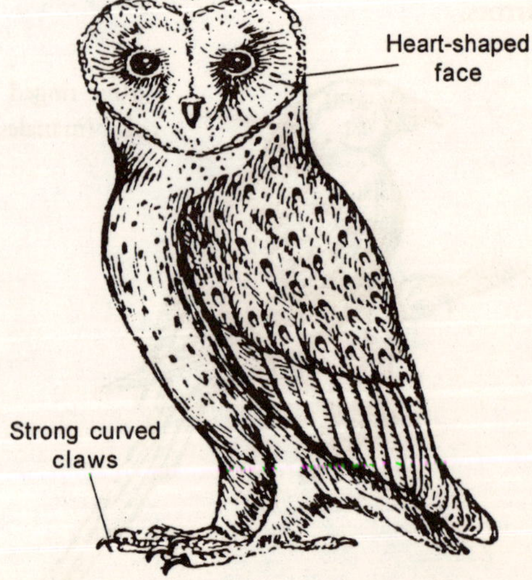

Fig. 26.16: Barn Owl

Food: These birds are predatory on rats and mice, and hence, economically important to mankind but they themselves cannot escape hunting by man.

Habits and Nesting: The bird nests in hollows in tree-trunks, rock-cavities, towers, cracks and crevices in ruined buildings, etc. Occasionally they use some twigs for preparing the nests.

The eggs are laid in one clutch or sometimes in two clutches of 3-11 eggs (usually 4-7) at intervals of few days. Incubation is done by the female for over one month. The nidicolous young are fed by both the parents.

17. Little Swift (*Apus affinis*) (B. name: Batashi)

Note: This species has been described by Salim Ali (1987) as the House Swift. But in recent checklist, this species has been named Little Swift, and the House Swift as *Apus nipalensis.*

Classification: Order Apodiformes; Fam. Apodidae.

This family is also moderately large containing 19 genera and 103 species and 11 subspecies. The genus *Apus* contains 17 species and 1 subspecies.

Morphological Characteristics: It is a small greyish-black bird having a white throat and a white rump. The tail is short, rectangular and slightly forked at the tip. The long, distally narrowing wings are spectacular in the flapping flight of this bird. The high aspect ratio of the wing (length/width ratio of the wing) is very effective in their untiring sustained flight.

Distribution and Habitat: The bird is distributed in Africa and Southern Eurasia. The bird covers a wide range of Indian subregion up to an altitude of about 6,000 feet (1,800 m) in the Himalayas.

Fig. 26.17: Little Swift (House Swift)

The natural home of the bird, so to say, is in the air. It spends most part of its life in flight, flying in small groups, screeching shrill and noisy calls or 'songs' in the sky and covering long distances.

Food: The food chiefly consists of dipterous insects like flies and midges that are captured in aerial sallies in the wide gape of the mouth.

Habits and Nesting: The swifts have all their four-clawed toes directed forward; they cannot perch, but can climb rough surfaces or walls or may look like hanging from the branch of a tree.

The nest is a cup-shaped depression formed of twigs, straw, feathers and odd materials bound together with the saliva of the bird. The nest is built in the cracks and crevices in the walls of buildings, as well as in the rocks and beneath the roof-tiles. While building the nest, an opening for the entrance/exit is always kept. The bird lays one clutch of 2-4 eggs (usually 3). Incubation is shared by both the parents and it takes about 14-20 days; the nidicolous hatchlings are also fed by both the parents.

18. Hoopoe (*Upupa epops*) (B. name : Mohonchura)

Classification: Order Coraciiformes; Fam. Upupidae.

In Monore and Sibley's Checklist of Birds (1993), the hornbills and the hoopoes have been taken out of the Order coraciiformes and placed in two separate Orders, Bucerotiformes and Upupiformes, respectively. The family Upupidae contains only 2 or 3 species.

Morphological Characteristics: The bird is about 11" (28 cm) long and is very distinctive for its plumage colouration. In general, the plumage is fawn coloured or brown mixed with pink and with black and white zebra markings on the back, wings and tail. The black and white markings are transversely arranged. An erectile fan-shaped feathery crest is present on the top of the head. The crest has also black and white bands. The long slender bill has a gradual downward curve.

Distribution and Habitat: This species is distributed in Europe, Central and Southern Asia and Africa. It is found throughout India and adjacent countries. It is mainly ground-dweller and found in open fields or thinly wooded countryside or in the hilly tracts up to 4,000-5,000 feet altitude.

Food: Hoopoe feeds on insects and their caterpillars and pupae.

Habits and Nesting: Hoopoes prefer wet grounds for probing into the soil to catch insects and their larvae. It usually moves singly or in pairs. Their calls sound like 'hoo-po' repeated several times. While digging in the soil in search of insects, the crest is contracted and folded back. The bird nests in tree-holes, cracks and ceiling in buildings with sticks, straw and rubbish.

Single clutch of 5-8 eggs (sometimes more) are incubated by the female for more than 2 weeks; hatchlings nidicolous and looked after by both the parents.

Fig. 26.18: Hoopoe

19. Little Green Bee-eater (*Merops orientalis*) (B. name: Banspati)

Classification: Order Coraciiformes; Fam. Meropidae.

The family contains 3 genera and 26 species of which the genus *Merops* contains 23 species.

Morphological Characteristics: The bird is larger than the House Sparrow (about 8"-10" (20-25 cm) long) and has the body plumage in general a grass-green colour. The plumage on the top of the head and neck is reddish-brown. The bill is long, gradually narrower rostrally and slightly curved downward. Ventrally, around the neck, the colour of the plumage forms a black band. The central rectrices in the tail are much longer and pointed, growing considerably beyond the length of other rectrices.

Distribution and Habitat: The bird is distributed in Africa, South of the Sahara desert, south and south-east Asia. It is distributed throughout India from about 5,000 feet (1,500 m) in the Himalayas, Bangladesh, Pakistan, Sri Lanka, Burma (Myanmar), etc.

The usual habitats are the open terrain, cultivate land, forest edges, large gardens, etc.

Food: The bea-eaters mainly feed on insects, particularly of the orders Diptera and Hymenoptera.

Habits and Nesting: This bird is in the habit of catching bees and other insects in air. After grasping the insects in the beaks, the bird returns to the perch, kills the prey and swallow it. They roost in large numbers in nearby trees.

This bird nests in colonies at the end of a tunnel of about 2.5 feet (75 cm) to 8 feet (2.5 m) long, dug out with the help of its beaks by the side of an earth-cutting, undulated sandy ground or the like. It has a single clutch of 4-7 eggs once in a year. Incubation and feeding of the nidicolous young are performed by both the parents.

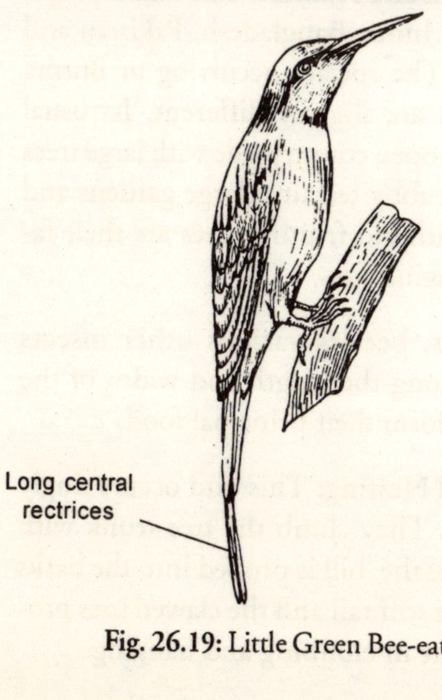

Fig. 26.19: Little Green Bee-eater

20. White-throated (White-breasted) Kingfisher (*Halcyon smyrnensis*) (B. name: Machhranga)

Classification: Order Coraciiformes; Fam. Halcyonidae.

The family contains 12 genera and 61 species; the genus *Halcyon* contains 11 or 12 species.

Morphological Characteristics: The bird is the commonest West Bengal Kingfisher. It is of the size of a woodpecker and has three very distinctive colour patterns in the body. While the upper plumage and that of the tail is bright blue, the entire throat and breast are white and the head, neck and lower part of the body are chocolate-brown. The long thick and sturdy bill as well as the legs are red. The powerful feet and toes are provided with cureved claws. A white patch occurs in the wings.

Distribution and Habitat: The bird is distributed in South Asia, and Philippines. It is found throughout India, both in plains and hills (lower altitudes). It is also found in Bangladesh, Pakistan, Sri Lanka and Burma (Myanmar).

Though for hunting fishes they prefer to watch from overhanging branches of trees on the banks of rivers, ponds, lakes, marshes, jheels or any waterbody, they may live away from these sites to feed on insects, lizards, mice and even small birds.

Food: Largely fishes but the food also comprises tadpoles, grasshoppers and various other insects, mice, lizards, etc.

Habits and Nesting: These birds have acute vision and from the watch-point on tree-branches, they can locate fish in water and with admirable speed are able to swoop on, catch and carry it to another tree-branch where the fish is battered into pieces and swallowed.

Like other Kingfishers, the nest is built in the form of a horizontal tunnel on the banks of water bodies or in an earth cutting. Eggs 4-7; digging of the tunnel, incubation and rearing of the nidicolous youngs are done by both male and female.

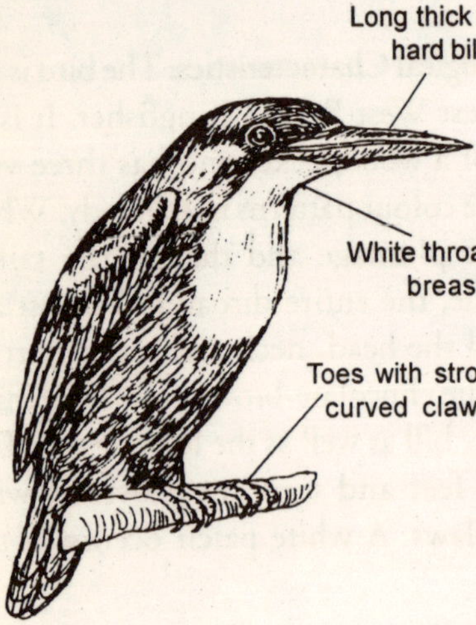

Fig. 26.20: White-throated Kingfisher

21. Golden-backed Woodpecker (*Dinopium benglalense*) (B. name: Kaththokra)

Classification: Order Piciformes; Fam. Picidae.

This species has also been named Black-rumped Flameback.

This large family contains 28 genera and 216 species; the genus *Dinopium* contains only 4 species.

Morphological Characteristics: The bird has a size between the Myna and the Bulbul. Its upper plumage on the rump is black and that on the sides and wings is golden-yellow. Ventrally the plumage is dull-white with thin black and wavy streaks. The crest on the top of the head is deep red, cheek white and the tail black. The bill is very stout, chisel-shaped and gradually pointed. Of the 4 clawed toes, 2 are directed forward and 2 backward; tongue long and protrusible.

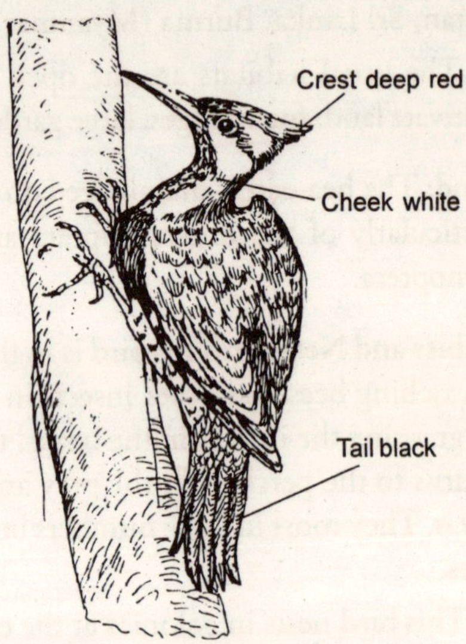

Fig. 26.21: Golden-backed Woodpecker

Distribution and Habitat: The bird is found throughout India, Bangladesh, Pakistan and Sri Lanka. The species occurring in Burma (Myanmar) are slightly different. Its usual habitats are open country side with large trees or thick shrubby terrains; large gardens and orchards with old fruiting trees are their favourite foraging sites.

Food: Ants, beetles various other insects occuring along the length and width of the tree-trunk form their principal food.

Habits and Nesting: This bird occurs singly or in pairs. They climb the tree-trunk with jerks. While the bill is probed into the barks of trees, the stiff tail and the clawed toes provide support in climbing and hanging.

The nest is a hole dug out by the bird in the tree-trunk or a tree-branch much higher up from the ground. Eggs 3 to 4, incubation and rearing of the nidicolous nestlings are done by both the parents.

22. Red-vented Bulbul (*Pycnonotus cafer*) (B. name : Bulbul)

Classification: Order Passeriformes; Fam. Pycnonotidae.

This too is a large family comprising about 21 genera and 138 species; the genus *Pycnonotus* contains about 40 or 44 species.

Morphological Characteristics: The bird is smaller than the pigeon and its head with the small crest, chin, throat and the bill are all black with a tinge of blue; the nape and the body, in general, are brown, while the rump and sides of abdomen are white. There is a distinct reddish patch at the base of the tail.

Distribution and Habitat: Its distribution zone is southern Asia; it occurs throughout India and also in Bangladesh, Sri Lanka and Burma (Myanmar); may be present up to lower altitudes in hills.

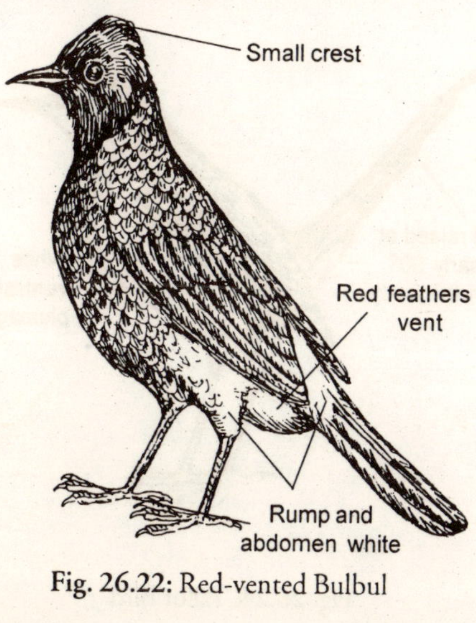

Fig. 26.22: Red-vented Bulbul

Their favourable habitat is the open countryside with lightly wooded forest, large gardens with variety of trees, etc.

Food: They feed on a variety of insects, fruits of the nature of berries and figs and other succulent fruits and vegetable matter, as well as on flower nectar.

Habits and Nesting: They are found usually in pairs or in small groups and sometimes come close to human habitation. The nest is a cup or saucer-shaped depression built with slender roots, twigs, fibres, etc. in a bush or higher branches of trees. Eggs 2-3; both incubation and caring of the nidicolous nestlings are done by both the parents.

23. Magpie-Robin (*Copsychus saularis*) (B. name: Doyale)

Classification: Order Passeriformes; Fam. Muscicapidae.

It is a large family; the genus *Copsychus* contains 8 species.

Morphological Characteristics: The bird is of the size of a Bulbul or a Wagtail and is generally a black bird with a beautiful combination of white plumage. The plumage on the upper part of the body from head to tail and on the wings is black, while that on the lower

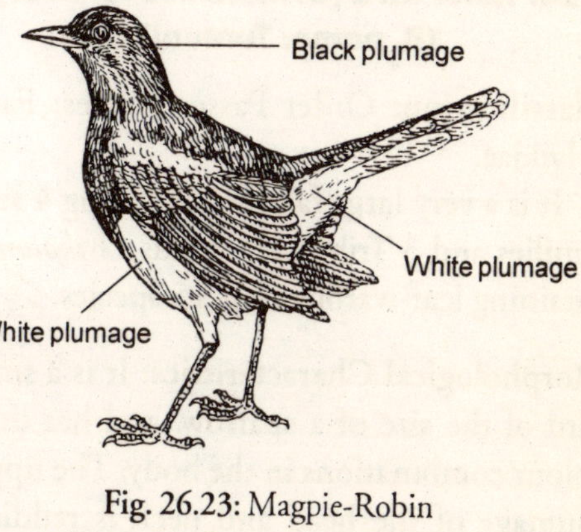

Fig. 26.23: Magpie-Robin

part including the tail is white. The wing too has two white patches on either side. The eyes are reddish. The tail is raised upward at an angle.

Distribution and Habitat: The bird has a wide distribution in the south and south-east Asia and Philippines. It is well-distributed throughout India, Bangladesh, Pakistan, Sri Lanka and Burma (Myanmar) in both plains and hills.

They prefer large branching trees around parks, fields, etc. in towns, villages and countryside, close to human habitation.

Food: This bird is mainly insectiovorous, picking insects from the ground and also feeds on small fruits and flower nectar.

Habits and Nesting: These birds are usually found in pairs. During summer, these birds emit charming notes and so are considerd as songbirds in our country.

The nest is a hole or a hollow in tree-trunks, cracked walls or in the depression of a wooden structure with the help of grass, thin roots, threads, hairs etc. formed into a pad. Eggs 3-5; incubation is done by the female and the nidicolous young is mainly reared by the male.

24. Tailor Bird (*Orthotomus sutorius*) (B. name: Tuntuni)

Classification: Order Passeriformes; Fam. Sylviidae.

It is a very large family containing 4 subfamilies and 3 Tribes; the genus *Orthotomus* cotaining leaf-warblers has 15 species.

Morphological Characteristics: It is a small bird of the size of a sparrow, and has three colour combinations in the body. The upper plumage of the head and neck is reddish-brown, that of the trunk, wings and tail is olive-green and the chin, throat and the entire underparts and a part of the tail are white. The tail is raised upward almost at a right angle to the body.

Distribution and Habitat: This species is distributed in the south and south-east Asia. It is found throughout India up to 5,000 feet (1,500 m) in the Himalayas (Salim Ali, 1987) and in Bangladesh, Pakistan, Sri Lanka and Burma (Myanmar).

The bird prefers to live in branching trees, bushes and shurbs.

Food: The bird largely feeds on insects and their eggs and larvae. It also feeds on flower-nectar.

Habits and Nesting: It is an agile bird found singly or in pairs, close to human habitation and also in bushes and groves on the outskirts of towns and villages. Though the bird is small, its voice is quite loud.

Usually, the nesting season is summer or the rainy season. The nest is built first by stitching two leaves like a skilful tailor into a cup or funnel-shaped structure within which

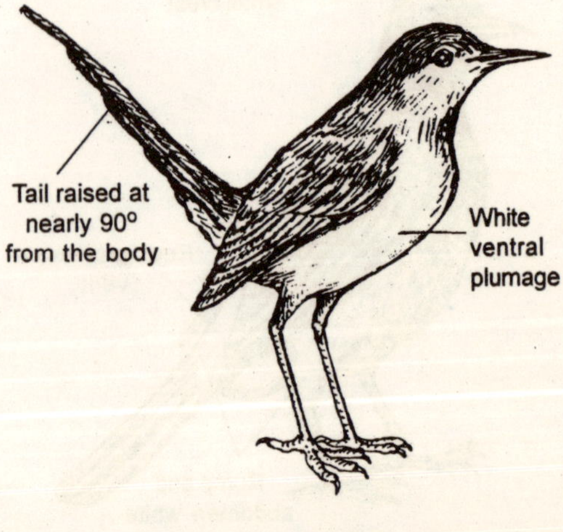

Fig. 26.24: Tailor Bird

a soft pad is formed with the help of cotton, hairs, fine twigs, vegetable matter, etc. The nest is built only a few feet higher from the ground. Eggs 3-4; incubation done by the female but the nidicolous hatchlings are cared by both the parents.

25. Purple Sunbird (*Nectarinia asiatica*) (B. name: Durga tuntuni)

Classification: Order Passeriformes; Fam. Nectariniidae.

It is a large family containing 8 genera and 170 species of flower peckers and Sunbirds.

Morphological Characteristics: The body of this bird is somewhat metallic violet which, in sunshine, reflects bluish-green colour. In the non-breeding season, both the male and female birds have the plumage on the back deep olive-brown and pale yellow below. The beaks are long, slender and rostrally curved; the tongue long, extendable. The thin, long tarsus is provided with curved claws.

Distribution and Habitat: The bird has a wide range of distribution in the south and southeast Asia. This Sunbird is found all throughout India, Bangladesh, Pakistan, Sri Lanka and Burma (Myanmar). In West Bengal and in other parts of India, it is a resident bird. The bird inhabits thinly-wooded part of the land with flowering plants.

Food: The food mainly consists of flower nectar which is sucked in by the slender bill and the tongue suitably adapted for this purpose. These birds are considerd as Asian counterpart of American Humming birds. However, these birds also feed on insects and spiders caught in flight.

Habits and Nesting: The Sunbirds sometimes exhibit hovering flight—like humming birds—while feeding on flower nectar. They are frequent visitors to flower gardens, orchards, growing cultivation, shrubs, etc.

The nest is a typical hanging pouch from the branch of a tree. The nest is built with twigs, straw, grass and other wastes and cobwebs. A clutch of 2-3 eggs are laid. Only the female incubates and the male assists in feeding the young.

Fig. 26.25: Purple Sunbird

26. House Sparrow (*Passer domesticus*) (B. name: Charai)

Classification: Order Passeriformes; Fam. Ploceidae. [In Monroe and Sibley's (1993) classification, the sparrows have been placed under the Fam. Passeridae and the Subfamily Passerinae, whereas the Weaver birds have been placed under the Subfamily Ploceinae of the same family.]

Subfamily Passerinae contains 4 genera and 36 species; the genus *Passer* contains 23 species and 3 subspecies (considered by others as species).

Morphological Characteristics: It is one of our most well-known birds that hops, jumps, flies and makes all sorts of noises in our buildings, corridors, window-cornices, etc. throughout the day. The bird is slightly smaller than the Bulbul and is about 5.5"-6" inches (13-15 cm) long. The bird shows sexual dimorphism. The male's head is dark-grey, nape ashy-brown, throat black, and the cheek and part of the breast are greyish-white. The upper plumage is brownish-grey with whitish streaks forming a concentric design. The female has a more uniform greyish-brown plumage above and ashy-white below. The female lacks black colour on throat. Beaks short but stout and conical from base to tip.

Fig. 26.26: House Sparrow

Distribution and Habitat: This species has almost worldwide distribution. As a resident bird, it is present throughout India in both plains and hills, as also in Bangladesh, Pakistan, Sri Lanka and Burma (Myanmar). They inhabit cultivation, dwelling houses, open fields, trees, bushes around countryside and almost every nook and corner in human habitation.

Food: This bird is omnivorous, feeding on grains and cereals, insects, fruits, fruit-buds, vegetable matters, flower nectar, etc. though it has preference for seeds, grains and cereals.

Habits and Nesting: The House sparrow is a very agile, chirping and noisy bird that is found in close association with human beings and their dwelling houses. In large flocks, they also invade crop fields and destroy huge quantities of growing crops. Again, they do some good to farmers by foraging on insect-pests of crops.

The nest is an untidy and shallow cup-like depression built with fine twigs, straw, feathers and rubbish, etc. usually in holes and cracks of buildings underneath the ceiling. Eggs 3-5; female incubates while the nidicolous hatchlings are reared by both the parents.

27. Baya Weaver Bird (*Ploceus philippinus*) (B. name: Babui)

Classification: Order Passeriformes; Fam. Ploiceidae (**see the note under classification at Srl. No. 26**); the genus*Ploceus* contains 63 species and 13 subspecies which have been considered as distinct species by others.

Morphological Characteristics: This bird is about the size of a sparrow and it shows sexual dimorphism, both in size and colour. The upper plumage of the head, nape and abdomen in the male is yellowish-white, while rest of the body and the wings have a golden-browen plumage, streaked with white lines. The legs are stout and the feet are provided with 4-clawed toes of which the hallux is directed backward. The female has a rather uniform golden-brown plumage above and a yellowish dull-white plumage below. The tail in both has rectrices arranged symmetrically to form a rectangle.

Distribution and Habitat: This species is widely distributed in the south and south-east Asia. It is found throughout India, Bangladesh, Sri Lanka and Burma (Myanmar). It is a resident bird of India, though some migrate locally. They inhabit lightly wooded forest, bushes around jheels and ponds, bushes around cultivation in countryside.

Food: The Weaver birds too, like sparrows, attack ripening crop fields in large flocks and destroy paddy and cereals. Thus they feed on crops, buds of crop plants, insects, etc.

Habits and Nesting: These birds are considered as pests of agricultural crops; like Reed Warblers, they are also found in thickets of reeds on the banks of large tanks.

Fig. 26.27: Baya Weaver Bird

These birds are famous for exceptionally skilful building of their nests. The nesting season coincides with the advent of monsoon in India. During breeding season the male bird builds several nests, each of an inverted funnel-shaped structure on single or separate tree branches. The nest is built with the help of straw, rough grasses, leaves, twigs in two chambers of which the egg chamber remains deeper. The inner wall of the dome near the egg chamber is plastered with wet mud. The male bird sits on the top of one of the nests that are surveyed by the female, one by one, as she comes out of the nest. The nest from which she does not come out is a signal to the male for her mate choice. Each male may have several females at the same time. Eggs 2-4; incubation done by the female and the nidicolous young are reared by both the parents.

28. Common Myna (Indian Myna) (*Acridotheres tristis*) (B. name: Shalik)

Classification: Order Passeriformes; Fam. sturnidae.

It is a large family containing about 38 genera and 148 species; the genus *Acridotheres* contains 7 species.

Morphological Characteristics: The bird is smaller than the pigeon and has an overall brown plumage in the body; that of head, neck and upper wing is dark-brown, brown on the back, lower wing and lower parts of the breast and abdomen. The ends of tail and wings are streaked blackish. The legs, beaks and back of the eye (including supercilium)

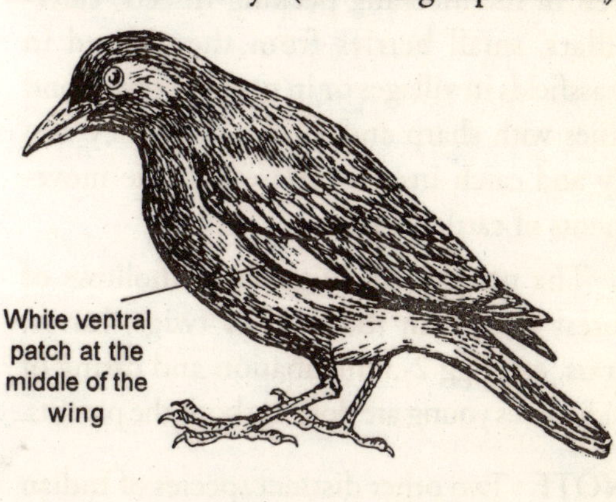

Fig. 26.28A: Common Myna

Fig. 26.28B: Hill Myna

are yellow. A white patch at the middle of the wing ventrally is very distinct.

Distribution and Habitat: It is common resident bird found all throughout India in both plains and hills; also common in Bangladesh, Pakistan, Sri Lanka and Burma (Myanmar) (with some racial differences).

Their habitats contain open fields, parks, cultivation around towns, villages and country side.

Food: These birds feed on a variety of insects, caterpillars, earthworms, small fruits, etc. and on anything found near human habitation.

Habits and Nesting: This bird is commonly seen in the morning pecking insects, caterpillars, small berries from the ground in grassfields in villages or in parks in towns and cities with sharp and loud noise. They also fly and catch insects exposed by the movements of cattle in grass.

The nest is built high up in hollows of forest trees with the help of twigs, leaves, grass, etc. Egg 2-3; incubation and caring of nidicolous young are done by both the parents.

NOTE : Two other distinct species of Indian Mynas are worth mentioning here. One is the **Hill Myna** *(Gracula religiosa),* which is shiny black, has a white patch on each wing and an orange-yellow patch of featherless skin behind each eye; this species is almost solely arboreal and has a restricted distribution, mostly in hilly tracts. The other species, the **Pied Myna** *(Sturnus contra),* described next (29) has a black and white plumage, while the skin around the eye, the legs and beaks are orange-yellow.

29. Pied Myna or Asian Pied Starling *(Sturnus contra)* (B. name: Goshalik)

Classification: Order Passeriformes; Fam. Sturnidae. The Family Sturnidae is quite large containing 38 genera and 148 species. The geneus *sturnus* contains 16 species.

Morphological Characteristics: The bird has the size of a common Myna. The lateral side of the head caudal to the eye bears a conspicuous patch of white while the orbital skin and the beaks are orange-yellow in colour. The upper breast, a long narrow patch along the wing, the entire belly and the ventral tail are all white or buffy white; the rest of the body is black. The feet are yellowish with curved claws.

Distribution and Habitat: The bird is distributed in the south and south-east Asia. In India, it is more common in the northern, central and eastern states and less common in the western and southern India. This bird is a permanent resident in West Bengal. They inhabit lightly wooded countryside with open fields as well as urban areas.

Food: They frequently visit the heaps of debris lying in the vicinity of human habitation and collect any edible veg.- or non-veg. items from the debris, as well as insects which form bulk of their daily dish.

Habits and Nesting: These birds, usually in a group, feed on insects around the body of grazing cattle in the vicinity of ponds, marshes or cultivation (much in the same way as the Cattle Egrets which may accompany them).

One pair or several pairs of birds may build nests in a single large tree. The nest is round or slightly elongated, built irregularly with the help of grass, leaves, fibres and waste materials high up in the tree (about 20-30 feet/ 6-9 m above the ground). The entrance/ exit opening in the nest is lateral in position. According to Salim Ali and Ripley (1987), 4-5 eggs are laid in a clutch. The eggs are glossy blue and both the parents share in building the nest, incubating the eggs and rearing the hatchlings.

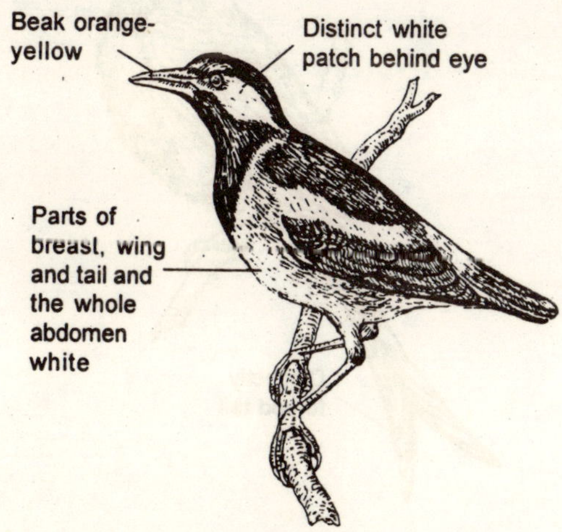

Fig. 26.29: Pied Myna

30. Black-headed Oriole (*Oriolus xanthornus*) (B. name: Benebou)

Classification: Order Passeriformes; Fam. Oriolidae.

NOTE: In Monroe and Sibley's A World Checklist of Birds (1993), the orioles have been placed in the Tribe Oriolini, Subfamily Corvinae and Family Corvidae. Monroe and Sibley's Checklist mentions the bird as '**Black-headed Oriole**'.

The orioles belong to two genera—*Oriolus* and *Icterus*, containing 27 species and 25 species, respectively, the latter containing 8 subspecies.

Morphological Characteristics: This bird, slightly smaller than or approximately of the size of a Myna (about 10"/25 cm), has its head, nape, neck, throat and rostral part of the breast covered with shiny black feathers. The rest of the body colour is bright yellow except margins of the wings and tail that are also black. The beaks are rosy and the eyes red.

Distribution and Habitat: The bird is distributed throughout India up to 4,000 ft (1,200 m) in the Himalayas and in Bangladesh, Sri Lanka and Burma (Myanmar). This bird, as well as the Golden Oriole *(Oriolus oriolus)*, inhabit leafy trees in both well-wooded countryside and in towns and cities.

Food: Their food chiefly consists of fruits, berries and flower nectar. They also feed on insects.

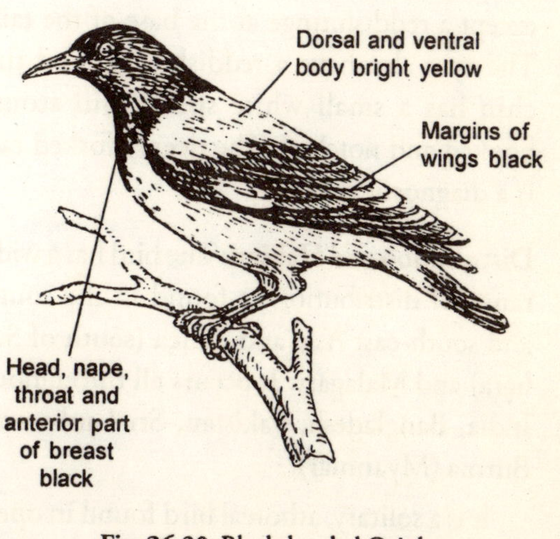

Fig. 26.30: Black-headed Oriole

Habits and Nesting: This bird is a good singer. In the breeding season, the male–sitting on the lofty branch of a tree—sings to attract attention of a female bird who reciprocates by singing.

The nest is typical in all the orioles–a deep cup-shaped depression neatly woven with grass, fibres and other materials bound with cobweb and the whole structure suspended hammock-wise at several points from the forked, horizontal leafy branch of a tree (Salim Ali & Ripley, 1987). Usually, 2-3 eggs are laid in a clutch (rarely 4 or more) and the incubation as well as rearing of the nidicolous nestlings are attended by both the parents

31. Black Drongo or King Crow (*Dicrurus adsimilis*) (B. name: Phinge)

Classification: Order Passeriformes; Fam. Dicruridae This family contains 2 genera and about 24 species. It is also known as Fork-tailed Drongo.

Morphological Characteristics: The bird has a length of about 10"-15" inches/25-38 cm (smaller than the Racket-tailed Drongo) and has a shiny black plumage all over the body except a reddish tinge at the base of the tail. The eyes also have a reddish tinge and the chin has a small white speck. Bill stout, hooked and notched. The deeply forked tail is a diagnostic feature.

Distribution and Habitat: The bird has a wide range of distribution as found in the south and south-east Asia and Africa (south of Sahara) and Malagasy. It occurs all throughout India, Bangladesh, Pakistan, Sri Lanka and Burma (Myanmar).

It is a solitary, arboreal bird found in open countryside and cultivation.

Food: The bird feeds on various kinds of insects and also on flower nectar.

Habits and Nesting: This is one of our well-known birds and found very often sitting on telegraph wires or on top of fencing wall with eyes fixed on flying insects disturbed by the movements of cattle through grass or for swooping on flying insects. As these birds destroy many harmful insects for crops, they are beneficial to agriculturists.

A cup or saucer-like nest is built by twigs and fibres usually at the horizontal fork of tree branch or at the end of a branch. Eggs 2 to 4 or 5; incubation and rearing of the nidicolous hatchlings are done by both the parents.

Fig. 26.31: Black Drongo

32. House Crow (*Corvus splendens*) (B. name: Kak)

Classification: Order Passeriformes; Fam. Corvidae. [In the new classification by Monroe and Sibley (1993), it has been considered as a very large Family divided into 7 Subfamilies of which Subfamily Corvinae has 4 tribes. The tribe Corvini, corresponding to the Fam. Corvidae of older classification, includes Crows, Magpies, Jays and Nutcrackers and has about 25 genera and 118 species.]

Morphological Characteristics: It is an unmistakably well-known bird observed nearest to our habitation. It is larger in size than the pigeon and has a wing span of about 2.5 to 3 feet/76-90 cm. The plumage colour is black except that from the back of the head and around the nape and breast–the colour is grey. Feet strong and toes provided with curved claws.

Distribution and Habitat: This species is widely distributed in south-central and southern Asia, parts of Africa and India, Bangladesh, Pakistan, Sri Lanka, and Burma (Myanmar). In India, it is found both in plains and up to certain altitude in hills.

The bird occurs in the vicinity of human habitations—in towns and villages, cities and their parks and market places.

Food: An omnivorous bird that feeds on anything left out of human habitations, marketplaces, fish and meat shops, carrions, mice or rotten scraps, insects, fruits, grains and almost anything edible.

Habits and Nesting: These birds are gregarious, very cunning and intelligent. They are scavengers of the wastes and filths of human habitation. The crows feed on almost anything eatable. They are sometimes attacking and may be a menace to weaker birds. Crows in large number may roost at night in large trees around parks, cities and towns.

The nest is a twig platfrom that is depressed in the middle and lined with wires, fibres, etc. Eggs 4 or 5; incubation and care of nidicolous youngs are done by both the parents.

Fig. 26.32: House Crow

27

Economic Importance of Birds

Salim Ali (1979, 1987) has rightly stated that without birds mankind would have perished from the fair face of the globe. To amplify in a broader sense, this truism appears a prophetic truth, because the insects, birds, trees and mankind are interdependent components of the ecosystem. It is unthinkable indeed whether trees would have survived without birds and mankind without plants–in the long run. The inexplicable balance of Nature has been established through millions of years through a meticulously woven network in the living world of plants and animals. As a part of this network, birds are one such essential link that has a direct or indirect effect on the economy and thinking of mankind.

From the dawn of human civilization, the birds have been regarded as either useful or harmful to mankind due to their close proximity with human societies. The domesticated varieties of poultry birds, for example, have largely compensated for the overall protein deficiency in many countries like ours. Conversely, the list of damages caused by some species of birds to our economy is sometimes quite alarming.

Thus, from the point of view of economic importance, birds' relation to mankind has been described under two heads—considering their activities—either **useful** or **harmful.**

27.1.A. BIRDS BENEFICIAL TO MANKIND

The birds, by their beauty, plumage, calls, songs, dialects and graceful movements, are a source of immense joy and beauty to mankind—'thing of beauty is a joy for ever'.

They also serve a number of useful purposes:

1. Birds as flower-pollinators

Just as the butterflies, bees and many insects are well-known agents for pollination of flowers, quite a number of species of birds—which are exclusively, frequently or occasionally **nectar-feeders**—act as pollinators of flowers. Such birds like Magpie-Robin (*Copsychus* sp.), Tailor Bird (*Orthotomus* sp.), Babblers (*Turdoides* sp.), Starlings (*Sturnus* sp.), Orioles (*Oriolus* sp.), Black Drongos (*Dicrurus* sp.), Purple Sunbird (*Nectarinia* sp.), White-eyes (*Zosterops* sp.)—while probing flowers for nectar–are smeared with pollens in their head, neck and body. Pollination inevitably occurs when these birds visit other flowers for nectar. Among the flowering trees, a few like the silk cotton tree or the coral tree having beautifully coloured and attractive flowers are the most favoured by birds.

2. Birds as seed-dispersers

Dispersal of seeds by birds from the sites of

the fruity trees to nearby or distant places, even covering hundreds of miles, is an usual phenomenon in the life of the birds. The birds are considered to be highly effective agents for the dispersal of seeds, and such dispersal is of great help towards afforestation of many barren areas. Of course, growth of many undesirable plants (e.g., *Lantana* weed, *Loranthus* tree-parasite) in our forest environment through random dispersal by birds cannot be avoided. On the other hand, seeds of sandalwood, mulberry, etc. dispersed by birds are of great economic importance. Among the tropical frugivores of Australia and New Guinea, the Birds of Paradise (Family: Paradiseidae) are the most prominent and effective seed-dispersers. In Australasian islands and in other parts of the Tropics, the fruit-pigeons, flower peckers, cuckoos, cuckoo-shrikes, orioles, thrushes, starlings, parrots and parakeets and the mynas are good examples of seed-dispersers.

3. Birds as destroyers of insect pests of crops

Of all the checks and balances drastically imposed by nature, the biological control of the insect pests of crops, cereals and fruits by birds is probably of great economic importance to mankind. The inconceivably prodigal rate of multiplication of insect pests would have been the greatest challenge to the very existence of human beings, had there been no check in the swarming population of insects pests through predation by birds. Insects form a common item of food for most birds. From the list of numerous insect-eating birds (some partly and some exclusively) mention may be made of egrets, ibises, snipes, coursers, cuckoos, swifts, hoopoes, shrikes, drongos, orioles, babblers, flycatchers, woodpeckers, and most passerine birds which feed on billions of insect pests of crops, cereals and their growing shoots throughout the year. The falconiform birds, on the other hand, destroy a large number of field rats and mice that also destroy enormous quantities of field crops as well as stored grains. Hence all these birds are of great economic importance to mankind.

4. Birds as scavengers

When we trail the source of an obnoxious smell coming from the vicinity of our villages, countryside or from the outskirts of township, we are sure to find that an assemblage of vultures, kites, hawks, eagles, etc. are dishing out the carrion of dead cattle or any other beast with amazing speed. The carcasses of animals, without being devoured by these birds, would have polluted the atmosphere unworthy of human habitation.

Within the premises of our localities, however, the crows are the most common scavengers. They are omnivorous and they render tremendous service from dawn to dusk in scavenging all filth and dirts from the heaps of our household garbage.

5. Birds as destroyers of rats and mice

The fecundity in the population of rats and mice–that are also very destructive mammalian pests of field crops and stored grains–is largely checked through predation by several species of birds like the merlins, kestrels, hawks, owls (particularly the Barn owl and the Great horned owl). These birds, therefore, are a great friend to agriculturists. Hundred thousand tonnes of crops and cereals are destroyed by different species of rats and mice annually. However, these falconiform birds very often prey on some beneficial birds, poultry farms, game birds, etc. that are directly concerned with our economy. But this loss is largely compensated by their beneficial activities.

6. Birds as food for human beings

Several species of anseriform and galliform birds are extensibly used as favoured food dish all over the world. Such birds comprise several species of ducks, swans and geese, several species of partridges, fowls, wild turkeys, wild fowls, guinea fowls, several species of water birds, and also quails, pheasants and pigeons. Modern hybridization experiments in poultry and duck farming have successfully augmented the quality and yield of both flesh and eggs. In recent times, the new improved poultry breeds have largely compensated for the overall protein deficiency in the country. Salim Ali (1987) reports that, in the earlier part of the twentieth century, some population of our country, living in the neighbourhood of 'jheels' and lakes, were used to subsist on (at least for several months in a year) wild fowls and different kinds of water birds collected through netting operations. The major portion of the collection was, no doubt, transported to market for earning livelihood.

7. Economic importance of bird feathers

Until recently, feathers of egrets, peafowls, ostriches, rheas and many other wild birds of our country were collected on commercial scale and transported to Europe and other countries for household decoration and ornamentation. In classical days, the native people using bow and arrows placed feathers on their arrows for speed, balance and direction, and the cushions, pillows, and mattresses made out of soft, heat-retaining downy plumage of birds had a great demand for the inhabitants in the European, American and Arctic cold.

8. Bird-nest as food-delicacy and aesthetic beauty

In older days, nests of Indian Edible-nest Swiftlet (*Collocalia unicolor*) and Andaman Grey rumped or White-nest Swiftlet (*Collocalia fuciphaga*) were collected in huge numbers and exported to China. These nests—built and glued with the help of copius secretion of saliva secreted by the sublingual salivary glands of these birds—were considered as dainty dishes in China and Burma (Myanmar). Though initially—in the early twentieth century—the trend in export of the birds nests was not encouraging (earning only several thousand rupees a year), it became a profitable business later, earning several lacs of rupees in a year. Apart from considering edible nests as a table delicacy, these are also used by many as decorative articles in houses.

9. Guano as farmers's boon

Many sea birds like the gannets (*Sula* sp. Europe and North America; migratory), cormorants, pelicans and others predating mostly on fishes have their excreta called guano which is rich in phosphorus and nitrogen; it also contains some amount of calcium and iron. The guano is dropped enormously like rainfall in certain places like island off the Chilean coasts and along coasts of West Africa, as also in our country. This excreta, either mixed with occasional rain water or simply dropped in drought areas, has the quality of a rich fertiliser for the growing crops. Thus, it is considered a friendly boon to the farmers.

10. Birds in other services of mankind

Quite a number of species of birds and their domesticated varieties have, in some way or other, come closer to human beings. The carrier pigeons or the messenger pigeons are well-known since long past to carry small letters with messages almost unmistakably to their destinations. Important sky-messages sent through 'carrier-pigeons' saved thousands of lives since time immemorial, even till the First World War (1914-1918). This was possible due to their learning through training,. The Hill Myna (*Gracula religiosa*) is a costly cage bird that learns to mimic and 'talk' human conversation almost distinctly. Same is also practised by the parrots, parakeets and many other birds. The song birds have soothing rhythmic tunes that fill our hearts with heavenly blissful joy.

Of many other beneficial effects by birds, their warning calls as signals and the use of their bodily parts as constituents of some medicines are specially noteworthy. Some birds send noisy warning calls to hunters or human beings treading forest tracks on the sight of tiger or poisonous snake in the area. The bones, muscles, feathers and eggs of many birds are used differently in the preparation of several indigenous medicines.

27.1.B. BIRDS HARMFUL TO MANKIND

Although the benefits contributed by birds to our economy are far greater than the damages done by them, yet in a number of ways their adverse effects on our economy are not very negligible. Some of their harmful effects are:

1. Damages to agricultural crops

In the growing season of crops, pulses and cereals, many birds damage them either by destroying the soft and succulent growing shoots of crop plants or simply by destroying the growing crops. Also, indirectly, the insect-eating birds, voraciously feeding on insects, destroy both beneficial and harmful insects and thus their predation on beneficial insects is to be looked upon as their harmful effects.

2. Damages to fruits and stored grains

Just as many species of birds habitually destroy field crops and vegetables, certain birds like the parrots and parakeets, the Green Imperial Pigeons, the Common Green Pigeons, the Drongos, the Koels, the Cuckoos, the Barbets, etc., that are either exclusively or primarily fruit-eaters, do considerable damage to fruit- gardens. This has an adverse effect on our economy.

Birds like sparrow and most granivorous birds are a menace to stored grains and they cause considerable economic loss to mankind.

3. Destroyer of honey-bees

Certain birds like the honeyguides (*Indicator* sp.; Order Pisciformes), the bee-eaters (*Merops* sp.; Order Piciformes) and a few other birds devour large number of honey-bees that are beneficial to mankind. At least 6 species of **honey-guides** are parasitic. These birds guide the short-legged **badgers** (*Meles* sp.; Fam. Mustelidae; Order Carnivora) to the combs of wild honey-bees. The badgers feed on the wax and combs, while the remnants of combs are eaten by the honey-guides. The bee-eaters too, in their aerial sallies, feed on a large number of honey-bees, wasps and other insects.

4. Birds as destroyer of poultry and game birds

Many falconiform and strigiform birds are a menace to poultry birds, game birds and a number of smaller birds. Thus some predatory birds become harmful to our economy when they destroy the birds useful to human society.

5. Extent of damages caused by birds to crops, fruits, etc. due to food preferences

Workers in the field of Agricultural Ornithology have surveyed the extent of damages caused by a number of birds to field crops, fruits, stored grains, vegetables, etc. in relation to agricultural economy. Of these birds—which comprise only 2-3% of the total bird population of India—a few birds such as the pigeons and doves, parrots and parakeets, sparrows, crows, weaver birds cause greater damage to human economy than others. However, as mentioned earlier, the beneficial role of birds in destroying many serious insect-pests and their larvae as well as rodent pests of field crops is more helpful to our economy—considering sum total effects.

The field crops generally comprise groundnut, maize, mustard, pearl millet, peas, pulses, rice, sorghum, sunflower, wheat, etc., while the fruits comprise almonds, ber, datepalm, grapes, guava, peach, etc.

Different birds show variable percentage of food preferences. For example, the parrots, parakeets and weaver birds consume 10-100 percent millets; pigeons, doves, parakeets and parrots consume up to 66% pulses and peas (both sprouting and ripe); pigeons and doves up to 12-85% of sorghum, and crows and parakeets up to 12-32% of fruits like grapes, guava, peach. The parakeets alone have the preference of consuming up to 63% ripening mustards (Dhindsa and Saini, 1994). As regards consumption of field crops by birds, it may be mentioned here that the birds in flocks feed much more plentifully than isolated birds of any species.

27.2. PROFIT & LOSS

It considering the bulk of benefit derived from various activities of birds, the damages caused by them and overall losses incurred are negligible. The role of birds as scavengers, flower pollinators, seed-dispersers, insect—and rodent—pest destroyers, etc. is of great economic importance to mankind. New growth and spread of forests on earth are appreciably contributed by birds. The final balance sheet of profit and loss to human economy reflects that birds' beneficial contribution to mankind is far more on the credit side compared to their harmful activities.

Birds are dictated by nature in their foraging activities. Human beings would do well not to meddle much in this delicate balance

Fig. 27.1: Both beneficial and harmful role of birds shown in a growing paddy field, bordered by fruiting trees. Birds' predation on field rats and harmful insects overshadows their destruction of fruits and honey-bees

of environment. To go against this fine–tuned mysterious balance of Mother Nature often invites disaster. A prime example can be cited.

In 1965-66, during the 'Cultural Revolution' in China, the leaders gave a clarion-call—amidst others—to eradicate sparrows from China. Statistics showed that sparrows ate-up about 5% food-crops.

Thousands of farmer-villagers, fully aided by the Government began the mass-killing of sparrows. Millions were killed—20-30—probably 40 millions (4 crores)!!

The resultant crop-yield was highly satisfactory—at first. But, soon, they fell to an alarming figure. The insects and pests the sparrows ate-up multiplied abnormally. These feasted on crops, unimpaired.

The ultimate result was that the yearly loss of crop—which was 5% due to sparrows—rose to 15% due to the sparrows being exterminated!! Overall, the loss to the national exchequer was in billions, and near-famine conditions prevailed for quite a few years.

And this huge loss continued for a few more years–until sparrows from other parts of the country had to be brought in—by the farmer-villages, again fully aided by the Government and reared!! They multiplied and restored nature's balance.

28

Agricultural Ornithology

Dhindsa and Saini (1994) of Punjab Agricultural University define Agricultural Ornithology as 'the science of birds in relation to agriculture' which, in other words, may be stated as 'ecology and management of birds in **agroecosystems**'. As our country primarily depends on agroeconomics, bird management is considered to be the focal point in Agricultural Ornithology. Such management includes (i) assesment of species-wise bird populations destroying agricultural crops, insect-pests of field crops, bird pests harmful to other kinds of farms related to agriculture such as sericulture, poultry, aquaculture, fisheries, apiculture, and (ii) conservation of bird species that are beneficial to agriculture. The new science has opened-up new horizons.

Like any other discipline of science, Ornithology—the science of birds—has also been expanding in all its branches, both basic and applied. Agricultural Ornithology is one such applied branch which is closely interlinked which a few basic tenets of Ornithology—food and feeding diversity of birds, community structure, feeding ecology and breeding ecology of birds that mainly depend for their nutrition on agricultural crops.

Agricultural resources affecting economy of countries like India include rice, wheat, maize, jawar, various oilseeds, pulses and cereals, both as field crops and stored grains. Varieties of fruits and nuts are also added to this list. Several hundred thousand tonnes of foodgrains are destroyed annually by bird pests. The birds either feed on them or simply destroy them right from the sapling stage up to ripening of the crops. The varieties of insects and other arthropods, their larvae and eggs, as well as quite a number of species of annelids and molluscs form favourite food niches for a wide variety of birds. On the other hand carnivory—on some vertebrates like fishes, amphibians and reptiles; and frugivory—on both succulent and dry fruits, are common food-habits of a large number of bird species, besides a few smaller birds known as sap-suckers and nectar-feeders.

Over 100 years ago, Oates (1889-91) and Blanford (1895-97) described briefly the food and food-habits of different birds. In early twentieth century, McAtee (1908) studied the food-habits of the grosbeaks. Subsequently, Manson and Maxwell-Lefroy (1912) and Baker (1922-27) studied the food-habits of birds in India. 1930 onward, information relating to various aspects of the life of birds–with particular reference to the biological role—of their organ-systems began to pile up. The food-habits and the diverse feeding adaptations in birds were being explained in the light of the functional morphology of their feeding apparatus for understanding the adaptation and evolution of the structures. However, the behavioural aspect of the feeding ecology of birds of agricultural importance received comparatively less attention in the past (Bhattacharyya, 1997).

28.1. FOOD AND FEEDING ECOLOGY OF SOME COMMON BIRDS IN RELATION TO AGRICULTURAL ECONOMY

While working on the feeding apparatus of certain columbid and passerine birds, Bhattacharyya (1997) observed the nature of food consumed and principal food-habits of 5 passerine and 5 columbid birds. He worked with the species that mainly comprised insectivorous, granivorous and frugivorous birds–with occasional divergence to other food-niches. Dhindsa and Saini (1994) report that the community structure of birds in agricultural lands is characterised by overdominance of only a few granivorous and omnivorous species while the rest of the species are represented in very small numbers. Toor et al (1986) mentioned that, from their survey of 26 species of birds frequenting a farm of stored grains in Punjab, the House Sparrow, *Passer domesticus*, was the most dominant species, covering 77% of the community. MacArthur (1972) reported (In: Dhindsa and Saini, 1994) that dominance by a few very common species is typical of disturbed habitats like agricultural areas.

Since the 1970s, excellent works have been done in the field of Agricultural Ornithology. To mention a few, the works of Ramzan and Toor (1972, 1973) on the damage to guava fruit (1972) and to maize crop (1973) by the Roseringed Parakeet, *Psittacula krameri* in Punjab; Dhindsa and Toor (1980) on the extent of bird damage to rice nurseries in Punjab and its control; Mathew et al (1980) on a comparative account of the food habits of some species of birds affecting agriculture; Dhindsa (1986) on the ecology of Indian Weaverbirds; Dhindsa et al (1991) on the House Crow damage to sprouting sunflower, and Saini and Toor (1991) on the feeding ecology and damage potential of feral pigeons *Columba livia* in agricultural habitats are specially noteworthy.

The insectivorous birds feed on varieties of insects—some beneficial and others pests of agricultural crops. Obviously, the balance in nature is superbly maintained without much impact on agricultural economy. If, however, this balance is disturbed by a gradual or abrupt change in the physical and/or biotic factors affecting insect populations or the type of vegetation amenable to the birds, the crop production will also suffer proportionately. Or, there may be considerable loss in the stored foodgrains. For example, both Jungle•Babbler (*Turdoides striatus*) and Magpie-Robin (*Copsychus saularis*) feed on—besides varieties of insects and caterpillars—appreciable number of earthworms each day. If the bird populations are increased and the earthworms are eaten up beyond the threshhold number, the natural fertility of the crop field is affected. This principle also holds true for insects whose number increases (thereby causing greater damage to crops), if the number of insectivorous birds in a locality abruptly decreases. Most of the insects-eating birds help increase annual yield of crops by feeding on insect pests and their caterpillars.

Among the columbid birds, *Columba* and both the species of *Streptopelia* feed on enormous quantities of millets, sorghum and wheat, besides grains of rice, cereals, and pulses, and are thus regarded as bird-pests when the extent of damage to the crops becomes alarming. Rana (1991) has reported that the Ring Dove (*Strepopelia decaocto*) alone causes considerable damage to millets, sorghum and wheat with variable percentage during pre-monsoon and post-monsoon periods. The Nicobar Green Imperial Pigeon (*Ducula aenea nicobarica*) and the Common Green Pigeon (*Treron phoenicoptera*)—which are almost exclusively frugivorous—act as bird-pests, either consuming huge quantites of nutmegs (*Myristica* sp; *Ducula*) or feeding on ripening fruits in orchards (*Treron*). At the same time, these birds also help dispersal of seeds and fruits.

Dhindsa and Saini (1991) have also given an account of the works of their colleagues (Toor et al, 1986; Saini and Toor, 1991, Dhindsa and Toor, 1990; Saini et al, 1993, 1994), stating the food preferences of captive or wild birds, dietary overlap of co-existing species, and many such points which are important aspects of feeding ecology.

28.2. EXTENT OF DAMAGES TO CROPS AND FRUITS BY BIRDS DUE TO FOOD PREFERENCES

Workers in the field of agricultural ornithology have also surveyed the extent of damages caused by a number of birds to fieldcrops, fruits, stored grains, vegetables, etc. in relation to agricultural economy. Of these birds–which comprise only 2-3% of the total bird population of India–a few (e.g., pigeons and doves, parrots and parakeets, sparrows, crows, weaverbirds, etc.) cause greater damage to human economy than the others. On the contrary, as already pointed out in the previous chapter, the beneficial role of birds in destroying many serious insect-pests and their larvae as well as rodent pests seems to be appreciably greater. The final balance sheet of profit and loss to human economy reflects that birds' beneficial contribution to mankind far exceeds their harmful activities.

Different birds show variable percentage of food preferences in the consumption of food. For example, the parrots, parakeets and weaverbirds consume 10-100% of millets; pigeons, doves, parrots and parakeets consume up to 66% of pulses and peas (both sprouting and ripe); pigeons and doves from 12% to 85% of sorghum, and crows and parakeets from 12 to 32% of fruits like grapes, guava, pitch, etc. The parakeets alone have the preference of consuming up to 63% of ripening mustards (Dhindsa and Saini, 1994).

28.3. CONSERVATION OF SPECIES BENEFICIAL TO AGRICULTURAL ECONOMY

It is alarming for the conservationists that inspite of their increasing efforts to conserve and sustain bird species beneficial to agricultural economy, such bird populations are gradually dwindling due to various factors of which deforestation around agricultural habitats is the main. Insectivorous birds like the Babblers, Munias, Magpie-Robins, Wagtails, Bulbuls, Shrikes, Thrushes, Mynas, Rosy pastors, etc., among many others, should be given their congenial habitats to thrive successfully for the benefit of agricultural crops. On the other hand, **the management and control of bird-pests is an important area of research in the agro-economic system.**

28.4. AGRICULTURAL ORNITHOLOGY : PROBLEMS AND PROSPECTS

First of all, it should be clarified that Agricultural Ornithology—often claimed to be a comparatively new subject—may not be considered as such. This is because the feeding ecology, adaptation and behaviour of birds, their beneficial and harmful activities affecting agro-economy, their breeding ecology, etc. are all included in the subject matter of ornithology. So agricultural ornithology, though not treated as a specialised branch in the past, received attention of the classical ornithologists as well. Recently, however, much more weightage has been given to the role of birds in relation to agriculture.

The problems necessitating expansion of researches in agricultural ornithology are quite a few—rapid increase in human population, planned reduction of forest habitats, contraction of agricultural lands and early rotation of crops through use of fertilizers, use of pesticides, etc. are the main. Research in all these aspects are expected to be carried out by an adequate number of scientists, which, in reality, has not materialised, especially in India and other Third World countries. This may be mainly due to paucity of funds for various projects in this field and partly to lack of enthusiasm in young researchers to work in this applied field of ornithology.

Dolbeer (1988), while discussing the role, impact and control of bird-pests in agriculture, has quoted a number of authors working in this field and commented that killing birds to protect agricultural crops is often viewed negatively today by the general public, particularly when (i) the agricultural losses are within a very small limit such as less than 1%, and (ii) many pest species also have attributes beneficial to man. Dolbeer (1988) has further commented that losses are not equally distributed; a small percentage of farmers receive significant losses (i.e., more than 5% crop) and for them the bird control measures, whether lethal or non-lethal, need to be considered. The lethal control methods should be applied most judiciously so that the killing of other birds not involved in agricultural damages is minimised.

Dhindsa and Saini (1994) have pointed out that 'the Roseringed Parakeet (*Psittacula krameri*) is the most harmful bird-pest for agricultural crops' although a few other species–the weaverbirds, the sparrows and doves—are also considerably harmful. For all these birds, especially for the Roseringed Parakeets, specific management techniques are to be evolved and applied carefully to control damages caused by such bird populations attacking a particular agricultural area.

The prospects of agricultural ornithology in India are interlinked with the problems outlined above. Such prospects are envisaged in several parameters: (i) improved research techniques to investigate and assess beneficial and harmful role of birds in agricultural lands on zonal basis; (ii) enthusiastic young researchers must be motivated to work in this field of applied ornithology, and (iii) both Goverment and NGOs should patronise more and more coordinated projects to study ecology and behavior of birds related to agriculture and to segregate beneficial birds from harmful bird-pests and to study the mammalian pests damaging agricultural crops, granaries and fruit gardens.

29

Threatened Birds of India and Conservation Strategies to Protect Them

Our rich biodiversity is fast declining. Our immense treasure of flora and fauna is in peril, not merely due to gradually changing biomes but also due to some abrupt changes of environment most carelessly brought about by so-called 'civilised' mankind in their march of Industrial Civilisation.

A recent assessment from the publication on 'Threatened Birds of India' (Vol 7, Nos. 1 & 2, 2002), compiled and edited by M. Zafar-ul Islam and Asad R. Rahmani of Bombay Natural History Society (BNHS), reveals alarming statistics to show different categories of the endangered, vulnerable and threatened species of birds of our country. These birds are either in a critical state of survival or threatened with extinction in near or distant future–unless a highly well-planned conservational protection is rendered to them. Needless to mention that this dismal picture is not only restricted to India, it is widespread throughout the so-called 'civilized' world. If strict conservation strategies are not implemented by the Goverments and Non-Goverment Organisations (NGOs) right now, we cannot save our country and the continents from gradual disappearance of a very colourful avifauna.

'Threatened Birds of India' reveals that **out of a total 2,700 birds species in Asia** as recorded in 2001, **323 species are threatened with extinction**—12% of the total bird species of Asia. In India too, the number of threatened birds deserving conservation is not less discouaraging. Out of the total of 1,224 (1,228 in another assessment) species (excluding subspecies) of Indian birds, 130 species making a total of about 11% are the immediate concern for conservationists. Of the 130 species of Indian birds, 8 have been listed as **'Critically Endangered'**, 10 **'Endangered'**, 57 **'Vulnerable'** and 52 **'Near Threatened'**, 1 **'Conservation Dependent'** and 2 **'Data Deficient'** (i.e., nothing much is known about them).

Islam and Rahmani (2002) referred to the Red Data Book published by the International Union for Conservation of Nature and Natural Resource (IUCN) and have explicitly outlined the Red List categories and criteria for all conservable species in all the continents of the world. According to the Red Data Book, **more than 1,000 species of animals being threatened are gradually approaching extinction.** The categories mentioned in the above paragraph and their precise definition and criteria as published by the IUCN

SSC, 1994, in their revised official booklet (source: Islam and Rahmani; In: Threatened Birds of India, Vol. 7, Nos. 1& 2, 2002; BNHS) are given below in a condensed form. Of the few critically endangered species, one or two species may have already been extinct and from the list of endangered species, some limited bird populations are declining so fast that unless prompt measures are adopted for their conservation, they too may be listed as critical.

29.1. RED LIST CATEGORIES AND CRITERIA: DESCRIPTION IN OUTLINE (BASED ON AND MODIFIED FROM ISLAM AND RAHMANI, 2002)

1.Critically Endangered

Species that are facing a very high risk of extinction in the wild in immediate future as defined by any of the given criteria (A to E):

A. Population decrease such as:

i. An observed or estimated reduction in population of at least 80% over the last 10 years (approx).

ii. A reduction of at least 80% projected or suspected to be met within the next 10 years or 3 generations—whichever is longer.

B. Area of occurrence estimated to be less than 40sq. miles (100 sq. km) or area of occupancy estimated to be less than 4 sq. miles (10 sq. km) and estimates indicating any two of:

i. Severely fragmented or known to exist only at a single location.

ii. Continuing decline in any of: (a) extent of occurrence; (b) area of occupancy; (c) area, extent and /or quality of habitat; (d) number of locations; (e) number of mature individuals.

iii. Extreme fluctuations in any of (a) to (e) mentioned above.

C. Population containing an estimated number of less than 250 mature individuals and either:

i. Continuing decline of at least 25% within 3 years or one generation, whichever is longer, or

ii. A continued decline in the number of mature individuals and population structure in the form of either (a) severley fragmented i.e., a subpopulation containing not more than 50 mature individuals, or (b) all individuals in a single sub population.

D. Population estimated to number less than 50 mature individuals.

E. Quantitative analysis showing that the probability of extinction in the wild—is at least 50% within 10 years or 3 generations, whichever is longer.

There are 8 species of Indian birds that are critically endangered.

2. Endangered

Though not critically endangered, a species is listed 'Endangered' when it is running a very high risk of extinction in the wild in near future, as defined by any of the folliwing criteria (A to E):

A. Decrease in population in the form of either:

i. An observed or estimated reduction of at least 50% over the last 10 years or 3 generations—whichever is longer.

ii. A reduction of at least 20% estimated to occur within the next 10 years or 3 generations—whichever is longer, **based on any of (b), (c), (d) or (e) as mentioned above.**

B. Area of occurrence less than 7800 sq. miles (20,000 sq. km) or area of occupancy estimated to be less than 780 sq. miles (2,000 sq. km).

C. Population estimated to be less than 2,500 mature individuals and either:

i. An estimated continuing decline of at least 20% within 5 years or 2 generations, whichever is longer.

D. Population estimated to be less than 250 mature individuals.

E. Quantitative analysis showing probablity of extinction in the wild at least 20% within 20 years or generations—whichever is longer.

There are 10 species of Indian birds that are endangered.

3. Vulnerable

Species that are not listed in Srl. Nos. 1 & 2 but are running a high risk of extinction in the wild in not long distant future (i.e., in a mid-term future), as defined by any of the criteria (A to E):

A. Decrease in population in the form of either:

i. Reduction of at least 20% over the last 10 years or 3 generations—whichever is longer.

ii. A reduction of at least 20% species anticipated within the next 10 years or 3 generations—whichever is longer.

B. Area of occurrence is less than 7800 sq. miles (20,000 sq. km) and area of occupancy is less than 780 sq. miles (2,000 sq. km).

C. Population less than 10,000 mature individuals and either

i. Continuing decline of at least 10% within 10 years or 3 generations—whichever is longer, or

ii. An observed continuing decline in numbers of mature individuals and population structure.

D. Population very small or restricted in the form of either of:

i. Population containing less than 1,000 mature individuals;

ii. Population having acute restriction in its area of occupancy, i.e., less than 40 sq. miles (100 sq. km) or in the number of locations (typically less than 5).

E. Quantitative analysis showing probability of extinction in the wild is at least 10% within 100 years.

There are 57 species of Indian birds that are vulnerable.

The next few catagories have been labelled as those of 'Lower Risk' in which the species are not listed in either of 1, 2 or 3 (i.e., Critically Endangered, Endangered or Vulnerable).

The lower risk categories are:

29.2.1. Lower Risk Categories

1. Conservation Dependent

Species which are targeted to be covered by species-specific or habitat-specific conservation programme for their survival and propagation, failing which the species in question may be listed in one of the threatened categories 1, 2 and 3, as mentioned above, within a period of 5 years.

At present there is only 1 species, the Dalmation Pelican (*Pelecanus crispus*), a winter visitor to India, covered by this category.

2. Near Threatened

Species that are not yet threatened, nor conservation dependent, but are close to Vulnerable category. There are 52 species of Indian birds that are Near Threatened.

3. Least Concern

Species that are neither Conservation Dependent nor Near Threatened.

29.2.2. Data Deficient

Species in this category lack in detailed information about their biology, life-history, distribution, etc. Hence the risk factor of these species cannot be ascertained. Once full knowledge of these bird species is available, it is quite likely that they may be listed in one of the Threatened categories. 2 species have been listed in this group: **Andaman Crake** (*Rallina canningi*) and **Nicobar Scops-owl** (*Otus alius*).

In the book (Threatened Birds of India) referred to above, the 'Threatened' bird species numbering 78 have also been classified according to habitats (in the cage below):

(a) Forest and Woodland Threatened species	:	29	species
(b) Grassland Threatened species	:	14	species
(c) Desert Threatened species	:	1	species
(d) Rocky Areas (including cliffs)	:	1	species
(e) Scrub Threatened species	:	4	species
(f) Wetlands (including littoral habitat) Threatened species	:	29	species

29.3. PREMIER ORGANISATIONS IN INDIA FOR THE PROTECTION AND CONSERVATION OF BIRDS

Since the 1970s, quite an appreciable number of Goverment and Non-Goverment Organisations, both in India and globally, have been devoted much more seriously than in the past for the protection and conservation of bird-life.

The Birdlife International, its UK counterpart **The Royal Society for the Protection of Birds** and the **Bombay Natural History Society** in India are the three premier organisations which, together, have established the Indian Bird Conservation Network(IBCN). To this have joined many other Non-goverment Organisations (NGOs), local Bird-watching groups and many individuals interested in the conservation of bird-life.

One of the main objectives of the IBCN was to locate and identify Important Bird Areas (IBAs) throughout the country. The IBA project was first launched by the Bombay Natural History Society in 1999. The selection and identification of the IBAs are not ony restricted to the biodiversity of India – these are also applicable to that of the world.

The aim of the IBA programme in India is to identify and protect the sites covering all habitats and their species, particularly those that are threatened most. The principal objectives of the IBA programme are [from 'Threantened Birds of India; Ed. Islam and Rahmani, 2002]:

i. To cater for the development of national conservation strategies.
ii. To highlight sites which are threatened or inadequately protected.
iii. To build national and regional networks of Ornithologists and conservationists through Indian Bird Conservation Network.
iv. To guide the works of national NGOs.

29.4. RED DATA BOOK

The International Union for Conservation of Nature and Natural Resources (IUCN), the International Council for Bird Preservation (ICBP), and the Bird Life International are very important organisations devoted for the conservation and protection of birds globally—particularly those Threatened and Near-threanted.

The Red Data Book was first conceived by Sir Peter Scott in 1963 and later published by the ICBP in 1980. Several other publications followed threafter, such as 'Birds to watch–I' (ICBP, 1988); 'Birds to watch–II' (Collar et al, 1994, by Birdlife International; 'Threatened Birds of the World' (Birdlife International, 2000); 'Threatened Birds of Asia' (Birdlife International, 2001).

The number of Threatend birds of Asia and India have been mentioned separately. The third global checklist of Threatened Birds reveals that there are 1,186 species of Threatened Birds in global assessment and their detailed mapping and illustrations have been completed and the Near-threatened birds have also been recorded for the first time (Islam and Rahmani, 2002).

The Important Bird Areas (IBAs) established by the BNHS in India are the important bird distribution sites throughout India and, according to Red Data Book, **17 migratory** and 57 **resident birds** have been found in our country. The Bird Life International is a partnership of non-government conservation organisations (NGOs) working in more than 100 countries for the conservation of birds as well as other forms of biodiversity. The Royal Society for the Protection of Birds (RSPB) of UK is the largest wildlife conservation charity in Europe and an UK partner of the Bird Life International. This organisation offers technical training and financial support to bird conservation organisations all over the world.

The Wildlife Preservation Society of India, Dehradun, established in 1958, is another important non-government organisation which has been doing commendable job in the management and conservation of wildlife of India. The World Wildlife Fund (WWF), established in 1961 in Switzerland, expanded its activities in India after the XIIth General Assembly of the IUCN held in Delhi in 1969 for preservation of the bird-population of India.

At the government level, several Acts have been promulgated since 1873. A few such Acts mentioned below are quite relevant for the conservation and protection of birds:

(i) The Wild Bird and Animal Protection Act, 1912.

(ii) Indian Board for Wildlife (IBWL), 1952.

(iii) Wildlife (Protection) Act, 1972 (Amended in 1982, 1986 and 1991).

(iv) Establishment of National Parks, Sanctuaries and Zoological Gardens.

29.5. SANCTUARIES AND NATIONAL PARKS IN INDIA

India has at present **92 National Parks** and **500 Sanctuaries** covering an area of 15.67 million hectares (38.7 million acres). Even then, it must be admitted that all these protected areas are not enough for the conservation of our rich avifuna and other forms of biodiversity. Many of these protected areas are not ideal sites for the conservation of Threatened or Near-threatened birds, they need to be covered by species-specific appropriate protection sites and effective management. This is because very few of our protected areas have been exclusively designed for the conservation of birds. In this regard, BNHS has a progressive programme for the conservation and protection of such birds under care of conservationists drawn from all sources like the central and State Governments, NGOs and various other small-scale voluntary organisations.

Examples of a few Sanctuaries and National Parks, States to which they belong and characteristic avifauna present in each are appended.

29.5.1. Important Sanctuaries in India

Table 29.1: Few Important Sanctuaries in India with Characteristic Avifauna

States	Sanctuaries	Characteristic Avifauna
Assam	Manas	Common Grey Hornbill, Spotbilled or Grey Pelican, White-necked Stork, Lorikeet, Yellow-headed Wagtail, Blackheaded Yellow Bulbul, Paradise Flycatcher, Peafowl, Sparrow Hawk, Crested Serpent Eagle, Cormorant
	Orang	Pelican, Blacknecked Stork, Adjutant Stork, Openbilled Stork.
Andhra Pradesh	Kolleru	Cormorant, Moorhen, Jacana, Migratory Ducks
Andaman and Nicobar Islands	Norkondam Island	Narkondam Hornbill
	North Reef Island	Andaman Teal, Nicobar Pigeon
	South Sentinel	White-billed Sea Eagle
Bihar	Valmiki	Paradise Flycatcher, Blue Magpie, Common Grey Hornbill, Himalayan Pheasants
	Mahuadar	Racket-tailed Drongo, Magpie Robin, Hawk Cuckoo, Spotted Munia, Jungle Fowl
	Udaypur	Famous for winter migratory birds

(*contd.*)

States	Sanctuaries	Characteristic Avifauna
Gujarat	Nal Sarobar	During winter Flamingo and different species of Ducks and Geese
Haryana	Sultanpur Jheel	Common and Purple Moorhen, Bronzewinged Jacana, Redwattled Lapwing, Little Ringed Plover, Wren Warbler, Baya Weaver Bird, White Eye, and in winter many migratory species like Lesser Whistling Teal, Mallard Duck, Shoveller, Greylag Goose, etc.
Himachal Pradesh	Govindasagar	Different species of Pheasant and winter migratory birds
Madhya Pradesh	Achanakmar	Blossom-headed Parakeet, Green Pigeon, Common Grey Hornbill
	Panchmari	Shama, Magpie-Robin, Blue Rock Thrush, Blackheaded Yellow Bulbul and several other species, Hawk-Eagle, Falcon, etc.
	Karera Bird	Great Indian Bustard, Red Junglefowl, Flamingo, etc.
Nagaland	Famie	Tragopan, Pheasant, etc.
Orissa	Chilka	Pochard, Teal, Shoveller, Mallard Duck, Grey Heron, Pelican, Darter, Cormorant.
Tamil Nadu	Mudumalai	Southern Trogon, Malabar Grey Hornbill, Crested Hawk-Eagle, Crested Serpent-Eagle, Myna, etc.
West Bengal	Narendrapur Wild Animal	Black-headed Oriole, Munia, Tailor Bird, Babbler, etc.
	Sajinakhali Wild Animal	Openbilled Stork, Cormorant, Egret, Night Heron, Lesser Whistling Teal, Brahminy Kite, etc.
	Bibhutibhusan Wild Animal	Little Grebe, Brahminy Kite, Openbilled Stork, Green Pigeon, Blossom-headed Parakeet, Hawk-Cuckoo, Koel, Blue-throated Barbet, Coppersmith, Redvented Bulbul, etc.
	Ballavpur Wild Animal	Black-headed Oriole, Magpie Robin, Shama, Redvented Bulbul, Iora, Indian Tree Pie, Spotted Dove, Indian Roller, Koel, Bronzewinged Jacana

29.5.2. Important National Parks in India

From the viewpoint of the richness of biodiversity India is one of the top 12 countries in the world. Besides the sanctuaries and the National Parks, India has also about a dozen of Biosphere Reserves for the protection, conservation and growth of animal resources.

The 92 National Parks in India have been designed such that their inner roads, water bodies, beautiful gardens, rest houses and secruity arrangements have promoted these sites as tourist spots. The names of a few National Parks in India are given. Besides the conservation and protection of mainly the mammalian fauna in the National Parks, these also contain characteristic avifauna.

Table 29.2. Important National Parks in India with Characteristic Avifauna

States	National Parks	Avifauna
Arunachal Pradesh	Namdafa	Different species of Colourful birds
Assam	Kajiranga	Different species of Colourful birds
Andaman and Nicobar Islands	Saddle Peak	Andaman Hill Myna, Nicobar Imperial Pigeon
Gujarat	Gir	Eagle, Sahi Falcon, Fish Owl, Crested Swift, Indian Roller, Coppersmith, Grey Partridge, etc.
Jammu and Kashmir	Dachigram	Different species of both resident and migratory birds
Karnatak National Park	Nagarhole	Commom Grey Hornbill, Malabar Pied Hornbill. Malabar Trogon, Fairy Bluebird, Thrush, Flycatcher, Warbler, Brahminy Duck, Lesser Whistling Teal, Pintail, etc.
Kerala	Periar	Oriole, Myna, Lesser Racket-tailed Drongo, Darter, Osprey
Madhya Pradesh	Kanha	Different species of birds, common with those of the sanctuaries of Madhya Pradesh
	Panna	Paradise Flycatcher, Peafowl, Junglefowl, Warbler, Adjutant stork, Serpent Eagle
Manipur	Sirohi	Full of many distinctive species of birds
Orissa	Simlipal	At present under Tiger Project, but varieties of birds are also present

(contd.)

States	National Parks	Avifauna
Rajasthan	Sariska	Peafowl, Red Junglefowl, Grey Partridge, White Scavenger Vulture, Tawny Eagle
	Desert	Famous for Great Indian Bustard
Uttar Pradesh	Jim Corbett	First National Park, then known as 'Ram Ganga National Park'. At present, it is under 'Tiger Project'. Besides varieties of mammals, different species of birds are considered as treasure of our country
	Valley of Flower	Famous for beautifully coloured flowers and bird species
West Bengal	Sundarbans, Gorumara, Shingalila, Neoraveli and Baxa	Besides different kinds of mammals in each, the varieties of birds are also their animal resources

NOTES :

1. It is important to note that both Sanctuaries and National Parks in our country are highly protected areas for the conservation of forests, water-bodies and wild life in their natural environment delimited and lawfully declared by the State and/ or Central Government through legislature.
2. A National Park, dedicated to the cause of the nation by the Central Government, may be declared so after considering various criteria such as scenic beauty of the area, luxuriant growth of plants with vast stretches of greenery, clean pools of water, flower gardens, orchards, provision for extensive road-links, rest-house and with amenities for attracting tourists. Grazing animals outside the boundaries of the National Parks are allowed, not within.
3. Fulfilling the above criteria, a wild life sanctuary or a similar vast conservable area may be declared as a National Park.

29.6. EXAMPLES OF A FEW THREATENED BIRDS OF INDIA AND CONSERVATION STRATEGIES TO PROTECT THEM

Of the total of about 130 threatened birds of India, listed under different categories, a few birds are briefly described:

A. Critically Endangered

8 species of birds have been declared critically endangered of which the first two described below are probably extinct by now (though unconfirmed as yet):

1. Pink-headed Duck (*Rhodonessa caryophyllacea*)

This bird, once appeared to be quite common in Bengal as described by earlier authors, has perhaps left no survivor in the wild by now.

It is known from records that the maximum number of this species occurred in India, less so in Burma (Myanmar), Bhutan, Bangladesh and Nepal. Decrease in habitat and increase in hunting through prolonged period might have reduced the number of species to a critical stage.

Since 1956, its killing, capture or egg collection has been totally banned by law. Reduction in numbers was noticed as early as 1878!

2. Himalayan Quail (*Ophrysia superciliosa*)

This bird is also likely to be extinct by now; if not, its population must be much reduced.

The bird was known only from the Western Himalayas between the altitudes 6,000 and 7,000 ft (1800-2,100 m) in the areas covering Mussoorie and Nainital. There are conflicting reports that a small party of 5 or 6 of these birds could be traced in West Kumaon and Nepal between 1945 and 1950, as also from the Mussoorie hills even in 1984.

Loss of habitat due to increase in human population in the restricted areas of these birds seems to be the main reason for decline of the species.

For conservation and protection, the species, though untraceable at present, has been kept fully restricted within the probable site of the Western Himalayas Endemic Bird Area, along with a few other threatend birds. The Badraj, Banog and Jharipani Sanctuaries of Uttaranchal are the probable sites (Ref: Threatened Birds of India, 2002).

3. Whitebacked Vulture (*Gyps bengalensis*)

This species is also known as **Whiterumped** or **Bengal Vulture**. This species, together with two other species of vultures, namely Long billed vulture (*Gyps indicus*) and slenderbilled vulture (*Gyps tenuirostris*), have been declared critically endangered because of their fast decline in numbers.

Once abundantly represented throughout India, Bangladesh, Pakistan and Myanmar, the Whitebacked Vulture has fast declined in numbers since 1990s, possibly due to some infectious viral or bacterial diseases or toxicity and indiscriminate use of pesticides in carcasses.

Worldwide awareness for the decline in population of this species has been effectively campaigned by the BNHS and other organisations in India and, in June 2000, the 'Asian Vulture Conservation Programme' has been established on emergency basis. It has now been found that the infectious disease is possibly caused by a viral attack.

Important Bird Areas (IBAs) for this species are Haryana, Uttar Pradesh, Rajasthan, Madhya Pradesh, Maharashtra, Tamil Nadu, West Bengal, etc.

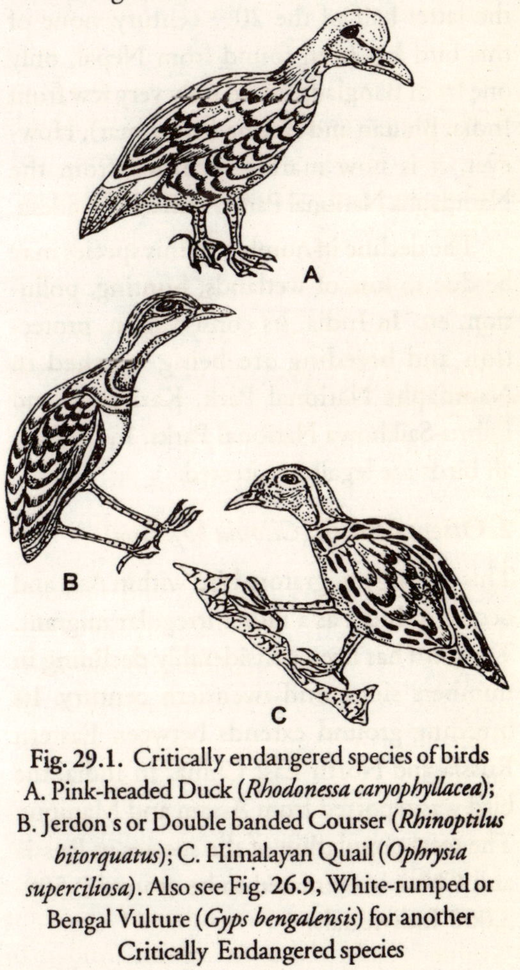

Fig. 29.1. Critically endangered species of birds A. Pink-headed Duck (*Rhodonessa caryophyllacea*); B. Jerdon's or Double banded Courser (*Rhinoptilus bitorquatus*); C. Himalayan Quail (*Ophrysia superciliosa*). Also see Fig. 26.9, White-rumped or Bengal Vulture (*Gyps bengalensis*) for another Critically Endangered species

NOTE: Besides the above-mentioned species, the Siberian White Crane (*Grus leucogeranus*), the Jerdon's Courser (*Rhinoptilus bitorquatus*), the Forest Owlet (*Athene blewitti*) are also critically endangered.

B. Endangered

10 species of Indian birds have been declared 'Endangered' which have been facing threats of extinction next to 'Crically Endangered'. 3 such species are briefly described:

1.White-bellied Heron (*Ardea insignis*)

Once distributed throughout Eastern India, Nepal, Bhutan, Bangladesh and Burma (Myanmar), Islam and Rahmani (In: Threatened Birds of India, 2002) report that from the latter half of the 20th century, none of this bird has been found from Nepal, only one from Bangladesh and only very few from India, Bhutan and Burma (Myanmar). However, it is now mainly recorded from the Namdapha National Park, Arunachal Pradesh.

The decline in number of this species may be due to loss of wetlands, hunting, pollution, etc. In India, its conservation, protection and breeding are being watched in Namdapha National Park, Kaziranga and Dibru-Saikhowa National Parks. In Bhutan, all birds are legally protected.

2. Oriental Stork (*Ciconia boyciana*)

This is a large migratory bird within Asia and occurs in India as a rather irregular migrant. This bird has been considerably declining in numbers since mid-twentieth century. Its breeding ground extends between Eastern Russia and North-East China. In India, the bird was reported from Assam and Manipur. The total population of the species in Russia and China is estimated to be around 2,500-3,000 individuals.

Gradual disappearance of its habitats, pollution, hunting and overfishing from the waterbodies are the main causes for the reduction in the number of this species.

The species is covered by the Indian Wildlife Protection Act, 1972, in Schedule IV.

3. Indian Bustard (*Ardeotis nigriceps*)

This resident bird, once present in large numbers throughout India, is now 'Endangered' due to fast-reducing population of the species.

This bird population occupied vast areas in India, extending from Western Orissa to northern and central India, Punjab, Rajastan up to border of Pakistan and in the south up to Tamilnadu. The species started declining in numbers since 1950. It was declared as globally threatened in 1966. It has dwindled from many earlier habitations and its total number may now be reduced to only 500.

The gradual reduction in number of the species may be attributed to various causes, mainly loss of habitats, extension of irrigation and cultivable lands, change of crop types, use of pesticides, pollution, hunting, etc.

The conservation strategies confirm legal protection imposing drastic action against killing of the species. The BNHS has adopted necessary conservation measures for eliminating threats to this species.

NOTE: Besides the above 3 species, the **Greater Adjutant Stork** (*Leptoptilos dubius*), the **White-headed Duck** (*Oxyura leucocephala*), the **Whitewinged Duck** (*Cairina scutulata*), the **Bengal Florician** [*Eupodotis (Houbaropsis) bengalensis*], the **Lesser Florican** [*Eupodotis (Sypheotides) indica*], the **Nordmann's Greenshank** (*Tringa guttifer*) and **Rufous-breasted Laughing Thrush** (*Garrulax cachinnans*) are also declared as **'Endangered'** species.

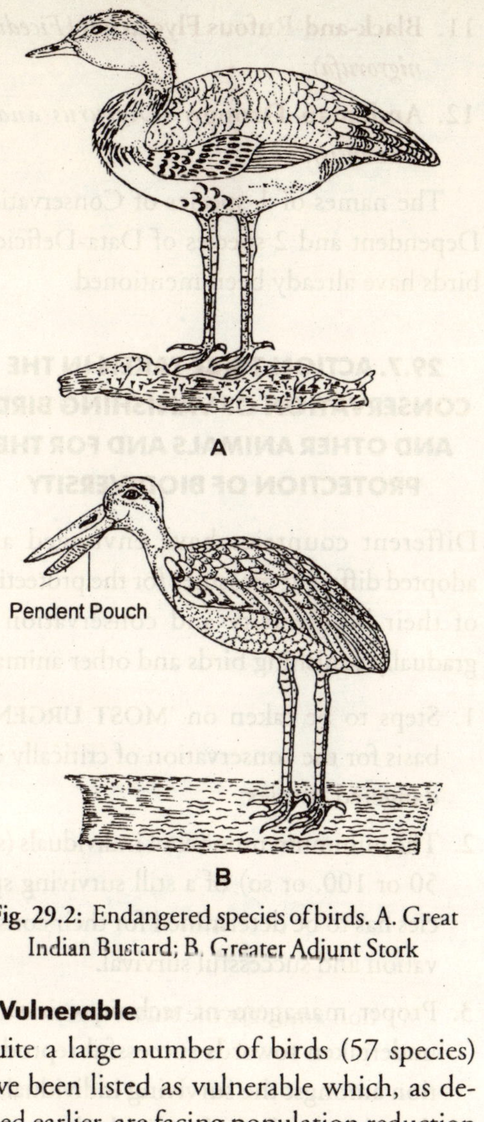

Fig. 29.2: Endangered species of birds. A. Great Indian Bustard; B. Greater Adjunt Stork

C. Vulnerable

Quite a large number of birds (57 species) have been listed as vulnerable which, as defined earlier, are facing population reduction or fragmentation at a slow or rapid rate. If this is allowed to continue, the birds under this category may be listed as 'Endangered' in near or not too distant future.

From the long list of vulnerable species, a dozen names are mentioned below as representative examples. All the vulnerable species in general have their populations reduced by about 20% since 1995, their areas of occupancy much contracted, sometimes limited around 780 sq. miles (2,000 sq. km) or less. In most cases such decline in population of birds has been due to increase in human population and decrease in forest tracts, change in crop types in cultivation, reduction in wetlands due to indiscriminate filling or geophysical changes due to climate. Species-wise conservation measures have been adopted by the Government and NGO establishments globally and, in India, these bird species have been covered under Schedule I to IV, as the case may be, of Indian Wildlife (Protection) Act, 1972.

Some Vulnerable Species

1. **Spot-billed Pelican** (*Pelecanus philippensis*)
2. **Nicobar sparrowhawk** (*Accipiter butleri*)
3. **Lesser White-fronted Goose** (*Anser erythropus*)
4. **Pallas's Sea-Eagle** (*Pallas's or Ringtailed Fishing Eagle; Haliaeetus leucoryphus*)
5. **Lesser Kestrel** (*Falco naumanni*)
6. **Manipur Bush-Quail** (*Perdicula manipurensis*)
7. **Western Tragopan** (*Tragopan melanocephalus*)
8. **Green Peafowl** (*Pavo muticus*)
9. **Sarus Crane** (*Grus antigone*)
10. **Pale-backed Pigeon** (*Columba eversmanni*)
11. **Tawny-breasted Wren-Babbler** (*Spelaeornis longicaudatus*)
12. **Nicobar Megapode** (*Megapodius nicobariensis*).

D. Near-threatened

Under the lower risk categories (next to Vulnerable), are included 52 species of 'Near-threatened' species which have their populations declined by about 10% or so since 1995, and their areas of occupancy have also been reduced—though not to the extent of vulnerable species. Most of the Near-Threatened species were enjoying a wide distribution in the past and even now, though reduced in population or rarely seen in many parts of the world, they occur in sizeable populations in others, ranging between 10,000 and 20,000 birds or more for each species.

In all cases, the reduction in population may be due to any one or a combination of factors like loss of sustainable habitats, increase in human populations, hunting of adult birds, pollution, deforestation, egg destruction, etc.

Many of these bird species inhabit protected areas now and are under close watch by the ornithologists.

From the long list of 52 species of Near-Threatened birds, names of a dozen birds are mentioned as respresentative examples:

Some examples of Near-Threatened birds

1. **Oriental Darter** (*Anhinga melanogaster*)
2. **Painted Stork** (*Mycteria leucocephala*)
3. **Black-headed Ibis** (*Threskiornis melanocephalus*)
4. **Lesser Flamingo** (*Phoenicopterus minor*)
5. **Lesser Fish-Eagle** (*Ichthyophaga humilis*)
6. **Black-bellied Tern** (*Sterna acuticauda*)
7. **Andaman Wood-Pigeon**(*Columba palumboides*)
8. **Malabar Pied-Hornbill** (*Anthracoceros coronatus*)
9. **Great Hornbill** (*Buceros bicornis*)
10. **Rufous-throated Wren-Babbler** (*Spelaeornis caudatus*)
11. **Black-and-Rufous Flycatcher** (*Ficedula nigrorufa*)
12. **Andaman Drongo** (*Dicrurus andamanensis*).

The names of 1 species of Conservation Dependent and 2 species of Data-Deficient birds have already been mentioned.

29.7. ACTION TO BE TAKEN IN THE CONSERVATION OF VANISHING BIRDS AND OTHER ANIMALS AND FOR THE PROTECTION OF BIODIVERSITY

Different countries have envisaged and adopted different measures for the protection of their biodiversity and conservation of gradually vanishing birds and other animals:

1. Steps to be taken on 'MOST URGENT' basis for the conservation of critically endangered species.
2. The minimum number of individuals (say, 50 or 100, or so) of a still surviving species has to be determined for their conservation and successful survival.
3. Proper management technique is to be undertaken towards successful reproduction amongst the surviving individuals so that the number of individuals remains well above the minimum.
4. To allocate causes for the gradual decline in number of a species and to take steps to restore natural habitat of a species.
5. In case the natural habitat of the species is found depleted due to dwindling of forest, shortage of food supply, pollution in the area of occupancy of the species or any other damage to the ecosystem, the threatened individuals may be conserved in a

Sanctuary or a Biosphere Reserve Forest. Biosphere Reserves are areas of terrestrial and coastal ecosystems. They are internationally recognised within the framework of UNESCO's Man and Biosphere (MAB) Programme. India has 13 Biosphere Reserves—Nilgiri, Manas, Sunderbans, Simlipal, Gulf of Mannar, Great Nicobar, Panchmari etc.

6. With the gift of modern computer technology, assessment in the number of threatened species (of all categories) may be monitered from time to time, keeping the individuals in improved habitat.

29.8. SHORT-TERM AND LONG-TERM CONSERVATION PROGRAMMES

With growing industrialisation in developing countries as ours, the ecosystems of the natural habitats for the birds, mammals and other animals have been largely eroded since the 1950s. To overcome such hazards, conservationists have to adopt both short-term and long term projects for the sustenance of our rich biodiversity.

For birds, the ideal programmes may be:

A. Short-term conservation programme need to be adopted for any species of birds fast declining. Once such an alarming decline in the number of individuals of a species is detected the habitat of such birds may be shifted to a zoo-garden or to any favourable site with the provision of adequate food suply. These birds may be rehabilitated to their original natural habitats on restoration and improvement of the same.

B. **Long-term conservation programme** for the threatened, vulnerable and near-threatened birds requires a well-thoughtout phase-wise policy congenial to the survival and sustenance of the species. The long-term conservation policy is generally aimed at restoring natural habitat of the species through modern management techniques. Alongside, specific conservation programme for a fast-declining species—such as 'The Asian Vulture Conservation Programme'–must be launched on an emergency basis. In other cases, extension of the overall 'Bird Areas' (as defined by the IBAs), restoration and preservation of wetlands, protection of forests, elimination of sources of pollution, etc. may be considered as important items of long-term conservation programme.

29.9. SOME COMMON INDIAN BIRDS BANNED IN TRADE

Out of a total of about 1,228 species of birds found in India (the number may be slightly more at present), two-third of them are found in West Bengal alone. As already mentioned, an alarming number of bird species are gradually dwindling within a span of 3, 5 or 10 years. The causes of decline may be attributed not only to the global change in climatology, topography, vegetation and the natural environment in general, but also to the man-made pollution of the environment, indiscriminate hunting and wilful disturbance in the ecological set-up. This has resultted in the loss of feeding and roosting habitats for the birds.

Out of several dozens of very common yet very interesting bird species that have been banned in trade under the **Wild life Protection Act of India, 1972: Schedules I-IV**, a few species are mentioned with the residential status of each:

Table 29.3: Species Banned in Trade

No.	Bird Species	Residential Status
1.	Shikra (*Accipiter badius*)	Resident bird of India SCH I
2.	Peregrine Falcon (*Falco peregrinus*)	Migratory; winter visitor throughout the subcontinent SCH I
3.	Osprey (*Pandion haliaetus*)	Migratory; winter visitor to India. SCH I
4.	Indian Peafowl (*Pavo cristatus*)	Resident and the National bird of India. SCH I
5.	Oriental Pied Hornbill (*Anthracoceros albirostris*)	Resident in India specially in West Bengal SCH I
6.	White-breasted Waterhen (*Amaurornis phoenicurus*)	Resident in India specially in West Bengal. SCH IV
7.	Hoopoe (*Upupa epops*)	Winter visitor throughout the Indian Subcontinent; abundant in West Bengal in the winter. SCH IV
8.	Magpie-Robin (*Copsychus saularis*)	Resident in India; abundant in West Bengal SCH IV
9.	Alexandrine Parakeet (*Psittacula eupatria*)	Largest Parakeet of India. SCH IV
10.	Roseringed Parakeet (*Psittacula krameri*)	Resident practically throughout India. SCH IV
11.	Blossomheaded Parakeet (*Psittacula cyanocephala*)	Resident throughout India and locally migratory. SCH IV
12.	Red Munia or Waxbill (*Estrilda amandava*)	Resident throughout India. SCH IV
13.	Scaly-breasted Munia (*Lonchura punctulata*)	Resident all over India. SCH IV
14.	Racket-tailed Drongo (*Dicrurus paradiseus*)	Resident; more or less throughout India. SCH IV
15.	Rosy Starling (*Sturnus roseus*)	Migratory; winter visitor throughout India. SCH IV

No.	Bird Species	Residential Status
16.	**Brahminy Myna (Brahminy starling;** *sturnus pagodarum***)**	Resident throughout Indian subcontinent. SCH IV
17.	**Asian Pied Starling** (*Sturnus contra*)	Resident in India; found abundantly in Kolkata and all over West Bengal. SCH IV
18.	**Red-whiskered Bulbul** (*Pycnonotus jocosus*)	Resident throughout India. SCH IV
19.	**Red-vented Bulbul** (*Pycnonotus cafer*)	Resident bird of India; abundant in West Bengal. SCH IV
20.	**Golden Oriole** (*Oriolus oriolus*)	Resident practically throughout India, except north east region. SCH IV
21.	**Black-headed Oriole** (*Oriolus xanthornus*)	Mostly resident bird of West Bengal; also **mentioned as Black-hooded** Oriole; inhabit other parts of India. SCH IV
22.	**Blue-throated Barbet** (*Megalaima asiatica*)	Resident of West Bengal; also found in other parts of India. SCH IV
23.	**Coppersmith Barbet** (*Megalaima haemacephala*)	Resident in West Bengal and throughout greater part of India. SCH IV
24.	**Baya Weaver Bird** (*Ploceus philippinus*)	Resident in West Bengal and throughout Indian subcontinent; Resident and locally migratory. SCH IV

By courtesy: Forest Department, Wildlife Wing, Government of West Bengal

NOTES:

1. It may be mentioned here that the Wildlife Protection Act of India, 1972, contains 6 Schedules which, till 2002, have been amended with new additions at least 6 times. Schedule I of the Act contains about 4 dozens of birds.

2. Hunting, poaching or wilful destruction of birds under Schedule I calls for highest punishment of 7 years' imprisonment and a fine of Rs. 10,000 or more.

PART II

BIOLOGY OF THE BLUE ROCK PIGEON, COLUMBA LIVIA Gmelin

30

Biology of the Blue Rock Pigeon, *Columba livia* Gmelin

30.1. INTRODUCTION

The pigeons are one of our nearest neighbours among birds and are found to roam about with their characteristic deep chirping notes almost fearlessly in the corners and corridors of our buildings in cities, towns and villages. They feel equally comfortable either in congested localities like market-places, railway stations, office buildings, residental complexes, as well as in the solitary corners of deserted buildings and farm-houses. They are found in large flocks near store-houses of rice, paddy, maize and cereals, a small quantity of which is always found scattered on the ground during transhipment. These grains are then pecked up by the pigeons from the ground as food.

In the strictly wild state, the Blue Rock Pigeon inhabits rocky sea coasts, cliffs and caves of the hills, in open country, as well as in the farms and suburbs of cities and towns. From the wild state this bird has mostly become semi-feral or feral through time and space as found today. This is a very old species that lived an arboreal life in the wild from time immemorial (recorded from ancient Egypt since 3000 BC and from the Middle East even earlier). Throughout centuries, the wild species came close to human habitation and gave rise to a few subspecies and an appreciable number of geographical races and domestic varieties. The pigeon has been considered as an ideal specimen for the study of structure, function and life-history of a carinate bird that is easily available.

30.2. SYSTEMATIC POSITION

Phylum	Chordata
Subphylum	Vertebrata
Superclass	Gnathostomata
Class	Aves
Subclass	Neornithes
Superorder	Neognathae
Order	Columbiformes
Family	Columbidae
Genus	*Columba*
Species	*C. livia*

The Family Columbidae is a large family comprising about 43 genera and 305 species (Salim Ali, 1987), or, in the more recent checklist by Monroe and Sibley (1993), the family contains 40 genera and 313 species and 25 subspecies or more. Goodwin (1967) also mentions about 40 genera in the family. The Genus *Columba* alone contains 54 species and 4 subspecies.

From the wild species quite a large number of domestic varieties of pigeons have developed. These are not to be regarded as subspecies, and, hence, these domesticated 'varieties' are not to be mentioned as *'Columba livia domestica'*, but, at most, the domestic variety may be indicated within a bracket after or below the species name. However, *Columba livia* has two distinct subspecies, *C. l. neglecta* and *C. l. indermedia.*

Some of the popular examples of domestic varieties are: **Pouters, Tumblers, Carriers, Fantail, Siraju, Jacobin** etc. The Indian wild species of Blue Rock Pigeon differs from that of Europe in having the lower posterior feathers of the rump and the abdomen ashy in colour, while the colour of those feathers in the wild European species is white.

30.3. GEOGRAPHICAL DISTRIBUTION

The Blue Rock Pigeon or the Rock Dove is widely distributed in Asia, Europe and North Africa wherefrom they have expanded their range in other countries. The two subspecies of *Columba livia* mentioned above have their origin in the Middle East and are now the resident birds of our country, though locally migratory. In Asia, the pigeons are distributed in the whole of India, Pakistan, Nepal, Bhutan, Bangladesh, Sri Lanka, Laccadive Island (now Lakshadweep), Burma (Myanmar) and Thailand up to 1000 ft (3,00 m) altitude in the hills.

30.4. HABIT AND HABITATS

The pigeons are very docile, harmless and peace-loving birds that are admired throughout the world as the messenger of peace. They are usually gregarious, but very often found to move and live in pairs. They are close companions of human beings, and, so, are least afraid of them.

Monogamy is a special characteristic of the pigeons. A pair of male and female pigeons live together for whole life. The pigeons have a reasonably good speed in flight (about 30-40 miles/50-65km per hour) and are quite alert on the ground while feeding.

In the wild, the Blue Rock Pigeon lives in colonies in cracks, crevices, ledges and cavities in rocky hills, in the holes of tree-trunks bordering countryside and in the forestry across foothills and hills, in the deserted buildings, ruins of ancient forts, etc. at a considerable distance or even far away from human habitation. From such roosting sites, they cover sometimes long distances to reach their feeding grounds and come back every evening. In such day-to-day foraging trips, these birds move in small, medium or large flocks to reach vast fields of cultivation and feed on all sorts of growing crops–paddy, pulses, cereals, groundnuts etc., thus causing appreciable damage to human economy. In contrast, the semi-feral and domestic varieties of pigeons living in association with human habitation have ample sites of feeding, including human charity and patronisation.

The habitats in domestic varieties have also changed considerably from those of the wild species. The domestic pigeons roost close to human habitation, with a wide choice from every nook and corner of buildings, cornices, cracks, holes in walls, junctions of ceilings and beams to gardens and trees around parks. They are equally comfortable in a noisy market place or in a solitary godown or a deserted building.

The male birds make a rhythmic 'gootrr.goo, gootrr.goo' call, keeping the head low and puffing up the throat. This call may be compared to a deep melancholic sound.

The pigeons breed almost throughout the year, although, in the higher altitudes in the Himalayas, the breeding season extends from May to July every year. The pigeons usually breed twice in a year.

Nest-building is an important function in the life-history of a bird. A general description about the selection of nesting sites and territoriality in different birds has been given earlier in this book. As already mentioned, the pigeons build their nest in the safe corner of a building such as in the cracks and holes in the wall, at the junction of ceilings and beams or in the solitary corners of a deserted building, but close to human habitation. Holes in tree-trunks are often used as nest-sites. The nest of a pigeon is an untidy and clumsy structure in the form of a shallow depression built by both the male and the female with fine twigs, threads, leaves, rubbish, etc. Each year there are 2 or 3 clutches (brood of young), each with 2 eggs which are incubated by both the parents for 16-19 days (usually 16 days in domestic pigeons). The nidicolous (young staying longer than average in the nest) nestlings are fed by both the parents with 'crop milk'. 'Crop milk' is a nutritious milky-white secretion from the crop glands, mixed with a portion of regurgitated and half-digested pulp of grains. The hatchlings grow and fly after about a month (30-35 days).

The description of the Common Pigeon given in this text is of the domestic variety of the Blue Rock Pigeon.

30.5. FOOD, FEEDING AND DRINKING IN PIGEONS

The pigeons are primarily granivorus but they occasionally invade other food-niches according to availability or being compelled by environmental hazards. Thus their food chiefly consists of the grains of rice, maize, wheat, millets, pulses and various other cereals, groundnuts, weed-seeds, soft shoots of growing crops, etc. Secondarily, they also feed on small fruits of the nature of berries and fruits of *Ficus* sp., insects, their eggs and larvae, and also worms.

The pigeons are very efficient ground-peckers. The **pecking** of grains from the ground is performed in quick succession with the help of their slightly curved bill-tip. Along with the grains of cereals and pulses, a few pebbles are also taken inside the stomach and thence to gizzard. These are called '**gizzard stones**' that help in churning of the food by the gizzard. The pigeons are not in the habit of taking large volume of food at a time, rather, they collect food several times a day. Before ingesting any food particle, a pigeon tests its acceptability by '**catch and throw**' method and then passes on the food particle to oropharyax for swallowing. The tongue-tip, the caudolateral processes of the tongue, the secretions of the mandibular glands and mucous—all help in this food-ingesting process (Zweers et al, 1994). These authors have extensively worked out the mechanism of food-getting devices of a number of bird families.

Bhattacharyya (1980, 1987), Zweers (1982a, 1982b) Bramble and Wake (1985) and Van Gennip (1986) have worked out the details of pecking, feeding system and feeding behaviour of the pigeons. Zweers, Berkhoudt and Vanden Berge (1994) state that all neognathous birds show pecking mechanism, irrespective of their specialised feeding habit and Zweers (1991a, b) assumes that pecking is the ancestral feeding design from which a wide range of trophic diversification has evolved through modifications and specialisations. McLelland (1979) has explained the role of the tongue in manipulating the food particles within the oral cavity before swallowing.

Zweers (1982) has worked exhaustively on the drinking mechanism of the pigeons. According to him, a **double suction method operates in the drinking process. First, the retraction** of the tongue causes decrease in air pressure in the buccal cavity which sucks in water. In the second phase, due to depressions of the pharyngeal floor, water from buccal cavity is forced into the pharynx and then swallowed.

In an earlier chapter (Chapter 9) of this book, a more detailed description on 'food, feeding and drinking' of birds, in general, has been provided.

30.6. EXTERNAL MORPHOLOGY

The Blue Rock Pigeon, *Columba livia* Gmelin, is a medium-sized, graceful flying bird, about 13 inches (33 cm) long and with a slaty or greyish colour traversed by a few darker bands. Some bright-green and magenta coloured feathers cover the entire neck and front part of the breast. The colour of the eyes and of the legs is somewhat pink. Two black cross-bands are found at the posterior aspects of each wing.

A pigeon's body is mainly divided into four parts: head, neck, trunk and tail. The pigeon can rotate its head 180° from the left to the right. The whole body is covered by feathers of various shapes and sizes.

The external morphological characteristics of different parts of the body of a pigeon are:

1. Head

(a) The head is almost rounded and much smaller in size in proportion to that of the body; the head is rostrally elongated to form the characteristic blackish pointed bill.

(b) The bill is rather medium-sized, unspecialised, and is actually formed by the upper and lower beaks which develop as a special type of hard, keratinised structures around extensions of the upper and lower jaws, respectively. The **mouth** is a slit-like aperture between the two beaks which, by their opening, form a wide gape.

(c) In general, this keratinised covering is called **rhamphotheca** of which the upper jaw covering is called **maxillary rhamphotheca** or **rhinotheca** and that of the lower jaw **mandibular rhamphotheca** or **gnathotheca**; the sharp rhamphothecal edges of the jaws are called **tomium**, while the dorsal ridge of the maxillary rhamphotheca is called the **culmen**.

(d) About the middle of the upper beak, on either side, lies an elongated slit-like aperture, the **external nare** or the **nostril**. This type of elongated nostril is called **schizorhinal nostril** which is characteristic of columbid and a few other families of birds.

(e) A small, round, whitish and featherless swollen part of the integument called '**cere**' (or cera) occurs medially at the base of the upper beak. Anteriorly, it overlies a small part of nostril. The function of cere is not well-defined, though some consider it as an integumentary receptor organ.

(f) Behind the cere, on either side of the head, lies a large rounded bright eye with one **upper eyelid**, one **lower eyelid** and a third transparent **nictitating membrane** which can be drawn as a curtain across the eye. Both the eyelids and the nictitating membrane are movable—they protect the eyes.

(g) Behind and below each eye lies the **auditory aperture**, which is covered by a few special type of **auricular feathers**. The aperture leads through a tubular canal—**the external auditory meatus** which terminates at the circular **tympanic membrane** or the **tympanum** at the entrance of the middle ear.

2. Neck

The neck can be easily identified from the rest of the body by its long cylindrical and flexible form. This flexibility is provided by the nature of cervical vertebrae and the adjoining muscles. The flexible neck thus works as an efficient pivot for a free rotation of the head through 180° degrees.

3. Trunk

(a) The rigid spindle-shaped body that becomes narrower rostro-caudally and extends between the base of the neck and beginning of the tail is called the trunk.

(b) The distinct bony ridge found midventrally in the trunk after removal of feathers is called the **keel** or **carina** which is characteristic of flying birds.

(c) Anteriorly, the two forelimbs have been modified as wings which, during rest, become folded and remain closely apposed to the body in the form of 'Z'. During flight, however, the two wings are extended at right angles to the trunk. Each wing is divided into three parts: arm, forearm and hand. Each hand terminates in three clawless fingers.

(d) On either side at the base of the wing, there is a membranous fold of skin called the patagium. In frontal aspect of the wing, the fold of skin located between the arm and the forearm is called the **prepatagium** and, along the posterior border of the wing, the fold of skin between the arm and the trunk is called the **postpatagium**. Both the patagia remain covered by the large flight feathers of the wing.

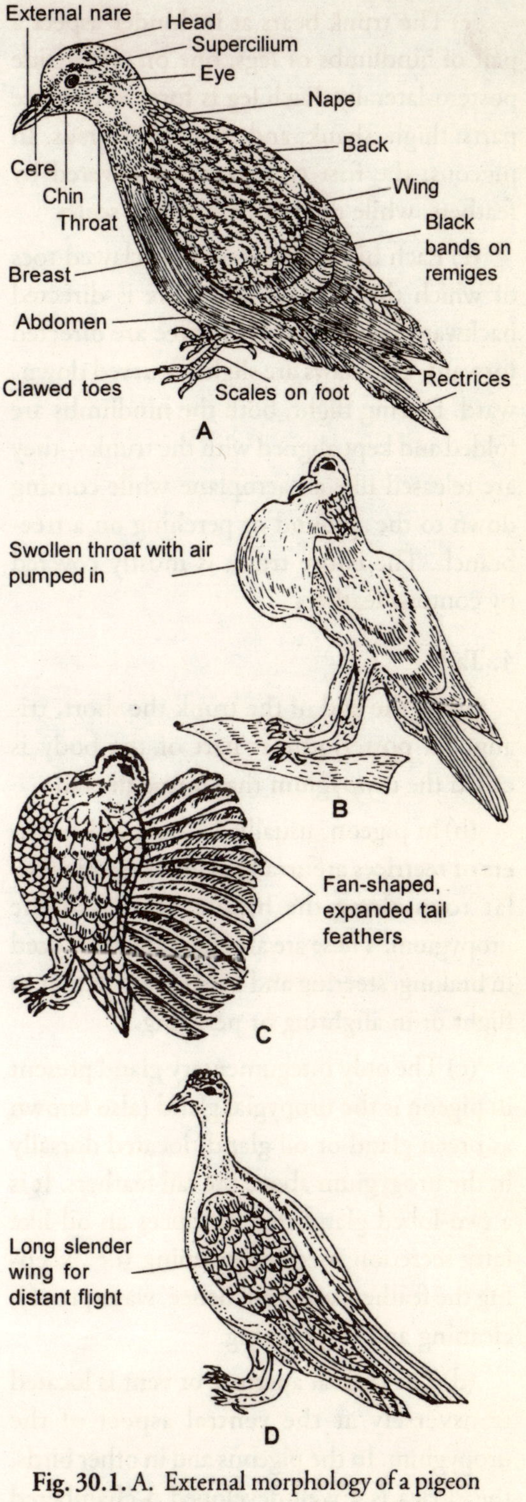

Fig. 30.1. A. External morphology of a pigeon (*Columba livia*); B. C. D. A few domesticated varieties; B. Pouter; C. Fantail; D. Carrier

(e) The trunk bears at its hinder aspect a pair of hindlimbs or legs, one on either side postero-laterally. Each leg is formed of three parts: thigh, shank, and tarso-metatarsus. In pigeons, the first two parts are covered by feathers, while the last part by the scales.

(f) Each hindlimb bears four clawed toes of which the first toe or hallux is directed backward, while the other three are directed forward. The claws are slightly curved downward. During flight, both the hindlimbs are folded and kept aligned with the trunk—they are released like an aeroplane while coming down to the ground or perching on a tree-branch. The entire trunk is mostly covered by contour feathers.

4. Tail

(a) At the end of the trunk the short, triangular, posteriormost part of the body is called the **uropygium** that forms the tail.

(b) In pigeon, usually 12 long **tail feathers** or **rectrices** are arranged in two semicircular rows along the hinder border of the uropygium. These are also flight feathers used in braking, steering and balancing—either in flight or in alighting or perching.

(c) The only integumentary gland present in pigeon is the uropygial gland (also known as preen gland or oil gland) located dorsally in the urogygium above the tail feathers. It is a two-lobed gland that produces an oil-like fatty secretion used for preening (i.e., keeping the feathers oily and, hence, waterproof), cleaning and brightening.

(d) The cloacal aperture or vent is located transversely at the ventral aspect of the uropygium. In the pigeons and in other birds, the cloaca is a well-developed 3-chambered organ.

30.7. INTEGUMENT

(a) The skin in the pigeon is devoid of skin glands; it is a thin dry loose and elastic covering.

(b) The skin is divided into two layers, the outer **epidermis** and the inner **dermis**. The epidermis is formed of several cell-layers of the which the outermost layer is the **epitrichium**. It is formed of a row of small flat cells. The middle layer, **stratum corneum**, has reached the peak of development in birds. It is formed of flat, keratinsed cells that are protective in nature. From this layer have arisen the rhamphotheca of beaks, the scales of feet, the claws in toes and a unique structure the feathers—which are the greatest endowment to birds for their flight. The innermost layer of epidermis is called stratum germinativum (Malpighian layer). This layer contains large columnar cells which lie perpendicular to the dermis and are able to divide mitotically.

(c) The **dermis** is comparatively thicker and is formed of mesodermal cells. It is formed of several layers of cells and contains muscle fibres, nerve fibres, blood capillaries, and connective tissues.

30.8. EXOSKELETAL SYSTEM

The exoskeletal structures in the pigeons, as in other birds, are derived from the epidermis. The stratum corneum of epidermis has undergone remarkable modifications giving rise to the hard keratinsed beaks, the scales, the claws, and the feathers.

1. Beaks, or the Bill: The beaks in pigeon are rather short, broad at base and narrower anteriorly. The upper beak-tip is slightly curved downward. This facilitates quick pecking of seeds and grains from the ground. Both the beaks are hardened by the keratinised coating called rhamphotheca. The mid-dorsal rhamphothecal ridge of the upper beak is called the **culmen,** while similar mid-ventral ridge is called the **gonys.** Chemically, the hard rhamphothecal coating is formed by the combination of phosphate compounds present in the stratum corneum, sufficient quantity of keratinised phospholipids, and calcium.

Functions: The beaks are primarily used as food-ingesting organs, and, alongside, these are used for nest-building and preening, i.e., spreading secretion of the oil gland all over the body feathers and for defence of the body.

2. Claws: In pigeon, each toe bears distally a downwardly curved claw which is also a hard keratinised structure derived from the epidermis. The development of claw is similar to that in reptiles.

Functions: Although claws are variously modified in structure and function in other birds, their main function in the pigeon are walking on the ground and perching on tree-branches.

3. Scales: In pigeons, the entire tarso-metatarsal region of the leg and the toes are not covered by feathers; instead, these parts are covered by hard, keratinised scales of reptilian size and structure.

The hard and sharp blade-like 'spur' that develops as a modification of scales on one side at the middle of the tarso-metatarsus in some galliform birds is not present in the pigeon.

4. Feathers: A detailed description about feathers in general has already been given in an earlier chapter (Chapter 18) of this book. Here only the structures and functions of feathers in *Columba livia* will be described briefly.

It has already been mentioned that the feathers are the unique evolutionary innovation of the epidermis in birds. The development of feathers has much similarity with that of reptilian scales (though not exactly the same).

Development of feathers: As to the origin of feathers, it is assumed that the feathers originated at some period of the Mesozoic era as an outstanding modification of the scales of some pro-avian group of reptiles.

Although the feather is epidermal in origin, its development is initiated by the dermis that pushes the epidermis outward in the form of elongated **dermal papilla** around which an annular groove and the **feather follicle** are formed. Within the follicle and the **periderm** that is formed around the thin cornified layer of epidermis, **barbs and barbules** as longitudinal ridges arise from the epidermis, the basal shaft forming the quill in a down feather. In a contour feather, the hollow quill extends as a solid axial shaft, the **rachis,** from which the branching system of barbs and barbules arise.

TYPES OF FEATHERS IN A PIGEON

Based on the shape, size and structure, **five** different types of feathers have been recognised, as described earlier in this book (Chapter 18) on the classification of feathers. In the pigeons, however, the following **four** types of feathers are present. The bristles are absent in the pigeons. The four types of feathers are: A. Contour feathers, including the flight feathers and coverts, B. Filoplume or hair feathers, C. Down feathers, including the nestling down and powder down, and D. Semiplume.

A. Contour feathers

The medium-sized and large-sized feathers

that offer general shape of the body of a pigeon are called the contour feathers. These contour feathers, based on their location and structure, have been named differently in different parts of the body. Hence, the large flight feathers of the wing and tail, known, respectively, as the remiges and rectrices, and the shorter coverts that overlie and protect the bases of the large flight feathers—are all modifications of the contour feathers.

In the pigeons, usually there are 23 remiges on each wing: **11 primaries** and **12 secondaries**. These are also known as wing quills. The tail quills or rectrices are 12, arranged semicircularly along the caudo-dorsal border of the uropygium. Among the primaries, 6 feathers attached to the metacarpal region are called **metacarpallaries** and 5 attached to the fingers are called **digitals**. As the secondaries are attached to the ulna of the forearm, these are called **ulnaris**. In a few other birds (not in the pigeon), a few feathers are also attached to the humerus. These are called **tertiaries**. A small but distinct bundle of feathers, attached to the first finger or the pollex, is called **'ala spuria'** or **'bastard wing'**. It acts as a slotting device, avoids turbulence and restores smooth flow of air.

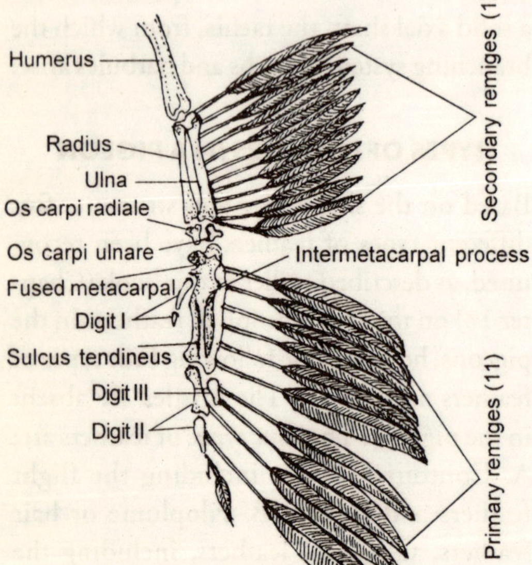

Fig. 30.2. Wing bones and remiges in a pigeon (*Columba* sp.)

A comparatively smaller type of modified contour feathers—called covert feathers—are arranged in a single row in pigeon along the bases of the remiges and rectrices at the upper and lower surfaces. Elsewhere, such wing and tail coverts may be arranged in more than one row at each of the upper and lower surfaces. The coverts close the gaps between the bases of the adjacent flight feathers, as also strengthen the bases of the flight feathers.

Just as remiges of the wing are the main organs in taking off and flight in birds, the rectrices act as a rudder in controlling direction during flight and in stalling, landing and perching.

Structure of a typical contour feather: In adult pigeon, a feather plucked from the body covering, wing or tail serves as an ideal specimen for the study of a contour feather.

The structural details of a contour feather in a pigeon are the same as described earlier under general description of the 'Types of feathers' (Chapter 18).

B. Filoplumes or Hair feathers

After plucking off the contour feathers from the body, some fine hair-like or needle-like feathers are found attached to the skin. These are called filoplumes.

The structure of the filoplume in a pigeon is the same as described earlier in this book (Chapter 18).

C. Down feathers

Extremely small, soft and whitish wooly feathers which provide natal covering on the body of the hatchling of a pigeon are called down feathers. These are also known as **'nestling down'**. The nestling down is replaced by contour feathers in the adult bird, but, beneath the contour feathers, the **'powder down'** exists in most birds.

The description of plumulae or down feathers, nestling down and powder down has already been provided under the heading 'Types of feathers' (Chapter 18).

D. Semiplumes

Most authors have recognised the semiplumes as halfway between the contour feathers and the down feathers. As in many other birds, these are also present in the pigeons.

The description of semiplumes has been given in an earlier chapter (Chapter 18) of this book under 'Types of feathers'.

Functions of feathers: The different functions of feathers in pigeons are almost the same as mentioned in Chapter 18.

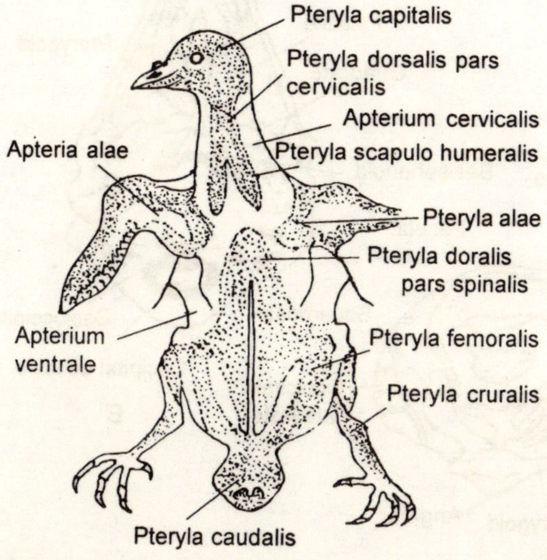

Fig. 30.3. Pterylae and Apteria in a Pigeon (Dorsal view)

30.9. PTERYLOSIS AND MOULTING IN PIGEONS

The arrangement of feathers in different parts of the body of a pigeon in definite feathered tracts (**Pterylae**) and the occurrence of featherless tracts (**Apteria**), as well as the progress of seasonal moulting in birds have already been described under the heads, **'Pterylosis'** Chapter 19 and 'Moulting' Chapter 7.

The nomenclature of the pterylae and apteria present in the pigeons remains the same.

30.10. ENDOSKELETON IN PIGEON

The description of the endoskeletal system in different birds, including subdivisions of the system, and structures and functions of different parts of the system has been provided in Chapter 21 of this book under the head, 'Endoskeleton and Muscles'.

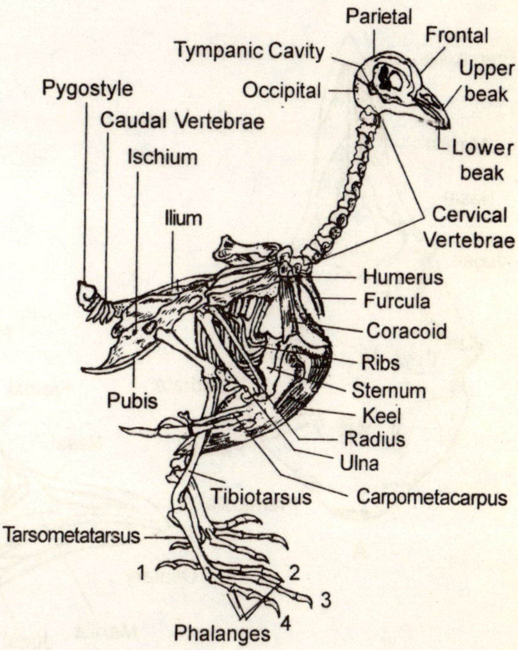

Fig. 30.4. Different parts of the skeleton (articulated) of a pigeon

Here only the endoskeleton of the Blue Rock Pigeon, *Columba livia,* is described.

The endoskeleton in pigeon is principally divided into two parts, — **A. Axial skeleton,** and, **B. Appendicular skeleton;** each again consists of several subdivisions.

A. Axial skeleton: The axial skeleton comprises the I. skull, II. the vertebral column, and III. the sternum and ribs. In pigeons, the endoskeleton is light, hollow and full of air-spaces. Frequent fusion of bones has rendered rigidity to the axial skeleton. Such morphological characteristics of the bones are directly concerned with volant adaption of the pigeons.

I. Skull

A few salient characteristics of the skull of a pigeon are:

i. The skull bones being pneumatic are very light and in the adult bird the sutures between the bones are absent.

ii. Both the jaws are toothless.

iii. The skull is joined to the vertebral column through a single occipital condyle.

iv. The vomer is absent and the maxillo palatines of both sides remain free; this type of skull is called **Schizognathous.**

v. The orbits are much larger compared to the size of the skull and the two orbits are separated by a thin interorbital septum.

vi. The mandibular articulation with the skull is **autostylic.**

The skull of a pigeon is subdivided into three heads: (a) Cranium, or the Brain case (b) Sense capsules, and (c) Visceral skeleton.

(a) Cranium, or the Brain case: The cranium is broad and rounded posteriorly and somewhat like a flat box anteriorly. Posteriorly, the large aperture of the cranium—the **Foramen magnum** through which the brain becomes continuous with the spinal cord—

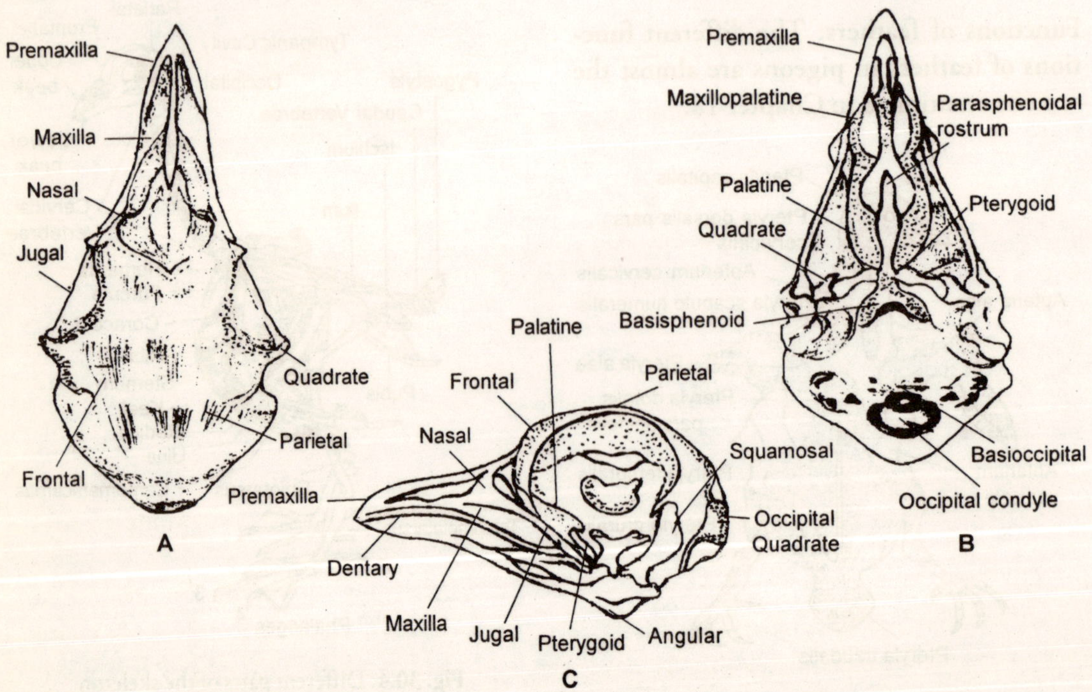

Fig. 30.5. Skull of a pigeon. A. Dorsal view; B. Ventral view; C. Lateral view

is bounded by the supraoccipital above, two exoccipitals, one on either side, and a single basioccipital below. The single occipital condyle is mainly formed by the basioccipital. Thus the cranium is roofed over from behind forward by a supraoccipital, a pair of parietals and a pair of frontal bones; its floor is formed by the basioccipital, basitemporal and basisphenoid bones, the last forming a rostrum in the pigeon (the vomer being absent).

In the nuchal region of the skull, dorsal to the foramen magnum the external median convexity of the supraoccipital and parietal bones is called '**Prominentia cerebellaris**'. A distinct arched crest separating the nuchal plane of the supraoccipital bone from the smoother part of the parietal bone is called the '**Crista nuchalis transversa**'. The sides of the skull are formed by the exoccipitals, squamosal, prootic and epiotic bones. The sides of the parietal cranium are formed by the squamosal and alisphenoid bones. Inwardly, the squamosal forms a small **Zygomatic process**. The sidewalls of the frontal region are mainly formed by the **Orbitosphenoid** bones.

(b) Sense capsules: Along with the development of the skull, three pairs of sense capsules—auditory capsules, optic capsules or orbits and nasal capsules—develop as part of the skull to protect the sensory organs lodged within.

i. **Auditory capsules:** On either of the occipital bone, an auditory or otic capsule is formed mainly by the prootic bone. Within this part the inner ear is protected all-round. Outwardly, the cup-like tympanic cavity of the middle ear is bounded by the squamosal above, basitemporal below and the sides are formed partly by the prootic and partly by the epiotic bones. A **tympanic membrane** (**Tympanum**) present within the tympanic cavity joins the ear ossicle of the middle ear called '**Columella**' (also called **Columella auris**). It is of hyoid arch origin and extends from the tympanic membrane to the vestibular window. It consists of a tripartite extra-columellar cartilage apposed to the tympanic or distal end, a middle rod-shaped bone and an expanded foot-plate proximally, occluding the vestibular window (NAA, 1993; Müller, 1963). The tympanic cavity communicates with the mouth cavity through an Eustachian Canal on either side. The movement of the columella is regulated by a muscle (M. tensor tympani; M. stapedius) attached to the extra-columella.

ii. **Optic capsule or the Orbit:** The orbits, one on either side of the skull housing each eyeball, are quite large cavities in the skull and are more or less round and deep. Medially, between the two orbits, lies a thin, vertical interorbital septum which is mainly formed by the mesethmoid bone and partly by the presphenoid, rostrum and orbitosphenoid bones. The frontal and lachrymal bones are located rostro-dorsally and dorso-ventrally, and a vertical plate of bone called Ectethmoid forms rostral wall of the orbit, separating it from the nasal cavity. At the caudo-dorsal aspect of the orbits, a part of the frontal and the alisphenoid bones form a downward pointed projection called the '**Postorbital process**' (**Pr. postorbitalis**). A tough broad band of '**Postorbital ligament**' (**POL**) attached dorso-ventrally by one end to the postorbital process and by the other end to the caudal mandible has a major role in the kinesis of the jaws.

iii. **Olfactory or Nasal Capsules:** The nasal capsule which encloses the olfactory organ is small in size. The cartilagenous 'Internasal septum' is a rostral extension of the interorbital septum and it divides the two nasal chambers medially. A pair of nasal bones on the roof and sides of the nasal chamber unite and form the nasal process which joins the frontal, lachrymal and the premaxilla. The **nasal conchae** occurring in the caudal part of the nasal cavity are cartilagenous in the pigeon.

(c) **Visceral skeleton:** The visceral skeleton comprises the bones of the upper and lower jaws and bones and cartilages of the hyoid apparatus.

i. **Upper jaw or the Premaxillo-maxillo-palatine apparatus:** The upper jaw bones are arranged in two series: an outer series comprising the Premaxilla, Maxilla, Jugal and quadratojugal, and an inner series comprising three bones—palatine, pterygoid and quadrate. Though both the series consist of separate bones, they are joined rostrally and caudally through their processes.

In the composition of the upper jaw, rostrally and laterally, a pair of Premaxillae (Ossa premaxillae) unite mid-dorsally to form a single triangular premaxilla. Caudally, the premaxilla with its three pairs of processes — **nasal processes, maxillary processes** and **palatine processes**—establish connection with the bones of the nasal cavity, upper jaw and palate. The maxilla of the upper jaw with its rostral process (Pr. maxillaris) and a caudal process (Pr. maxillopalatinus) form the sides of the palate. The rod-shaped slightly curved jugal bone (Arcus Jugalis) is connected to the maxilla in front and to the quadrato-jugal behind. The quadrato-jugal is a small rod-shaped structure, thicker and broader caudally, and narrower rostrally. This is the terminal segment of the outer series of bones of the upper jaw.

The quadrate is a stout, broad, tri-radiate bone whose cranial end is immovably articulated to the squamosal and otic process (**Pr. squamosus and Pr. oticus**) of the skull and the broad pillar-like foot is articulated (slight movably) by its lateral and median condyles (**condylus lateralis** and **condylus medialis**) to the articular mandible, whereas the rostral **orbital process of the quadrate** is obliquely directed forward and free to move by muscular action. The rod-shaped pterygoid bone is obliquely disposed in front of the quadrate to which it is movably articulated at one end and rostrally to the caudal end of the palatine.

The modern concept of **jaw kinesis** has been discussed in some details in an earlier chapter (Chapter 21) of this book. However, it may be mentioned here that according to de Beer (1954), Fisher (1955), Simonetta (1960), Manger Cats-Kuenen (1961), Zusi (1962), Bock (1964, 1974), and Bühler (1981), the upper jaw-cranium, the quadrate-pterygoid-palatine-maxillary complex, the jugal-quadrato-jugal and the lower jaw have all formed a kinetic relationship with one another in such a way that the upper jaw kinesis and the opening and closure of the two beaks depend on the synergistic force produced by these components.

ii. **Lower jaw (Ossa mandibulae):** The mandible consists of two slender mandibular rami which, in the pigeon, are slender rod-shaped bony components, slightly curved inwardly, broader posteriorly, and narrower anteriorly. Apically at the beak-tip the two rami are united by a pair of small bones called '**Mentomandibular**' (Romanoff, 1960; In: NAA, 1993) and the place of union is called '**Mandibular symphysis**'. The mandible appears 'V'-shaped and each rames consists of six bones of which the small coronoid bone (**Os coronoideum**) is usually absent in birds, but a small coronoid process (**Pr. coronoideus**) is present in most birds (also in the pigeon) at the dorsal edge of the caudal mandible. The strong, flat and tough aponeurosis of M. adductor mandibulae externus superficialis is attached to this coronoid process.

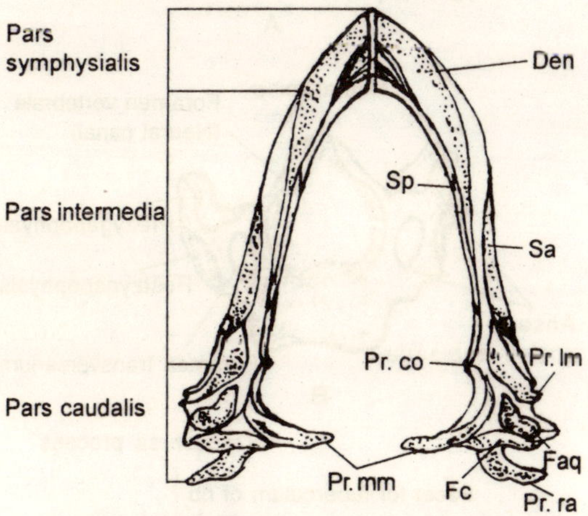

Fig. 30.6. Dorsal view of the mandible (Ossa mandibulae) in pigeon. Den, Dentary; Fc, Fossa caudalis; Faq, Fossa articularis quadratica; Pr. co, Processus coronoideus; Pr. lm, Processus mandibulae lateralis; Pr. mm, Processus mandibulae medialis; Pr. ra, Processus retroarticularis; Sa, Os supraangulare; Sp, Os spleniale

The other five bones of each mandibular ramus—Dentary, Articular, Angular, Suprangular (Surangular) and Splenial—have been discussed in Chapter 21 of this book.

iii. Hyoid apparatus: Ventral to the buccal cavity and lying posterior to the tongue,

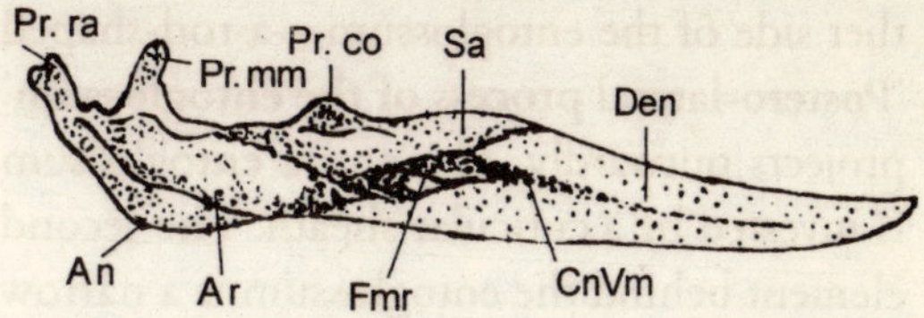

Fig. 30.7. Lateral view of the right ramus of mandible in pigeon. An, Os angulare; Ar, Os articulare; CnVm, Canalis neurovascularis mandibulae; Fmr, Fenestra mandibulae rostralis; Pr. co, Processus coronideus; Pr. mm, Processus mandibulae; Pr. ra, Processus retroarticularis

a few bones and cartilages combine to form the hyoid apparatus which supports the tongue and controls its movement.

The hyoid apparatus in pigeon consists of three median, unpaired bony elements and two lateral, paired and rod-shaped bony elements. Both the median and lateral elements have portions of cartilages. The tongue in pigeon consists of an unpaired, arrow-head-like **entoglossum** which has been formed by the complete fusion of a pair of paraglossalia present in other birds. The entoglossum is rostrally cartilagenous and caudally—from ei-

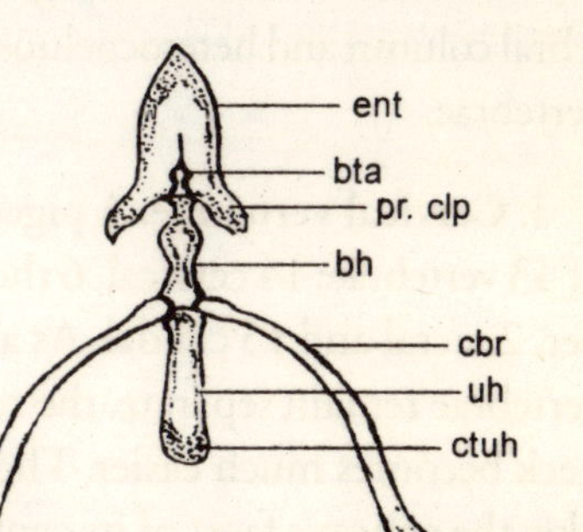

Fig. 30.8. Hyoid apparatus in pigeon (enlarged ventral view). bh, basihyale; bta, basihyale-tongue articulation; cbr, ceratobranchiale; ctuh, cartilagenous tip of urohyale; ent (pgl), entoglossum (Paraglossum); ebr, epibranchiale; pbr, pharyngobranchiale; pr. clp, caudo lateral process of paraglossum

ther side of the entoglossum—a rod-shaped '**Postero-lateral process of the entoglossum**' projects outwardly. The entire entoglossum is invested by a cuticular sheath. The second element behind the entoglossum is a narrow rod-shaped laterally compressed bone called the '**Basihyale**' (or Basibranchial), and the last caudal element is an elongated flat '**Urohyal**' whose terminal end of about 0.8 in (2mm) long is cartilagenous.

On either side of the basihyal copula are articulated a long bony '**ceratobranchial**', a comparatively shorter '**Epibranchial I**' and a terminal small cartilagenous '**Epibranchial II**'. These lateral branchial arches represent '**Posterior Cornu**' of the hyoid. These two cornua appear as an inverted 'V'.

II. Vertebral Column

The vertebral column of a pigeon has several specialities such as increase in the number of cervical vertebrae due to the presence of a long flexible neck which can be rotated freely, fusion of vertebrae in certain regions of the vertebral column and heterocoelous nature of the vertebrae.

i. Cervical vertebrae: A pigeon has a total of 43 vertebrae: 14 cervical, 6 thoracic, 6 lumber, 2 sacral and 15 caudal. As all the cervical vertebrae remain separate, the rotation of the neck becomes much easier. The atlas is ring-like; the concave facet of its centrum fits into the occipital condyle of the skull. All the cervical vertebrae have their rudimentary ribs fused to their respective short transverse processes forming '**Ansa costotransversaria**'. These serve for the attachment of tendons of lateral cervical muscles. The axis or the second cervical vertebra has an '**Odontoid process**' rostrally for fitting into the atlas, a pair of **pre-zygapophyses** and a pair of **post-zygapophyses**, and a **hypapophysis.** All the cervical vertebrae have each a pair of '**Proc. costalis**' laterally and a pair of "**Foramina transversaria**' (Transverse foramen), one on either side, at the base of the transverse process. The transverse foramen on either side form a continuous canal known as '**Canalis vertebrarterialis**' through which pass the vertebral artery and vein.

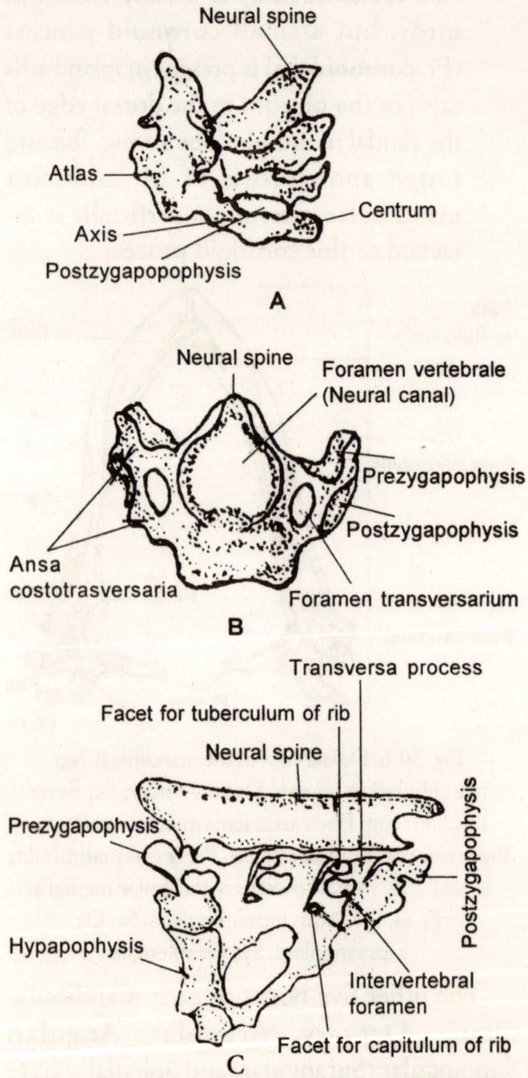

Fig. 30.9. Vertebrae of a pigeon. A. Lateral view of the atlas and axis (articulated); B. Anterior view of a cervical vertebra; C. Lateral view of a thoracic vertebra

The atlas is apneumatic (without air-space) in all the birds. The hypophyses provide origin-sites for the flexor muscles of the neck.

ii. Thoracic vertebrae: Details have been given in the earlier chapter under the head 'Endoskeleton and Muscles' (Chapter 21).

In pigeon, there are 6 thoracic vertebrae of which the first one is free and the last fused (may be free in some birds) with the synsacrum. The second to fifth thoracic vertebrae are fused to form the single, compact structure—the '**Notarium**'.

Each thoracic vertebra is characterised by a pair of double-headed vertebral ribs that, directly or indirectly, join the sternal ribs. The ventrally flat centrum forms the hypapophysis, a pair of pre- and a pair of postzygapophyses and the laterally-compressed neural spines of fused vertebrae forming a vertical blade-like structure.

The two heads of each vertebral rib, the '**tuberculum**' and the '**capitulum**' are joined to the base of the transverse process and to the centrum respectively. An **uncinate process** (**Proc. uncinatum**) arising from the caudal border of each vertebral rib joins the rib in front.

iii. Synsacrum: In pigeon, the last thoracic, 6 lumbar, 2 sacral and the proximal 5 caudal vertebrae fuse to form a rigid compact

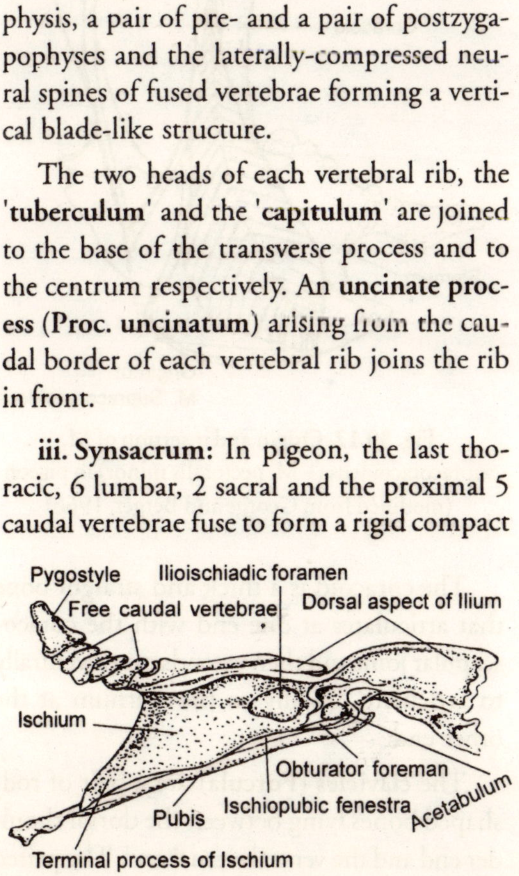

Fig. 30.10. Synsacrum, pelvis and caudal vertebrae (Lateral view)

and somewhat triangular bony plate known as the **synsacrum**. It bears fenestrae and smaller foramina traversed by nerves and blood vessels. The number of vertebrae forming the synsacrum varies in different birds. The synsacrum is ankylosed to the hip-bone (**Os coxae**) by synostosis.

iv. Caudal vertebrae: In pigeon, the intermediate series of 6 caudal vertebrae are free, amphicoelous and movable. Each of them has a pair of short transverse processes.

The last segment of vertebral column in the pigeon is a compound upturned fused element called the '**pygostyle**' (**Os pygostylus**) or the ploughshare bone, consisting of 4 terminal caudal vertebrae fused (may be 5 or 6 in others).

III. Sternum and Ribs

The sternum—breast-bone—with the characteristic vertical keel or carina is a distinctive feature in carinate birds. In pigeon, a broad flat and rigid plate, somewhat boat-shaped in appearance, is found along the ventral thoracic region. It has a concave inner and a convex outer surface. A vertical ridge—called **keel** or **carina**—arises along the median convexity of the sternum. The keel serves for an extensive origin site of the largest flight muscle, **M. pectoralis major**.

In the nomenclature of the parts of sternum used in NAA, 1993, a reference has been made to Fürbringer (1888) who called the cranial part of the sternum '**costosternum**' to which the ribs are attached and the caudal part as '**Xiphosternum**' (= **metasternum**).

The details of the parts of sternum have been described earlier under 'Sternum and ribs' in Chapter 21 on 'Endoskeleton'.

Of the seven pairs of thoracic ribs in pigeon, the first two pairs are joined to the cervical vertebrae, while the remaining five pairs are truly thoracic ribs, each consisting of two elements, vertebral rib and sternal rib–both fully ossified. The sternal ribs are articulated to distinct notches of the sternum at '**Margo costalis sterni**'. The double-headed vertebral ribs are attached to the vertebrae.

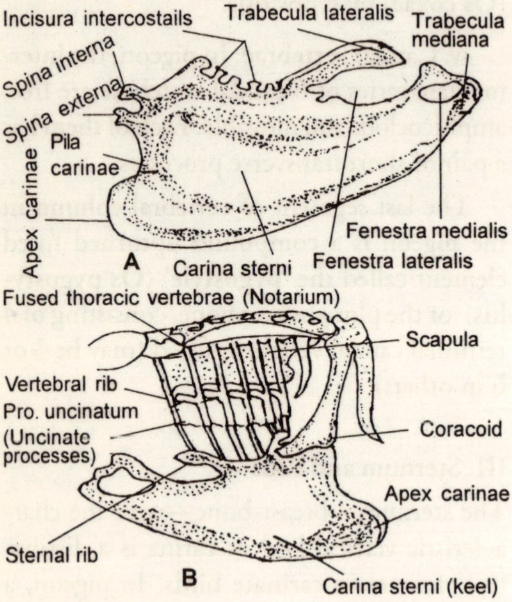

Fig. 30.11. Sternum and ribs of a pigeon. A. Left Lateral view of the sternum and B.vertebral column with the ribs attached

B. Appendicular skeleton: The pectoral girdle and the forelimb bones, and the pelvic girdle and hindlimbs bones are described as appendicular skeleton.

1. Pectoral girdle and wing skeleton: For a detailed description of the appendicular skeleton in general, Chapter 21 on 'Endoskeleton' may be referred to.

In pigeon, the pectoral girdle is a bilateral organisation of bones comprising on either side a **sabre-shaped scapula**, a broad, **pillar-like coracoid** and a **'V'-shaped clavicle** called **Furcula or Merrythought bone.** In all carinate birds, the coraco-scapular angle is less than 90° (greater than 90° in ratite birds). The main body of the sub-cylindrical and then flat sword-shaped scapula has a pointed caudal end, and on one side of the knob-like cranial end lies the pointed acromion. The glenoid process of the scapula fits into the cup-shaped depression of the coracoid, both thus forming the **coraco-scapular joint.**

Dorsally, the scapula, coracoid and the clavicle at their junction form a **'Foramen triosseum'** (Canalis triosseus; NAA, 1993). The powerful tendon of **M. supracoracoideus (= M. pectoralis minor)** passes through this foramen.

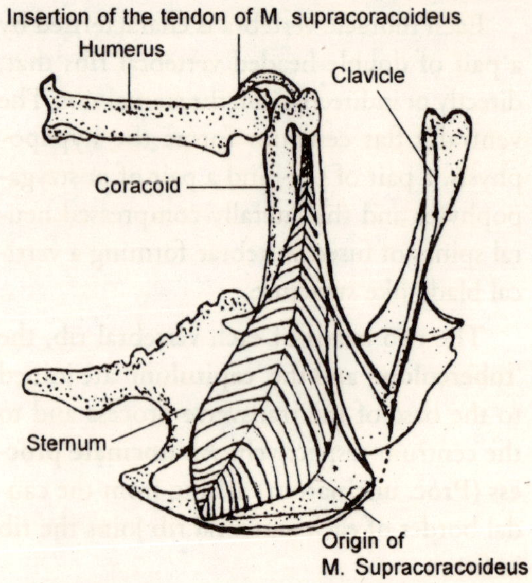

Fig. 30.12. Origin and insertion of M. supracoracoideus (= M. pectoralis minor) in pigeon (modified from George and Berger, 1966)

The coracoid is a thick and straight bone that articulates at one end with the coraco-scapular joint and then extends caudoventrally to articulate broadly to the sternum at the other end.

The **clavicles (Furcula)** are a pair of rod-shaped bones lying between the dorsal shoulder end and the ventral sternal end. The paired clavicular rami ankylose at their ventral ends and thus form the '**V'-shaped** furcula.

The wing skeleton or the forelimb bones in pigeon comprise the **humerus, radius, ulna, two carpals, carpo-metacarpus** and **three digits**. The head of the humerus articulates in the shallow **glenoid cavity**. The scapulo-

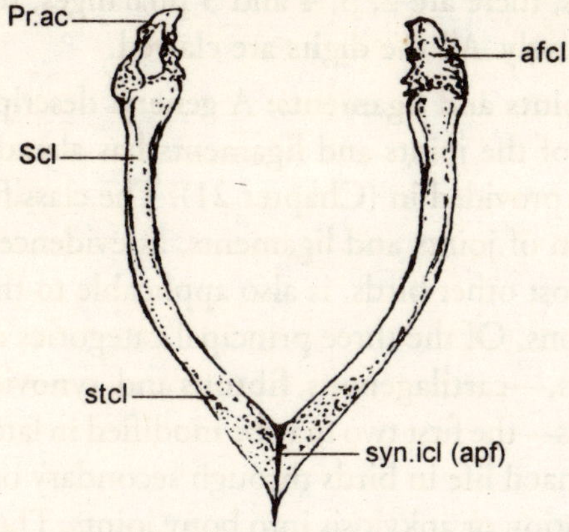

Fig. 30.13. Caudal view of the clavicula in pigeon. afcl, articular facet of clavicula; pr. ac, acromial process; Scl, scapus claviculae; Stcl, sternal end of claviculae; Syn. icl, synostosis interclaviculae (apophysis furculae)

coraco-humeral articulation is a ball-and-socket joint which admits free rotation of the humerus on either side. The humerus is a stout and straight bone whose head articulates in the shallow glenoid cavity. The articular surface at the head of humerus is called **'Caput humerus'**—it has a dorsal tubercle. The tendon of M.supracoracoideus is attached to it. On the caudal surface of the head of humerus lies the **'pneumatic foramen'** within a depression called **'pneumatic fossa'**.

The ulna is a thicker and longer bone than the radius. The pointed process at the proximal end of the ulna is called the **Olecranon process.** Proximally, the ulna bears two depressions at which the dorsal and ventral condyles of the humerus articulate. Distally, the ulna bears two condyles with a groove between them known as **'Trochlea carpalis'**.

In adult pigeons, of the four carpal bones, two named **'radiale'** and **'ulnare'**—persist, while the other two fuse with the compound bone **carpometacarpus.** The pigeon has three metacarpal bones and three digits. The metacarpals are fused at both proximal and distal ends. Of the three digits, digit I (Pollex or allular digit) and digit III each bear a single phalanx, while digit II bears two phalanges. Digital primaries (feathers) are attached to the dorsal surface of the carpometacarpus.

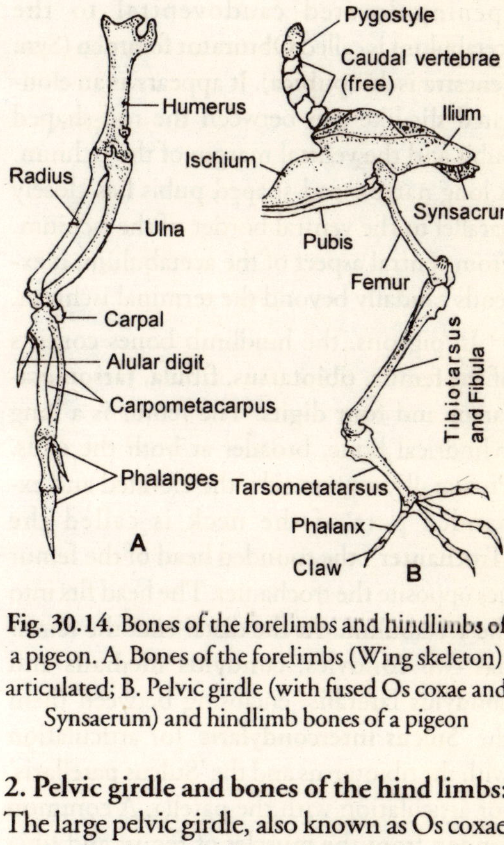

Fig. 30.14. Bones of the forelimbs and hindlimbs of a pigeon. A. Bones of the forelimbs (Wing skeleton) articulated; B. Pelvic girdle (with fused Os coxae and Synsaerum) and hindlimb bones of a pigeon

2. Pelvic girdle and bones of the hind limbs: The large pelvic girdle, also known as Os coxae or the hip bone, is formed on either side by three bones—the **Ilium, Ischium** and the **Pubis**. Each half of the pelvic girdle, on either side, is called **Os innominatum**, and the pelvic girdle is formed by joining of these two halves with the synsacrum. The ilium is a strongly developed bony plate which lies dorsal to the acetabulum and by its dorsal and dorso-lateral cristae remains connected to most of the synsacrum. The ischium, also a broad plate, lies caudoventral to the ilium.

Both the ilium and ischium articulate at the **acetabulum** which serves as the facet for articulation of the head of the femur. The major part of the ischium lies caudal and ventral to the large oval aperture called '**Foramen ilioischiadicum**'. The foramen, bounded dorsally by the ilium and ventrally by the ischium, serves as the passage for ischiadic nerves and blood vessels. An elongated oval opening located caudoventral to the acetabulum is called **Obturator foramen** (Syn: **Fenestra ischiopubica**). It appears as an elongated slit-like gap between the rod-shaped pubis and the ventral margin of the ischium. A long narrow rod-shaped pubis lies closely parallel to the ventral border of the ischium. From ventral aspect of the acetabulum, it extends caudally beyond the terminal ischium.

In pigeons, the hindlimb bones consists of the **femur, tibiotarsus, fibula, tarsometatarsus** and **four digits**. The femur is a long cylindrical bone, broader at both the ends. Proximally, on one side, the elevated and expanded part of the neck is called the '**Trochanter**'; the rounded head of the femur lies opposite the trochanter. The head fits into the acetabulum. At the distal end, the femur has two condyles, **condylus medialis** and **condylus lateralis**, enclosing between them the '**Sulcus intercondylaris**' for articulation with the tibiotarsus and the '**Sulcus patellaris**' for articulation with the patella. A common tendon from the muscles of femur and tibia gives rise to a sesamoid bone that forms the patella.

The tibiotarsus is straight, stout, longer and more slender than the femur. Along proximal one-third of this bone there is a distinct '**cnemial crest**'. The fibula is a very small narrow and reduced bone attached along the outer side of the tibiotarsus. The tarsometatarsus bone is formed by the union of the outer tarsal bones with the second, third and fourth metatarsal bones. This bone is shorter than the tibiotarsus and fibula. Of the four digits, the first one is the hallux which is directed backward and the other digits are directed forward. From the first to the fourth digits, there are 2, 3, 4 and 5 phalanges, respectively. All the digits are clawed.

Joints and ligaments: A general description of the joints and ligaments has already been provided in (Chapter 21). The classification of joints and ligaments, as evidenced in most other birds, is also applicable to the pigeons. Of the three principal categories of joints,—**cartilagenous, fibrous** and **synovial joints**—the first two may be modified in later postnatal life in birds through secondary ossification or ankylosis into bony joints. They may, however, persist in many birds as cartilagenous and fibrous joints. The synovial joints in adult birds are truly bony joints forming '**Synostoses**'.

The special branch of study dealing with the nature, location, structure and function of the joints and ligaments is designated as '**Arthrology**' which is closely interlinked with the study of '**Osteology**' and '**Myology**'.

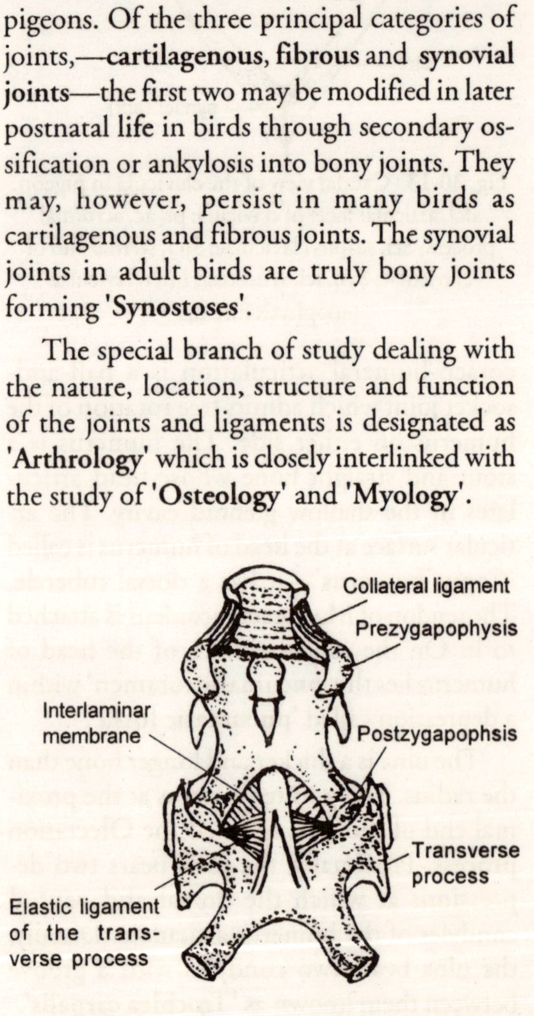

Fig. 30.15. Dorsal view of a small segment of the vertebral column of a pigeon showing a few ligaments and joints (based on J. J. Baumel, In: NAA, 1993)

As in most other birds, the joints and ligaments of the pigeons are well-developed and clearly definable.

30.11. FLIGHT MUSCLES AND PERCHING MECHANISM IN PIGEON

The wing movements are extremely significant in the flight of birds. Like other flying birds, the pigeons also possess a well-developed and complex system of muscles. The muscles that are directly or indirectly concerned with the flight of a pigeon are called the flight muscles.

Mention may be made here that a muscle is composed of a large number of contractile myofibrils. A skeletal muscle has the characteristic that it must have an **origin-end** wherefrom the muscle arises, and an **insertion-end** where the muscular action becomes effective. The two ends of a muscle are either directly attached to bones (in which case these are called fleshy attachments) or these two ends may be attached differently in different species, such as a bone at one end, and, at the other end, a stiff membrane to which are attached the fine thread-like, plate-like or flat aponeuroses or a thick tendon or a flat ligament which are connective tissue elements.

However, a few muscles may be attached at both the ends through such connective tissue elements. To differentiate between the origin and insertion of a skeletal muscle, a simple definition may be recalled that by muscular contraction, the end of the muscle which is nearly static or less mobile is the origin and the end that causes greater movement of the bone is the insertion of the muscle.

The nomenclature of avian muscles has been a complex subject due to frequent changes in the naming of muscles during the last 100 years, based on different criteria. However, the naming of the muscles is commonly based on their functions (e.g., M.depressor mandibulae), origin and insertion with respect to skeletal parts (e.g., M. scapulohumeralis), location with respect to an organ or skeletal part (e.g., M. supracoracoideus), or on the shape and size of the muscle (e.g., M. coracobrachialis longus).

A. **Flight muscles:** The flight muscles of pigeon are described in a tabular form:

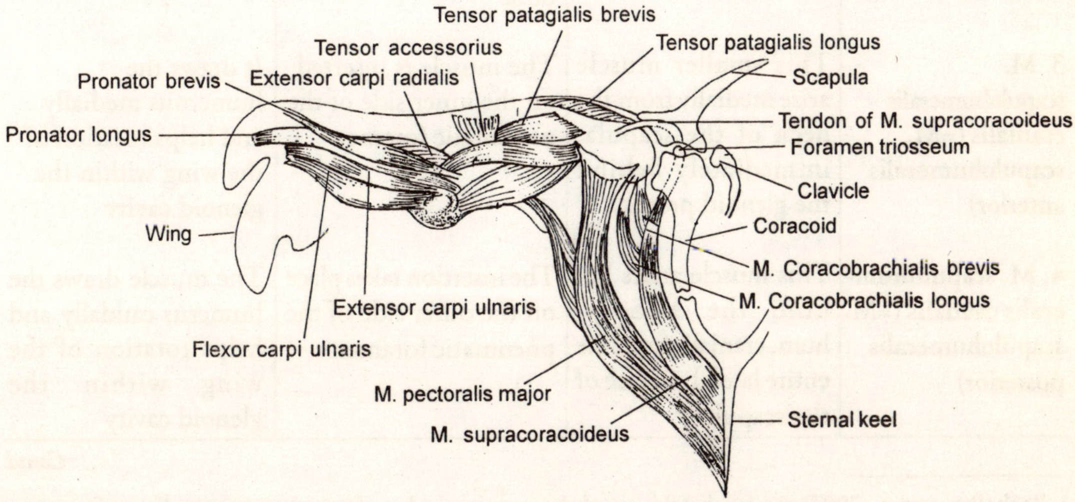

Fig. 30.16. Some important flight muscles of a pigeon (Ventral view)

Table 30.1 Names, Origins, Insertions and Functions of the Flight Muscles of Pigeon

Muscles	Origins	Insertions	Functions
1. M. pectoralis major (largest among the flight muscles)	This muscle makes up about 1/5th of the total volume of muscles of the body. The muscle arises from more than ventral half of the keeled sternum and from the clavicle	The muscle inserts by a very broad and flat tendon on the lateral and ventral aspect of the head of humerus[1]	This is the largest depressor muscle of the wing
2. M. pectoralis minor (= M. Supracoracoideus)	This muscle arises inner and dorsal to M. pectoralis major from the ventral aspect of the sternum	The long, stiff tendon of the muscle runs through rostral sternum, passes through the foramen triosseum (= Canalis triosseus) to reach the dorsal aspect of the head of humerus and inserts there on the dorsal tuberculum. In some birds, the canal is formed only by two bones—the coracoid and the scapula. In pigeons, it is formed entirely by the coracoid bone	It is the main elevator muscle of the wing
3. M. scapulohumeralis cranialis (=M. scapulohumeralis anterior)	This smaller muscle arise medially from the neck of the scapula, immediately behind the glenoid process	The muscle is inserted on the inner side of the pneumatic foramen	It draws the humerous medially and helps rotation of the wing within the glenoid cavity
4. M. scapulohumeralis caudalis (=M. scapulohumeralis posterior)	This muscle arises behind the M.Scap. hum. cranialis from the entire lateral surface of the scapula	The insertion takes place on the outer side of the pneumatic foramen	The muscle draws the humerus caudally and helps rotation of the wing within the glenoid cavity

Contd.

1. Bock (Pers.comm., 2007) reports that this muscle has a major tendon of insertion on its medial surface.

Muscles	Origins	Insertions	Functions
5. M. scapulohumeralis profundus	It arises deeper from the neck of the scapula	The muscle inserts on the frontal aspect of the distal humerus	The muscle draws the humerus medially and caudally towards the scapula
6. M. tensor patagialis longus (= M. tensor propatagialis; M. deltoideus, pars propatagialis, M. propatagialis) In recent nomen clature, this muscle has been considered as a sub division of the deltoid complex of muscles	The muscle arises from M. pectoralis major and extends along the outer border of the prepatagium	The muscle inserts on the patagial membrane on the frontal aspect of the head of humerus	This muscle completely extends the prepatagial membrane
7. M. tensor patagialis brevis (M. deltoideus, pars propatagialis brevis)	This muscle arises from the distal end of the scapula above the M.deltoideus major	The insertion takes place on the distal end of the humerus by the side of the elbow through a tendon	The muscle lifts the humerus and extends the prepatagial membrane
8. M. tensor accessorius patagialis	It arises from the connective tissue membrane of the M.biceps brachii	The muscle inserts broadly around the middle of the patagial membrane	This muscle too helps in the extension of the prepatagial membrane
9. M. coracobrachialis longus (=M.coracobrachialis cranialis)	The triangular muscle arises from the acrocoracoid process of the coracoid and lies narrowly below the M. pectoralis major	The narrow tendon of this muscle is inserted on a distinct 'impressio' at the intertubercular plane (**Planum intertuberculare**) of the proximal end of the humerus	This muscle assists the M. supracoracoideus in depressing the wing

Contd.

Muscles	Origins	Insertions	Functions
10. M. coracobrachialis brevis (= **M. coracobraclialis caudalis**)	The muscle arises from the inner side and attaches to the shaft of the coracoid	The insertion takes place at the proximal end of the humerus ventrally on the distinct '**Tuberculum ventrale**'	This muscle too, smaller in size than the 'longus' part, helps in depressing the wing
11. M. deltoideus major (= **M. deltoideus, pars major**) Brown et al (1989) distinguishes M. deltoideus as a complex which consists of three parts: M.deltoideus, pars propatagialis; M. deltoideus, pars major, and M. deltoideus, pars minor	The muscle arises from the rostro-dorsal border of the scapula	The muscle inserts on the Crista deltopectoralis (Crista deltoidea) of the humerus, opposite to the attachment of M. pectoralis major	The muscle lifts the humerus upward and causes full extension of the prepatagium
12. M. deltoideus minor (=**M. deltoideus, pars minor;** M. coracobrachialis anterior	The muscle arises by two slips: the dorsal slip from the head of scapula and the ventral slip from the sternocoracoclavicular membrane	Both the slips insert more or less in common on the proximal end of crista deltopectoralis of the humerus	It assists in the function of M.deltoideus major
13. M. triceps brachii. This large muscle complex exists in the form of three separate muscles — **M. scapulotriceps, M. humerotriceps, M. coracotriceps**	M. scapulotriceps arises from the scapula, superficial to M. scapulohumeralis cranialis and also from the humerus; M. humerotriceps arises from proximal humerus either by a	M. scapulotriceps inserts on the dorsal aspect of the Proc. cotylaris dorsalis of the ulna	The contraction of all the three muscles of M. triceps brachii complex extends the forearm. M.coracotriceps is not present in a few birds, but, in *Columba*, this muscle acts as a sensitive mechanoreceptor

Contd.

Muscles	Origins	Insertions	Functions
	single or two slips; M.coracotriceps (also known as M. **anconeus coracoideus;** Berger 1966) arises from the coracoid	M. humerotriceps inserts on the olecranon process of the ulna; The muscle is inserted on the proximal ulna	(NAA, 1993)
14. M. biceps brachii. This muscle corresponds to the same in mammals	In avian biceps the scapular head is absent; the muscle arises by two heads: from the coracoid and from the humerus. The humeral head that arises from a crista of the bicipital groove is larger; the smaller head arises from the coracoid	The two bellies and two tendons of insertion of the muscle are attached to a tubercle on the inner side of the radius and unla	This muscle acts as a powerful flexor of the forearm, folding it towards the arm
15. M. expansor secondariorum (Syn: M. dermoulnaris)	This non-striated muscle consists of a proximal tendon, a middle fan-shaped belly and a distal tendon. The proximal tendon of origin is attached to the fascia of M. scapulohumeralis (or directly to the scapula and coracoid)	At insertion, the distal tendon is attached to ventral epicondyle of the humerus and directly on the follicles of five secondary remiges (ie., 2nd to 6th) in the region of elbow (Berger, 1966; Vanden Berge, 1970)	The muscle extends the secondary remiges

In addition to the above-mentioned flight muscles of pigeon, other muscles are attached—either to the radius or the ulna or the carpo-metacorpus and digits—mainly as **flexors** and **extensors** (e.g., **M. flexor carpi ulnaris; M. flexor digitorum superficialis et profundus; M. extensor carpi radialis; M. extensor carpi ulnaris**, etc.). These also assist in the flight of a pigeon.

B. Perching Mechanism in Pigeon : Very often we see the pigeons and many other birds alighting and changing positions on tree branches, standing with a grip on a long wire or a rod with perfect balance. After flying for some time, the bird comes to the perch, often jumps and moves from one tree branch to the other, and then takes rest by gripping the support underneath with the help of toes adapted for the purpose. This kind of movement is called perching of the bird, and the process by which this adaptation has been achieved is called 'perching mechanism'.

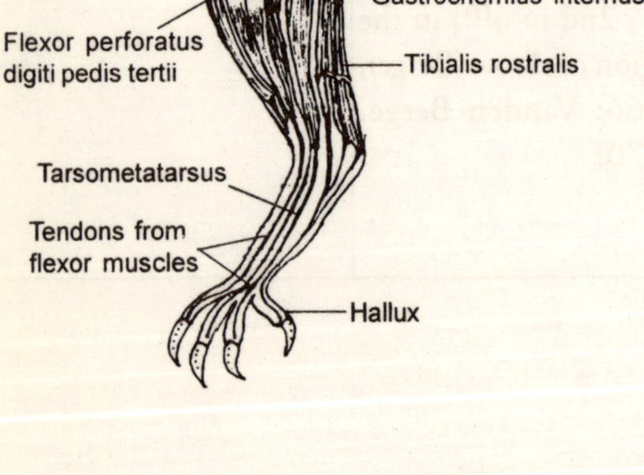

Fig. 30.17. Medial view of the right side of pigeon showing certain leg muscles and tendons used in perching mechanism

The handlimb muscles of the pigeons are quite well-developed. These muscles are used both for perching and walking on the ground. The muscles concerned with the perching mechanism in pigeons are mainly the flexors and the extensors.

These muscles are the same in the pigeons—as described in an earlier chapter (Chapter 20)—in general in the passerine and other birds under the head 'Modifications of Feet and Toes in Birds'.

M. ambiens is a small spindle-shaped muscle that arises from the preacetabular tuberculum of the ilium and inserts on the proximal aponeurosis of the digital flexor muscles. The distal tendon quite often perforates the patellar ligament before insertion.

30.12. DIGESTIVE SYSTEM

The digestive system of a pigeon mainly consists of an alimentary canal and its associated digestive glands. The digestive system in the pigeon has a few special characteristics due to its perfect volant adaptation and the habit of pecking cereals and pulses from the ground as food. The shape, size and form of the bill are directly correlated with the type of food and feeding habit. The high rate of metabolism in the pigeons, as in many other bird species, demands a constant feeding process unlike in other vertebrates.

The diversity in the alimentary adaptation in birds has reached great perfection. The pigeon is an ideal example of such alimentary adaptation among granivorous birds. In pigeon, the feeding apparatus which comprises the 'jaw apparatus, the tongue apparatus and the glottal apparatus, the salivary glands, the crop, the structure of the stomach and intestine, the digestive glands and the cloaca have all been suitably modified with the ultimate goal of a thorough and complete digestion.

After the description of the alimentary organs, the mechanism of feeding, digestion and absorption has been described in brief.

A. Alimentary canal

Although the alimentary canal extends from the mouth aperture to the cloacal aperture, it has several sub-divisions such as the feeding or ingesting organs, the digestive organs, the absorptive organs and the egesting organs. A short description of the organs is given:

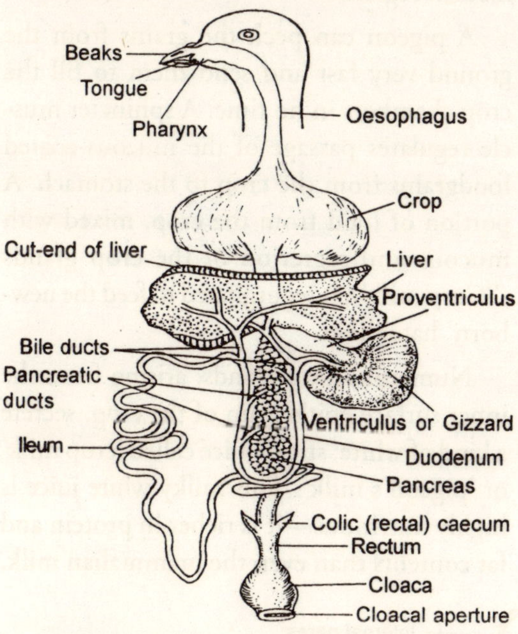

Fig. 30.18. Diagrammatic view of the digestive organs of a pigeon

(a) Feeding or ingesting organs: The feeding apparatus comprises certain essential parts of the organs of alimentation and is concerned with the selection—collection—orientation and transportation of food through the gullet to the digestive organs.

The feeding apparatus is formed of these structures:

i. Beaks: In pigeons, the toothless and elongated jaw bones invested with thick epidermal rhamphothecae form the upper and lower beaks. The upper jaw is schizorhinal (i.e., with an elongated external nare on either side) and rhynchokinetic (i.e., the upper jaw can be raised as a whole at the nasal-frontal Flexon zone). The upper jaw kinesis is reasonably well-developed in the pigeons, as they can select and peck the foodgrains from the ground in quick succession.

ii. Mouth: The widely open aperture bounded by the two beaks forms the mouth.
In fact, the rostral portion of the large cavity within the beaks constitutes the mouth which is also the beginning of the alimentary canal.
iii. Buccal cavity: Behind the mouth aperture, the spacious cavity that extends up to the pharynx is called the buccal cavity. The buccal cavity is lined by stratified squamous epithelial cells that are keratinised. In most birds, the buccal cavity and the pharyngeal cavity join to form the '**oropharynx**'. The rostral part of the roof of this cavity is dry, hard and devoid of mucous glands, the caudal part is somewhat softer.

Numerous horizontal folds called '**rugae palatinae**' are found over the surface of the entire palate. The narrow ducts from the mandibular, palatal and lingual salivary glands open into the cavity of the oropharynx. The internal narial apertures located at the mid-ventral palate communicate with the external nares through the median choanal slit of the palate.

In pigeon, a narrowly triangular tongue called **entoglossum** is situated on the floor of the buccal cavity. In fact, the paired paraglossal elements of the tongue have fused in the pigeon to form the single entoglossum. The caudal end of the tongue is slightly broader and it bears on either side a short and gradually tapering **posterior paraglossal process** (Processus paraglossalis caudalis). The folds of the oropharynx to which the base of the tongue is attached are called '**lingual frenulae**'. Posterior to the tongue, and on either side of the **glottis**, are found quite a good number of pores from the taste buds and mucous glands.

iv. Pharynx: A comparatively smaller part of the alimentary canal lying posterior to the buccal cavity is called the pharynx. It is also called the oropharynx. In this region, behind the internal narial apertures, the two eustachian canals unite and open by a single **eustachian aperture.**

v. Oesophagus: The pharynx leads through a food-intake aperture or the gullet to a long cylindrical and thick-walled canal called oesophagus. It extends along the entire neck region dorsal and parallel to the trachea and, before reaching the trunk, it gives off a thin-walled pouch called **crop** that lies bilaterally extended on either side of the oesophagus.

In pigeon, the length and diameter of the oesophagus is larger compared to the body size. The wall of the oesophagus is formed of rigid keratinised stratified squamous epithelial tissue. In pigeons, the longitudinal muscles of the mucous layer are only limited to the thoracic part of the oesophagus (Landolt, 1987).

vi. Crop: The oesophagus, near its caudal end, gives rise to a thin-walled and spacious sac-like structure—the crop. It serves as a temporary storage chamber for the cereals, pulses, seeds, grains and fruits that the pigeon takes in as foodgrains and transport them to the crop. The crop is commonly present in birds belonging to the Orders Columbiformes (e.g., pigeons, doves), Galliformes (e.g., Fowls), Psittaciformes (e.g., Parrots and Parakeets), etc. The crop is a single sac that lies in two main parts, one on either side of the oesophagus, and between the two there is a short median region.

A pigeon can peck the grains from the ground very fast and send them to fill the crop chambers in no time. A sphincter muscle regulates passage of the mucous-coated foodgrains from the crop to the stomach. A portion of food from the crop, mixed with mucous and secretion of the crop glands ("**Crop milk**'), is regurgitated to feed the newborn hatchlings.

Numerous crop glands, arising from the inner surface epithelium of the crop, secrete a kind of white, sticky juice called 'crop milk' or 'pigeon's milk'. This milky white juice is highly nutritious—it is richer in protein and fat contents than even the mammalian milk.

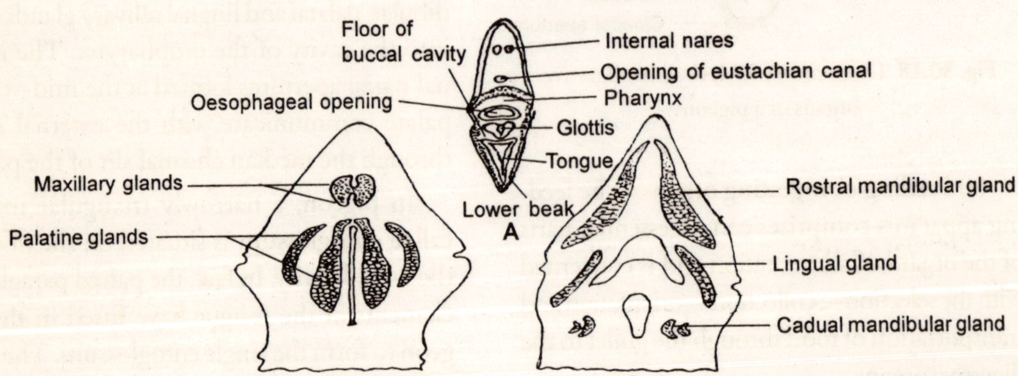

Fig. 30.19. Location of structures within the buccal cavity and pharynx of a pigeon. A. Interior view of the buccal cavity; B. Location of maxillary and palatine glands; C. Location of mandibular and lingual glands.

CROP-MILK COMPOSITION

Water......................65-80%

Protein....................15%

Fat..........................25-35%

Lecithin..................5%

Ash.........................1.5%

Lactose sugar............scanty

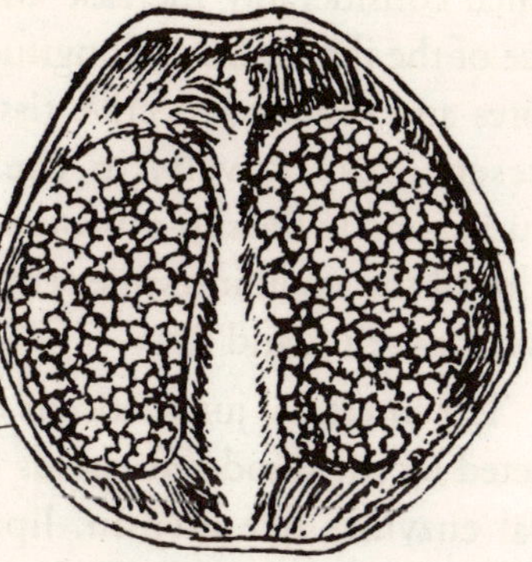

Fig. 30.20. Dissected view of the crop in pigeon showing crop glands

Prolactin hormone secreted by the anterior pituitary gland regulates secretion of the crop glands. Besides the milk secreting crop glands, the crop possesses adequate number of mucous-secreting glands.

(b). Digestive organs: The subdivisions of the stomach and duodenum of the small intestine are the sites for preliminary and final digestion of food particles, respectively.

vii. Stomach: The stomach in pigeon is divided into two main parts: **proventriculus gastris** and **ventricular gastris or gizzard**. A small tubular third part called **pars pylorica gastris** is also present between the gizzard and the duodenum.

The oesphagus, on entering the thoracic cavity slightly towards the left, forms the tubular proventriculus–the broad, spindle-shaped first part of the stomach. The inner wall of proventriculus gives rise to numerous tubular gastric glands, and, hence, by definition this part is the real stomach. A narrow tubular isthmus occurs between the proventriculus and the next part of the stomach—the gizzard. The mucous membrane of the inner wall of proventriculus contains numerous papillae called **'Papillae proventriculi'** which actually bear pores of the deep compound glands lying deeper within the proventriculus. Lying between the papillae on the inner surface of the proventriculus, there are large number of wavy folds and grooves called **'Rugae proventriculi'**. Both hydrochloric acid and pepsin are secreted from the same proventricular glands.

The second part of the stomach, called ventriculus or gizzard, appears like a rigid compactly muscular biconvex lens. Because of the covering of thick muscular layers, the lumen of the gizzard is very small. The surface epithelium of the inner wall of the gizzard(also known as '**Cuticula gastris**') is extremely thick, hard and folded. This epithelium also contains numerous tubular gizzard glands which secrete a substance of the nature of carbohydrate or protein, or both. This forms a kind of chemical compound called '**Koilin**'. Koilin is an extremely hard substance of the inner wall of the gizzard. Some amount of elastic fibres and collagen fibres are also present below the mucosa of the gizzard. Along with foodgrains, the pigeon also transports a few pebbles to the gizzard. These pebbles act as a grinder for churning of foodgrains transported by the proventriculus to the gizzard. Besides this, the foodgrains remain in the gizzard till they undergo preliminary digestion with the help of acidic and enzymatic secretions. The pH value of gizzard is sufficiently acidic.

The third part or the **pyloric part of the stomach** is small, tubular; its inner surface also secretes Koilin. The well-grinded, soft and chemically broken down food bolus is transferred to the small intestine through this pyloric part.

(c). **Absorptive organs:** The distal, longer and narrowly tubular ileum is the absorptive region of the essence of finally digested food.

viii. Small Intestine: Duodenum[1] **and Ileum:** The small intestine is divisible into two parts: a proximal tubular duodenum (that forms an 'U'-shaped loop by its descending and ascending limbs) and a distal narrow tubular ileum (which is closely coiled centrally). The inner space and the two limbs of the duodenum are attached by a fold of the mesentery. The duodenum arises from the inner side of the junction of the proventriculus and the gizzard, descends down adjacent to the wall of the gizzard, and then curves upward forming an 'U'-shaped loop. The ascending limb is as long as the descending one. The duodenum is about 30 in (75 cm) long, which is about one-third the length (7.5 ft/ 2.25 m) of the small intestine. Long ago, Beddard (1911) stated that in a few fruit-eating pigeons, the duodenal loop is not formed. The pancreas is located in the space between the two limbs of the duodenum. Two bile ducts from the livers and three pancreatic ducts from the pancreas open separately into the distal limb of the duodenum. Another bile duct opens into the proximal duodenum at the pylorus region. From the mucous membrane of the innermost surface of duodenum arises numerous **finger-like villi, crypts of Lieberkühn** and **goblet cells.**

The distal end of the duodenum slightly curves to the right and forms the long narrow coiled tubular (the tube with uniform diameter) ileum. The proximal and distal ends of the ileum are more or less straight or much less coiled than the central part which is closely coiled. From the mucous membrane of the ileum arise numerous long, narrow and closely-set finger—like projections called **villi** which considerably increase absorptive surface of the ileum. Some longitudinal muscle fibres and some connective tissue fibres are present, respectively, in the muscularis mucosa and in the sub-mucous layer. The pH value of the small intestine remains most active between 5 and 8.

The intestinal juice, **succus entericus**, secreted by the duodenal glands contains several enzymes like erepsin, lipase, maltase, invertase, etc. which also help in final digestion of the food absorbable by the ileum.

(d). **Egesting organs:** The non-absorbable residue of food substances together with excess water are collected in the region of the alimentary canal known as the large intestine. Here the excess water is almost completely reabsorbed.

ix. Large intenstine: The portion of the alimentary canal between the ileum and the cloaca is the large intestine which, in higher vertebrates, usually consists of two parts—the colon and rectum. In lower vertebrates, the ileum almost imperceptibly leads to the rectum, there being no distinctive part as colon. However, in the elasmobranch fishes, a pair of long **rectal glands** and in the reptiles, one or two **colic caeca** at the junction of the small and large intestine mark the begin-

1. The duodenum, although described here under 'Absorptive organs', forms an essential part of the 'Digestive organs'.

ning of the latter. In pigeons too, the colon is indistinguishable and both the colon and the rectum are almost of the same diameter as the ileum. A pair of **colic caeca** (also known as **rectal caeca**) are present at the junction of the small and large intestine. The two caeca open separately into the rectum as is the usual case in most birds. In *struthio* and *Pterocles* (Sandgrouse), the two caecal ducts unite to open into the rectum by a single aperture. In pigeon, the rectum is small and it leads into a spacious and compartmentalised cloacal chamber.

x. Cloaca: In the pigeons, as in other birds, the short rectum opens into a wide, three-chambered cloaca—the last part of the alimentary canal. In pigeons, the cloaca is very well-developed and divided into three compartments of which the proximal first compartment (following the rectum) is widely tubular and the two distal compartments are large, wide and sac-like. From the proximal to the distal end, the chambers are named as **Coprodaeum**, **Urodaeum** and **Proctodaeum**, respectively. The rectum opens into the first compartment, the ureters and the reproductive ducts open into the second and the

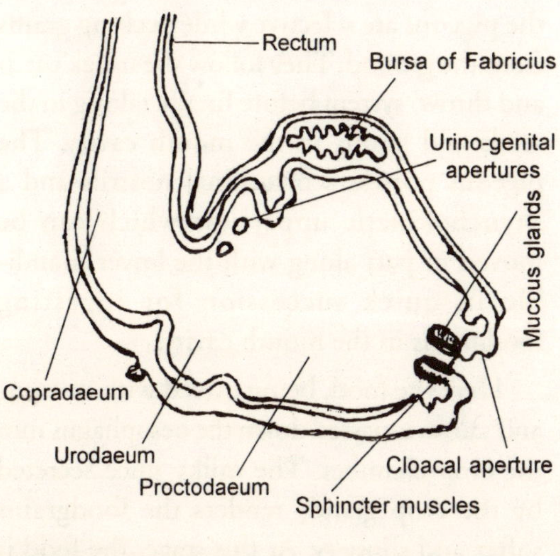

Fig. 30.21. Interior view of the cloaca of a pigeon

proctodaeum opens to the exterior through the cloacal aperture. Numerous goblet cells are observed in the inner wall of the cloaca.

Bursa of Fabricius: In many textbooks, this term has been commonly used as Bursa Fabricii. This sac-like or bursate unpaired organ arises as a lympho-epithelial structure from the cloacal dermis of hatchlings and young pigeons and later gets attached to one side of proctodoeum. According to Chang et al, 1955, and Epple and Stetson, 1980, it is a thymus-like lymphoid gland that forms antibodies in hatchlings and young pigeons for their protection against germs. The gland also participates in the production of White Blood Corpuscles (WBC).

B. Digestive glands

A number of digestive glands are associated with the alimentary canal and these open into the canal at different sites.

The principal digestive glands are:

i. **Salivary glands**: A pair of small **maxillary glands** (**Gl. maxillaris**) occur at the front end of the palate. Behind these, one or two elongated **palatine glands** (**Gl. palatinae**) are located on either side of the median line (Zweers, 1982). On the floor of the buccal cavity, on either side of the lower jaw, is found a spindle-shaped **sublingual gland** (**Gl. sublingualis**). This gland is also **known as rostral mandibular gland** (**Gl. mandibulares rostrales**). Behind these, on either side, occurs a narrower **caudal mandibular gland** (**Gl. mandibulares caudales**). Homberger and Meyer, 1989) have designated this gland as **mandibular gland** (**Gl. mandibularis**). Besides these the nomenclature of the rostral and cadual lingual glands (Gl. linguales) have been conceived differently by different authors. The salivary glands mainly secrete mucous, but the saliva also contains small quantities of enzymes.

ii. **Gastric glands:** Numerous tubular gastric glands are located in the mucous membrane of the fundic region of the stomach, whereas the cardiac and pyloric regions contain mainly the mucous-secreting glands. The gastric glands are of two types: the **chief cells** secreting pepsin (which is a proteolytic enzyme), and the **parietal** or **oxyntic cells** secreting hydrochloric acid–which kills germs and bacteria taken along with the food, and acidifies reaction in the stomach.

iii. **Livers:** In pigeons, these are the largest digestive glands situated on either side of a median space that accommodates the heart, duodenum and the gizzard. Deep-brown in colour, it is divided into two lobes—the right lobe is slightly larger than the left. Two bile ducts, arising separately from the right lobe and the left lobe, open directly into the distal and proximal arms of the 'U'-loop of the duodenum, respectively. The gall bladder is absent in the pigeons, as the bile is poured directly into the duodenum for a fast digestion of food.

iv. **Pancreas:** In the space between the two arms of the duodenal loop, the pancreas is located as a long flat and yellowish-red or pink-coloured gland. The pancreatic juice secreted by the three lobes of this gland—dorsal, ventral and splenic lobes—is conveyed by the three pancreatic ducts, one arising from each lobe, into the duodenum. The pancreas is both an exocrene and endocrine gland. The endocrine portion contains cells called **'Islets of Langerhans'** which secrete two important hormones called 'Insulin' and 'Glucagon'.

v. **Intestinal glands: Brünner's glands** are present in the sub-mucous layer of the duodenum and numerous intestinal glands, including the Crypts of Lieberkühn. Lymphoid tissues are present in the mucous membrane of the ileum at the bases of the closely packed long fingerlike villi. All these glands serete the intestinal juice called 'Succus entericus ' which contains certain digestive enzymes of the nature of enterokinase, peptidase, disaccharidase, nuclease, etc. The endocrinal epithelial cells of the intestine also secret certain hormones such as secretin, pancreozymin and cholecystokinin which activate the secretion of intestinal juice.

C. Mechanism of Feeding, Digestion and Absorption of Food in the Pigeons

The pigeons are mainly granivorous, feeding on rice, pulses, wheat, maize, millets and seeds of various fruits. They also feed on small fruits and insects and their larvae. Like other birds, the pigeons are selective while pecking grains from the ground. They follow the usual 'catch and throw' system before finally taking in the seeds and grains in the mouth cavity. The pigeons possess schizorhinal nostrils and a rhynchokinetic upper jaw which can be moved in part along with the lower mandible in quick succession for ingesting foodgrains in the mouth cavity.

Here the food, being mixed with mucous and saliva, is passed down the oesophagus into the crop chamber. The milky juice secreted by the crop glands renders the foodgrains softer and slippery. At this stage, the food is transferred to the proventriculus of stomach. A small amount of carbohydrate-breaking enzyme present in the saliva breaks down the carbohydrate food to a small extent.

The gastric glands present in the mucosa of proventriculus secrete gastric juice that contains hydrochloric acid and the protein-breaking enzyme pepsin. While the former acidifies reaction in the stomach and has mainly a germicidal action, the latter partially converts the proteinous food into intermediate products like proteoses and peptones. The carbohydrate-breaking enzymes like maltase and sucrase convert the complex carbohydrate food into double sugar maltose and sucrose. The gastric lipase converts the fatty food into fatty acid and glycerine.

In this condition the partially digested acid chyme reaches the gizzard where the food is thoroughly churned and converted into a half-digested food bolus. The food bolus, on entering the duodenum, immediately comes in contact with the bile, the pancreatic juice and the succus entericus. As the pigeons lack a gall bladder, the bile secreted by the livers is directly poured into the proximal and distal ends of the U-shaped loop of the duodenum through two principal bile ducts. Likewise, the pancreatic juice is conveyed to the ascending and descending loops of the duodenum through 2-3 shorter pancreatic ducts.

Bile neutralises the acidic food, kills germs and bacteria and emulsifies fat from the fatty food. The pancreatic enzyme trypsin and other aminopeptidases break down the proteoses and peptones first into polypeptides, then to oligopeptides (i.e., smaller peptides) and then, finally, into amino acids. The pancreatic carbohydrate-breaking enzymes amylase and maltase convert the disaccharides into monosaccharides or glucose and the pancreatic lipase act on the triglycerides and convert them into monoglycerides, fatty acids and glycerine.

Although, generally, the intestinal juice Succus entericus contains the protein-breaking enzyme erepsin and the carbohydrate-breaking enzymes sucrase and lactase, their presence in *Columba* has not been proved beyond doubt.

The food, being thoroughly digested in the duodenum, enters the ileum in a liquefied state. The essence of food is now absorbed by the mucous membranes of the ileum. The excess water is absorbed by the inner walls of the rectum. The vegetable fibres are further acted upon by the caecal juice and a small amount of excreta (food-residue after digestion and absorption)–mixed up with concentrated urine–is ejected to the exterior through the cloacal aperture.

30.13. CIRCULATORY SYSTEM

In the pigeons—as in all other vertebrates in general—the circulatory system comprises two elaborate systems of branching tubes of different diameters ramifying throughout the body. These are: (I) the blood vascular system formed of heart, arteries, arterioles, capillaries, venules and the veins; (II) the lymphatic system formed of lymph capillaries, lymph vessels, sinuses, nodes and lymph organs of different types.

1. Blood vascular system

Quite a few structural and functional changes have taken place in the blood vascular system of the pigeons owing to their flight adaptations. Such changes may be mentioned as larger size of the heart compared to that of the body, greater diameter of the large arteries and veins in the body, ventricles with a thick muscular wall, presence of papillary muscles for a more efficient contraction of the left ventricle and also the presence of a single-flapped muscular valve controlling blood flow at the right auriculo-ventricular aperture.

The blood vascular system in a pigeon includes blood, heart and the arteries and veins with their branches. The portal system forms an essential part of the venous system.

1.1. Blood: It is a liquid connective tissue in which different types of corpuscles are found floating in varying numbers. The straw-coloured liquid matrix of blood is called plasma. Plasma constitutes about 55 percent of blood and the rest 45 percent (approx.) is formed of the corpuscles. The plasma contains about 91 percent of water and about 7-8 percent of protein and lipids, besides small percentage of various inorgantic substances. Among the proteins, the albumin and globulin mainly act as antibodies, and prothrombin and fibrinogen take part in the clotting of blood. Besides these, sodium, potassium, chloride, bicarbonate, sulphate, phosphate ions are also found dissolved in the plasma. The plasma reaction is slightly alkaline (pH 7.3), as the proteins and bicarbonates react with one another to maintain the alkalinity. The plasma circulates glucose, amino acids, fats and fatty acids from the absorbed food; urea and uric acid from the excretory system, and also transport dissolved gases, hormones and vitamins.

In a pigeon, three types of corpuscles are found afloat in the blood: the **erythrocytes** or the **Red Blood Corpuscles (RBC)**, the **leucocytes** or the **White Blood Corpuscles (WBC)** and the **thrombocytes** that help in clotting of the blood. In the pigeons, the **Blood Platelets** as found in mammalian blood are absent. In a pigeon, the RBC (found in a much greater number than the WBC) are oval, nucleate and occur in about 2-3 million/mm^3 of blood. This number far exceeds that of the reptiles.

Each red blood corpuscle is formed of an iron-containing protein called haemoglobin in which 4 iron atoms are attached to a protein molecule. Each iron atom combines with an oxygen ion to form a complex **oxyhaemoglobin** compound. 95 percent of haemoglobin is formed of a complex protein called 'globin' and only 5 percent is formed of a red-coloured iron compound called 'haematin'. Thus the oxyheamoglobin of the RBC transports oxygen to all the cells and tissues of different organs in the body. Hence the oxyhaemoglobin is also known as the respiratory pigment. The RBC also help in the removal of carbondioxide from different organs along with the plasma.

In pigeons, as in other vertebrates, the white blood corpuscles or the leucocytes occur in different shapes and sizes, and, in general, these are larger than the RBC, but lesser in number. Each has a distinct nucleus in the cytoplasm and because of the absence of the respiratory pigment hemoglobin, the leucocytes appear whitish. Based on the structure of the nucleus and on the nature of the cytoplasm, whether granular or smooth, the leucocytes have been divided into two principal types: **Granular** and **Agranular**.

The granular leucocytes have their cytoplasm with numerous small granules and the nucleus in each is either sharply constricted or divided into segments. These cells, usually round, are capable of diffusing out of the endothelial tissues of the blood capillaries or enter into them through amoeboid movement. Examples of granular leucocytes are the basophils, eosinophils and the neutrophils (also known as polymorphonuclear leucocytes). The granular leucocytes are larger and have their nucleus undivided. The lymphocytes and the monocytes are of this category.

The leucocytes have a defensive role by killing bacteria and various germs causing any infection in the body.

The third category of cells that helps in the clotting of blood are the thrombocytes, each having a spindle-shaped or oval body with clear cytoplasm and a distinct nucleus. In the mammalian blood, the corresponding cells are called platelets that are devoid of nucleus.

The blood-forming or **haemopoietic tissues** are those that give rise to different kinds of blood corpuscles in different vertebrates. In general, from the anurans up to higher vertedrates, the red bone marrow is the main source that gives rise to the red blood cells, granular leucocytes and the spindle cells or the thrombocytes. The lymphocytes, monocytes and the agranular leucocytes are usually derived from the lymphoid tissue. In the embryo of higher vertebrates, before the development of spleen and bone marrow, the mesenchyme tissue and the liver function as the main source for the development of red blood cells.

1.2. Heart : The heart of a pigeon is proportionately larger compared to the body size of the bird. It is a broadly triangular structure located in the thorax, ventral to the lungs. The 4-chambered heart–that appeared in the crocodiles–reached perfection in the carinate birds. In birds, the right side of the heart is completely separated from the left through interauricular and interventricular septa that are devoid of any intercommunicating aperture.

The entire heart is enclosed by a transparent, slightly whitish and thin pericardial membrane. A kind of serous fluid circulating between the heart and the pericardium keeps the outer surface of the heart moist.

The heart is the central pumphouse of the blood vascular system. Being formed of a thick muscular wall, it is capable of rhythmic contraction and relaxation which ensure circulation of blood throughout the body through arterial, venous and lymphatic systems at species-specific pressure and temperature. Externally, as viewed from sides, the junction of the two auricles and the two ventricles on either side is marked by a depression called the '**coronary sulcus**'.

Development of heart: In vertebrate embryos in general, the initial stages in the development of heart have a close similarity. The heart arises from the heart-forming cells which are modified mesenchymal cells that originate from the **splanchnic mesoderm** at the ventral aspect of the primitive archenteron. The cells reorganise themselves into two endocardial tubes lying closely adherent. Soon they merge with one another and form a single, spacious endocardial tube.

The heart tube is formed of three layers: the outermost **epicardium**, the middle thick and muscular **mesocaridum** (also known as the myocardium) and the innermost **endocaridum** formed of endothelial tissues. The dorsal and ventral mesocarida disappear and a single pericardial cavity surrounds the heart. As the length of the heart tube increases in the mid-region, it twists in the form of an 'S', thus giving rise to chambers from behind forward –the atrium or the auricle, the ventricle and the conus arteriosus. Dorsal to and behind the atrium lies the sinus venosus into which the large veins drain the blood.

1.2.a. Structure of the heart: The two auricles are laterally broad, spacious and thin-walled, whereas the two ventricles have thick muscu-

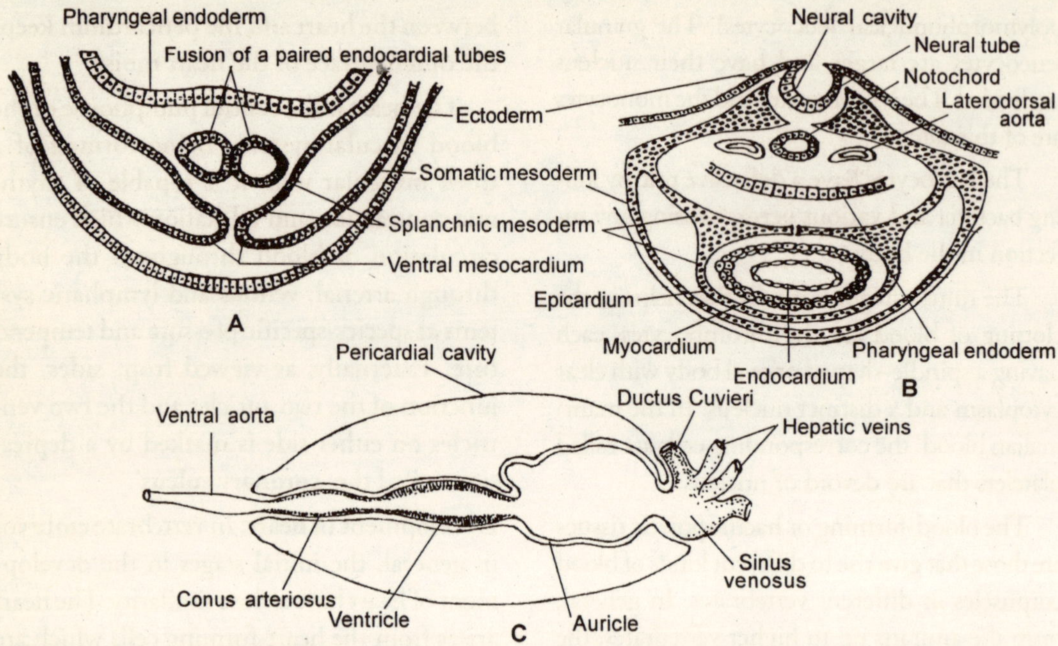

Fig. 30.22. Early stage in the development of a heart of a pigeon. A. Diagram showing paired endocardial tubes fusing; B. Formation of a single endocardial tube surrounded by the mesocardium and the epicardium; C. Early stage in the development of a primitive heart

lar walls and they converge posteriorly to form the broad apex of the triangular heart. In pigeons, the interauricular septum is shifted more towards the left, so that it divides the two auricles unequally and it is not in a line with the interventricular septum that also divides the two ventricles unequally. Due to complete longitudinal partitions between the auricles and the ventricles, the right side of the heart remains completely separated from the left side. Due to shifting in position of the septa, the right auricle is larger than the left, while the left ventricle is larger than the right, although the wall of the right ventricle overlaps semicircularly a portion of the wall of the left ventricle.

Fossa ovalis is an oval depression on the wall of the interauricular septum in the adult bird. This depression marks the presence of an oval aperture, the **Foramen ovale**, in the embryonic condition of the bird. This aperture–which served as the only point of communication between the right and left side of the heart—is completely closed in the adult. In the heart of a pigeon, the **Sinus venosus** and the **Conus arteriosus** are absent, as the Sinus venosus is incorporated in the walls of the right atrium and the conus arteriosus is modified to form the main arterial arches. Two large precaval veins and a single large postcaval vein collect unoxygenated (or least oxygenated) blood from the anterior and posterior parts of the body, respectively, and drain the blood directly into the large right auricle.

From the right auricle the venous blood enters into the right ventricle through the right auriculo-ventricular aperture guarded by a valve formed of a **single muscular flap (unique in birds)**. The blood from the right ventricle enters with the contraction of the muscular wall of the right ventricle into the large pulmonary aorta whose aperture is guarded by a set of three unidirectional **semilunar valves.**

The blood is transmitted to lungs for oxygenation. With the diastole of the auricles, the oxyygenated blood from the lungs reaches the left auricle through two pairs of pulmonary veins. With the contraction of the left auricle, the oxygenated blood enters the left ventricle through the large, round left auriculo-ventricular aperture guarded by a two-lobed membranous valve which corresponds to the mitral valve in the mammalian heart.

In both the ventricles, particularly in the left ventricle, tendon-like muscular chords span the ventricular cavities with their one end attached to the valves, thus keeping them adherent to the walls of the auriculo-ventricular apertures and the other end attached to the thick, muscular inner wall of the ventricles forming **Columnae carnae.** This layer develops from the myocardium. In the left ventricle, the greater part of the chordae tendinae is modified to from spindle-shaped muscles called **Papillary muscles** (Musculi papillares) which considerably increase contractility of the left ventricle.

The contraction of the left ventricle forces the oxygenated blood to enter the only persistent right systemic or aortic arch (the left aortic arch being abortive in birds). The opening of the right systemic arch into the left ventricle is also guarded by a set of three semilunar valves. The aortic arch supplies oxygenated blood to the anterior as well as posterior parts of the body through the arterial system. As soon as the systolic movement of the heart is over, it is followed by the diastole which brings in again unoxygenated blood in the right auricle and oxygenated blood in the left auricle. The cycle is thus rhythmically repeated.

The diastole-systole cycle and the heart beat are characteristic features in the heart of birds and mammals. The specific and rhythmic contractions of the heart are triggered by a special kind of tissues called the '**Junctional tissues of the heart or the Nodal system**'.

1.2.b. Junctional tissues of the heart (Nodal system; Pacemaker): This tissue system is comprised of the following components— The pacemaker which initiates the rhythmicity of the heart-beat is formed of a bundle of a typical muscle fibres:

i. Sinoatrial node (S-A node): In higher forms where the sinus venosus has been incorporated in the wall of the right auricle, the sinoatrial node lies embedded in the wall of the right atrium–upwardly.

ii. Atrioventricular node (A-V node): The other node that is also formed of the same typical mass of muscle fibres is called the Atrioventricular node that lies downwardly in the interatrial septum of the heart of a pigeon. If the S-A node becomes non-functional for some reason or other, the A-V node becomes fully functional as the pacemaker.

iii. Internodal tracts: Three internodal tracts, the anterior, middle and the posterior convey stimulus from the S-A node to the A-V node and the junctional tissues in the walls of the ventricles.

iv. Bundle of His and Bundle branches: From the A-V node arises a thick, thread-like Bundle of His that divides into two main bundle branches, the left bundle branch and the right bundle branch. The left bundle branch further subdivides into a superior and an inferior division, both terminating in delicate muscle fibres.

v. Purkinje fibres: A bunch of fine, thread-like delicate muscle fibres arise from the Bundle of His and are distributed in the walls of the ventricles. The rhythmic waves of con-

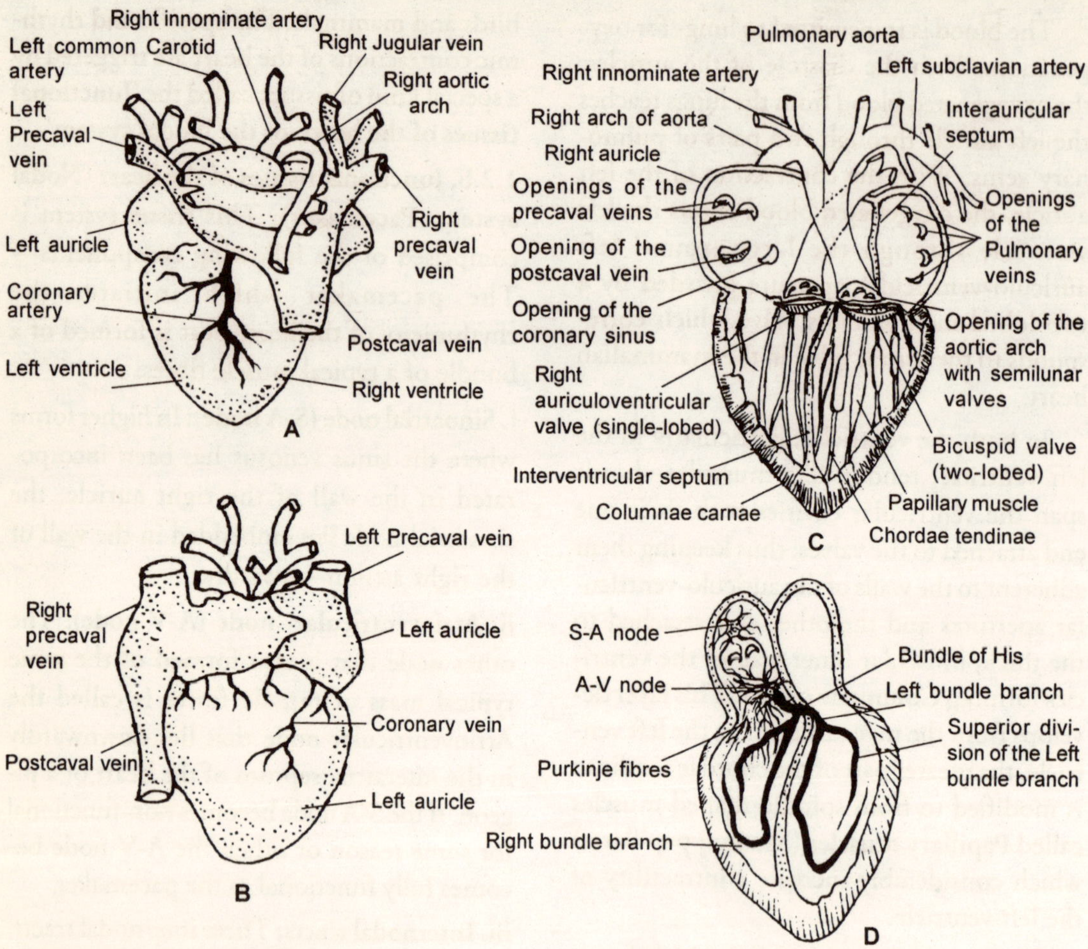

Fig. 30.23. Structure of the heart of a pigeon. A. Dorsal view of the heart; B. Ventral view of the heart; C. Ventral sectional view of the heart showing internal structures; D. Junctional tissues of the heart (Pacemaker system)

traction are finally distributed to the walls of the ventricles through Purkinje fibres.

The two nodes, the Bundle of His, the bundle branches and the Purkinge fibres constitute the nodal system of the heart and as this system initiates rhythmicity of the heart, it is also known as the '**Pacemaker**'. The pacemaker is obviously under the control of the nervous system.

For respiration and nutrition of the heart itself, the pigeons possess a well-developed coronary circulatory system. This system is formed of coronary arteries, veins and their branches and sub-branches forming a network.

1.3. Arterial system

The size of the heart and the diameter of the arteries and veins in pigeons are proportionately larger compared to those in other vertebrates. These are more muscular and are able to circulate more blood to various organs in the body. The arterial system in a pigeon is concerned with the supply of oxygen and nutrition to the cells and tissues of various organs in the body as fast as possible. From the opposite direction unoxyzenated blood (or blood with negligible amount of oxygen) reaches the heart from all over the body through the venous system. The only exceptions are the pulmonary arteries and the pulmonary veins. The former carry unoxygenated venous blood to the lungs for oxygenation and the latter bring back oxygenated blood from the lungs to the heart.

From the heart of a pigeon, two main aortae originate—the pulmonary aorta and the right systemic aorta. The arterial system is formed by their branches, sub-branches and their capillary networks.

In the arterial system of a pigeon, the arteries and their supply-organs are described:

1. Pulmonary aorta: In the evolution of aortic arches in vertebrates, it is the sixth embryonic aortic arch that is modified to form the pulmonary aorta. This aorta arises broadly from the right ventricle, shifts somewhat obliquely to the left as it crosses over the right aortic arch ventrally and then divides into the right and left pulmonary arteries entering into the right and left lung, respectively.

2. Right aortic arch (Right systemic arch or aorta): In embryonic condition, the fourth arch or the systemic arch–together with the portion of radix aorta on the left side in birds–disappear. So the right aortic arch is connected only to the dorsal aorta. In pigeons and in all birds, the ventral aorta splits into two portions: a systemic aorta that persists on the right side only, and a pulmonary aorta. These two aortae are connected to the left and right ventricles, respectively.

3. Brachiocephalic arteries (Innominate arteries): Two stout brachiocephalic arteries (Innominate arteries) arise as branches of the right aortic arch slightly ahead of the origin of the aorta from the left ventricle. From each brachiocephalic artery arise, on either side, a **common carotid artery** and a **subclavian artery**.

A. Common carotid artery: Each common carotid artery, proceeds rostro-ventrally along the lateral wall of the long neck and gives off branches:

i. **Internal carotid artery:** Arising very near the base of each common carotid artery, the two internal carotids come closer to each other, run parallel for a short distance, and then both the arteries enter ventrally into the hypophysial canal of a number of cervical vertebrae lying in the mid-region of the neck. While running through the hypophyseal canal, the right internal carotid artery lies dorsal to the left one and, further, rostrally the two internal carotids emerge from the hypophyseal canal and, on reaching the base of the skull, they enter the interior of the skull to supply the brain with oxygenated blood. A descending oesophageal artery arises from the base of the internal carotid and it curves downward to supply the upper and a portion of the middle oesophagus.

ii. **External carotid artery:** It arises almost collaterally from the base of the origin of internal carotid artery, and is distributed laterally to supply the head, face, oesophagus, palate and adjoining muscles.

iii. **Comes nervi vagi:** As this part of the common carotid artery on either side of the neck accompanies the vagus nerve and the jugular vein in close proximity and extends all along the neck up to the base of the skull, the artery has been so named. This artery joins the internal carotid artery at the rostral end of the neck. Caudally, from the comes nervi vagi, two small arteries arise inwardly: the upper one is the **ascending oesophageal** that runs forward to supply the caudal part of the oesophagus, and the lower named ingluvial artery supplies the crop chamber. Baumel et al (In: NAA, 1993) mention that in *Gallus* and *Columba*, the oesphageal artery is a branch of the Coeliac artery and it arises as an upward offshoot from very near the origin of the coeliac artery from the dorsal aorta. The **descending oesophageal arteries** supply rostral part of the oesophagus.

iv. **Vertebral artery:** It arises from the common carotid artery, just above and close to the origin of the internal carotid. The two vertebral arteries run inward and pass through the vertebrarterial foramen of the cervical vertebrae. The vertebral arteries supply blood to the cervical vertebrae and the brain.

v. **Syringio-bronchial artery:** This small artery, too, arises from the common carotid artery, close to and just below the origin of the internal carotid artery. This artery supplies blood to the syrinx and bronchus.

B. Subclavian artery: This is a very thick wide and stout artery that arises from the innominate artery in common with the origin of the common carotid artery and then enters the wing on either side. Each subclavian artery is mainly formed of three branches and their sub-branches, although a smaller fourth branch—the sternoclavicular artery—also arises from the subclavian.

i. **Axillary artery:** In fact, the subclavian artery is continued within the arm and shoulder as the axillary artery and within the forearm as the brachial artery. Both these arteries have a number of branches.

ii. **Pectoral artery:** This too is a thick and wide artery that supplies blood to the huge pectoral muscles. This artery is subdivided into several branches.

iii. **Internal mammary artery:** From the base of the subclavian artery on either side, two internal mammary arteries originate and, in the pigeon, they supply blood to the ribs and sternum.

iv. **Sternoclavicular artery:** From the other side of the origin of the internal mammary arteries, this small artery arises from the subclavian artery and soon divides into two branches to supply the sternum and the clavicle.

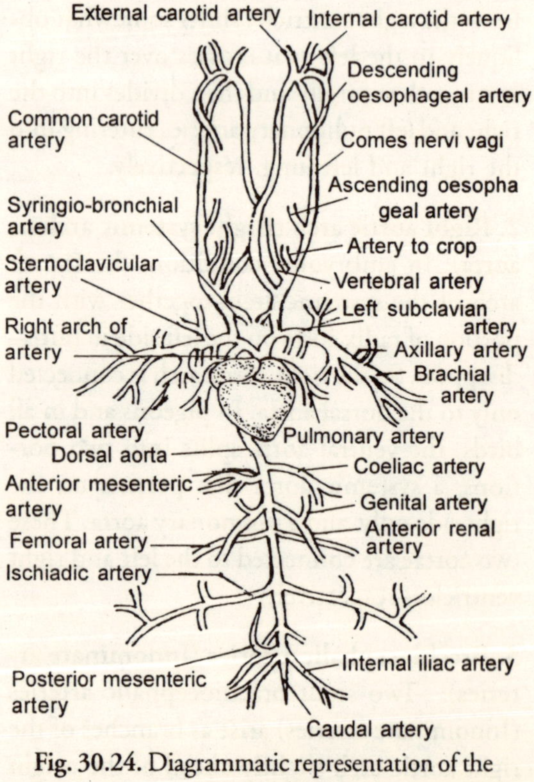

Fig. 30.24. Diagrammatic representation of the arterial system of a pigeon

C. Dorsal aorta

The right aortic arch curves round the heart dorsally and then extends straight along the mid-dorsal line of the body posteriorly as a thick and wide artery, giving off a number of arterial branches to different organs as:

i. **Dorsal intercostal arteries:** About 3-4 pairs of dorsal intercostal arteries (derived from the ventral rami of the segmental arteries), arising from the proximal part of the dorsal aorta, supply blood to the intercostal muscles.

ii. **Coeliac artery:** It is a thick, wide and long artery that arises from the proximal dorsal aorta and forms a complex system of branching arteries. It is formed of three main branches. The first one, a smaller and narrower branch, the **artery to the dorsal proventriculus**, arises from very near the origin of the coeliac artery. Then the thick and wide coeliac artery divides into two main branches: a sinister branch supplying the ventral proventriculus, stomach (muscular gizzard) and left lobe of the liver; a dextrous branch of the coeliac artery gives off arteries to the right lobe of the liver, duodenum, ileocaecal complex, a separate gastroduodenal artery to the stomach and duodenum, a distinct pancreaticoduodenal artery (which is the continuation of the dextrous branch of the coeliac artery) to the pancreas and duodenum and two or three short arterial branches to the spleen (splenic arteries).

iii. **Anterior mesenteric artery (Arteria mesentrica cranialis; NAA, 1993):** This artery arises from the proximal dorsal aorta, slightly caudo-dorsal to the origin of the coeliac artery. Distally, the artery divides into branches to supply the ileum and its mesenteric fold.

iv. **Genital artery:** A pair of spermatic or ovarian arteries, one on either side, arise from the middle of the dorsal aorta and supply blood to the genital organs and their ducts, the adrenal gland, and also a branch to the cranial lobes (anteriormost lobes) of the kidneys.

v. **Renal artery:** In pigeons the first pair of renal arteries arise directly from the genital arteries, one on either side or, less commonly, from the dorsal aorta, from the slightly caudo-dorsal to the origin of the genital arteries and supply blood to the first lobe of the kidney. The second and third pair of renal arteries supplying blood to the second and third pair of kidneys, respectively, arise from the ischiadic artery. Each renal artery subdivides into a number of branches within each kidney lobule. The first pair of renal arteries give off three or four primary branches within the first kidney lobule, the second renal artery sends only two branches while the third sends four or five branches within the third kidney lobule.

vi. **External iliac artery:** A common iliac artery is not present in birds; the external iliac artery and the internal iliac artery arise independently from the dorsal aorta. Arising from the dorsal aorta on either side, the external iliac artery courses between the first and second lobe of the kidney and—on entering the thigh—the artery is known as the femoral artery. A narrower pubic artery arises from the external iliac artery and runs along ventral border of the pubis.

The femoral artery is long and wide and supplies the thigh and pelvic muscles.

vii. **Ischiadic artery:** It is a pair of thick and wide arteries that arise from the dorsal aorta, one on either side, and each passes slightly caudolateral between the middle and caudal lobe of the kidney where the artery gives off—at the level of the kidney—an upward and a downward renal artery to the middle and caudal lobes of the kidney, respectively.

The ischiadic artery is mainly concerned with the supply of blood to the pelvic limbs.

viii. **Posterior mesenteric artery (caudal mesenteric artery):** This single median artery arises from the ventral aspect of the dorsal aorta and soon divides into two rami–the R. cranialis and R. caudalis supplying the rectum, the cloaca and the cloacal bursa. In fact, a bursocloacalis artery, arising from the R. caudalis of the posterior mesenteric artery, supplies the bursa and the cloaca in the pigeons.

ix. **Internal Iliac artery:** 'In birds and other higher vertebrates the A. iliaca internal is distributed to the terminal parts of the digestive tract, reproductive tract, and urinary system' (NAA, 1993).

However, in the pigeons, this artery arises in the pelvic region of the dorsal aorta, one on either side, slightly caudal to the origin of the posterior mesenteric artery. From the site of origin of the internal iliac arteries, the terminal segment of the dorsal aorta is known as the median caudal artery. Because, the internal iliac artery on either side divides into a lateral caudal artery distributed to the dorsolateral tail region and supply the bulb of the rectrices, the muscles and integument of the uropygium, and a cloacal artery (**A. pudenda**) supply the cloaca as well as the bursa.

x. **Caudal artery:** It has already been mentioned above that the narrower terminal segment of the dorsal aorta is called caudal artery or the median caudal artery, as there are a pair of lateral caudal arteries. All these arteries supply blood to different aspects of the cloaca and its derivatives such as the bursa, the bulbs of the rectrices, the uropygial gland, etc.

1.4. Venous system

In different parts of the body the arterial capillaries are in close contact with the venous capillaries in such a way that they form an almost inseparable winding network. Through this network oxygen and essential nutrients are transported on one hand (arterioles and arterial capillaries), and carbon dioxide as well as removable metabolites are transported on the other (venules and venous capillaries). The extensive system of venous capillaries and venules ultimately form larger venous vessels forming the venous system in the body. The blood—rich in carbon dioxide and containing scanty amount of oxygen—is collected from all over the body through the venous system. The venous blood is finally transported to the heart and, thence, to the lungs for O_2-CO_2 exchange. Thus the blood vessels in the pulmonary circuit act unlike other arteries and veins—as the pulmonary arteries carry unoxygenated blood to the lungs and the pulmonary veins carry oxygenated blood from the lungs to the heart.

As in other higher vertebrates, the venous system in pigeons may be studied under three heads: **A. Systemic veins, B. Pulmonary veins,** and **C. Portal veins.**

A. Systemic veins

Three large veins or caval veins—namely, a pair of precaval veins collecting blood from anterior parts of the body and a single wide postcaval vein from the posterior parts—drain the blood into the right auricle through separate openings. These veins and their branches are called the systemic veins as they are concerned with the collection of CO_2-rich blood from all over the body and return it to the heart.

1. Precaval veins: In pigeons, the right and left precaval veins return the blood directly into the atrium of the heart, there being no sinus venosus (which has been absorbed in the wall of the atrium). Each precaval vein is formed of the following veins:

i. **Jugular vein:** A long, straight and wide jugular vein runs along either side of the neck region accompanying the vagus nerve. At the top of the neck region, the two jugular veins are interconnected by a short, transverse vein known as '**Interjugular anastomosis**'. However, Baumel et al (NAA, 1979, 1993) con sider it more appropriate that this transverse connecting vein connects the two rostral cephalic veins rather than the two 'jugular veins'.

   Each jugular vein receives: (a) a facial vein from the skin and muscles of the head and face; (b) a tracheal vein from the trachea; (c) a cervical cutaneous vein from the skin of the neck region; (d) an oesophageal vein from the oesophagus; (e) a crop vein from the crop chamber, and (f) a vertebral vein from the vertebrae.

ii. **Brachial vein:** The venous capillaries of the wing and shoulder on either side unite to form the brachial vein.

iii. **Pectoral vein:** The extensive system of venules and capillaries within the large pectoralis muscles unite to form the large and wide pectoral vein. This vein opens at the base of the jugular vein.

iv. **Intercostal and thoracic veins:** The dorsal and ventral intercostal veins and the internal thoracic veins (the external thoracic veins are the same as pectoralis vein) return blood from the ribs and sternum.

2. Postcaval vein: In pigeons, venous blood from the entire posterior part of the body is collected by a thick and wide postcaval vein which opens by a large opening into the posterior part of the right auricle. The post-caval vein is formed by the union of a number of small and medium-sized veins from different parts of the body lying posterior to the heart.

The following veins return blood to the postcaval vein:

i. **Hepatic vein:** A pair of wide and large hepatic veins, left and right, together with a median and two or three smaller lateral hepatic veins open into the postcaval vein.

ii. **Renal veins:** The renal veins, arising from the kidney lobes, are in direct communication with the internal iliac and the renal portal vein on either side. They return the blood from the kidney.

iii. **Iliac veins (V. iliaca communis):** Two broad iliac veins—called Vena iliaca communis—unite medially at the cranial face of the kidneys to form the still broader postcaval vein. The iliac vein, before continuing outwardly as the external iliac vein, receives, on either side, a number of important veins, each with several branches. These veins are the genital veins, renal veins (both cranial and cadual divisions), the renal portal vein (both cranial and cadual divisions) etc.

iv. **Genital veins:** In adult pigeon, either the spermatic vein in males or the ovarian vein in the females return blood from the testis and ovary, respectively, to the iliac vein.

v. **Femoral vein (Vena femoralis):** In each pelvic limb of pigeon, as also in most birds, the femoral vein forms a very well-developed branching system of veins of which V. ischiadica, V. poplitea, V. tibialis, V. fibularis and, distally in the leg, V. metatarsalis—are the main. In fact, the external iliac vein continues distally as the V. femoralis through the Foramen iliofemoralis.

vi. **Ischiadic vein (V. ischiadica):** Behind the knee, the popliteal vein—being joined by the V. tibialis and V. fibularis—courses forward as the ischiadic vein. Hence, it is the proximal prolongation of the popliteal vein. The ischiadic vein forms an anastomosis with the V. femoralis and is known as **the 'Anastomosis ischiofemoralis'**. Most of the blood from distal to the hip joint is diverted from the ischiadic vein to the femoral vein through this anastomosis.

vii. **Coccygeomesenteric vein** (Syn: **V. mesenterica caudalis**): This vein, also known as the inferior mesenteric vein, connects the caudal vein with hepatic portal vein. This vein curves inwardly and caudally, and is connected at the confluence of the two renal portal veins and the single median caudal vein.

viii. **Caudal vein** : A straight, short, caudal vein (sometimes two) returning blood from the uropygium or the reduced tail drains it into the interiliac anastomosis (Anastomosis interiliaca). From the anastomosis this vein extends caudally along with the caudal artery.

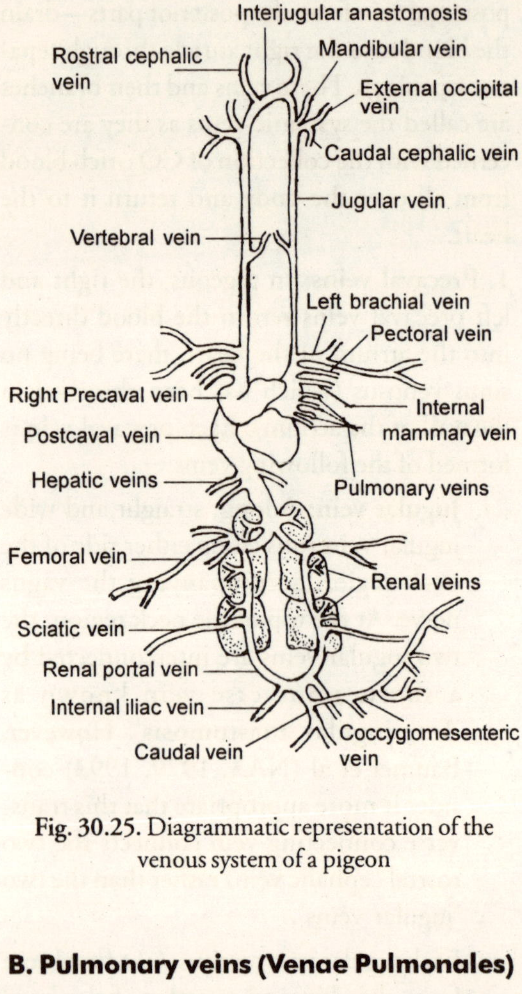

Fig. 30.25. Diagrammatic representation of the venous system of a pigeon

B. Pulmonary veins (Venae Pulmonales)

Two pairs of pulmonary veins, one pair from each lung of *Columba*, drain oxygenated blood into the left atrium. Each pulmonary vein (vena pulmonalis) is fed by at least three tributaries or radices called Rdx. cranialis, Rdx. caudomedialis and Rdx caudolaterials. These radices lie either medial or ventral to their corresponding pulmonary arteries. Deeper in the lungs, the interparabronchial veins are the tributaries of each of the three main radices of the pulmonary vein.

C. Portal veins

The system of veins which collect blood from different parts of the body posterior to the heart and, instead of returning it to the heart via the postcava., break-up into capillaries inside some intermediate organs (livers in hepatic portal system and kidneys in renal portal system). The capillaries reunite to form veins that join the postcaval.

Two such portal systems–**hepatic portal system** and the **renal portal system**—are commonly present throughout the vertebrate series, although the renal portal system may not be truly represented in all the vertebrate groups.

1. Hepatic portal vein: In the pigeons, the hepatic portal system of veins is very well-developed. The hepatic portal veins have been referred to as afferent veins conducting venous blood to the liver from the digestive tract and the spleen. The digestive tract covers a wide area from the proventriculus to rectum. Most birds, however, possess a major right hepatic portal vein and a lesser left hepatic portal vein. The more prominent right hepatic portal vein is formed by the union of the common mesenteric vein (which, in turn, is formed by the confluence of the superior and inferior mesenteric veins), gastro-pancreaticoduodenal vein and the proventriculo-splenic vein. The ileocaecal vein

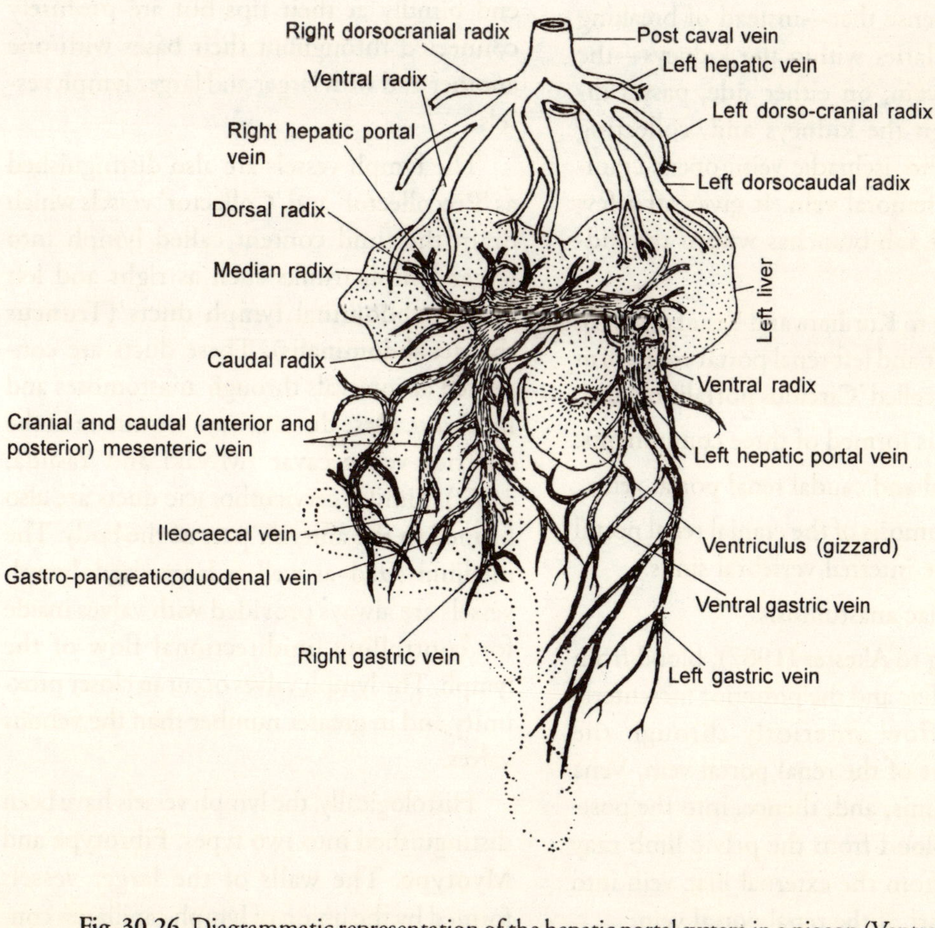

Fig. 30.26. Diagrammatic representation of the hepatic portal system in a pigeon (Ventral view)

opens near the base of the superior mesenteric vein. The left hepatic portal vein is formed by the union of the left gastric vein, ventral gastric vein, proventricular caudalis vein, and the pyloric vein.

All these verins form venous capillaries within the sinusoids of the liver; the other set or the efferent capillaries reunite to form the hepatic vein that directly connects to the postcava.

2. Renal portal vein: According to Johnson (1978), the presence of a functional renal portal system in birds is important in the excretion of uric acid by tubular secretion of the kidney. But in the pigeons and in other birds, the renal portal system is not very well-developed in the sense that—instead of breaking up into capillaries within the kidneys—the renal portal vein, on either side, passes directly through the kidneys and, collecting blood from the ischiadic vein, opens cranially into the femoral vein. It gives off a few branches and sub-branches within the kidneys.

According to Kurihara and Yasuda (1975), the main right and left renal portal veins form a venous ring called '**Circulus portalis renalis**'.

This ring is formed of three components:

(a) Cranial and caudal renal portal veins.

(b) Anastomosis of the cranial renal portal veins with the internal vertebral sinus.

(c) Interiliac anastomosis.

According to Akester (1967), blood from the internal iliac and the posterior mesenteric veins may flow anteriorly through the 'caudalis' part of the renal portal vein, Vena iliaca communis, and, thence, into the postcaval vein. Blood from the pelvic limb may be diverted from the external iliac vein into the caudal part of the renal portal vein.

2. Lymphatic system

Although extensive studies on the lymphatic system of lower vertebrates are rather lacking, the birds and mammals possess a well-developed lymphatic system formed of lymph vessels of different sizes—lymph capillaries and nodes, sinuses and lymph organs of different types.

After the heart and blood vessels are laid down, the lymphatic system develops independently at a later stage and the main lymphatic vessels establish connections with the large veins near the heart. The lymph capillaries are the finest branches of the lymphatic system. Tissue fluid from the blood capillaries diffuses into the lymph capillaries which end blindly at their tips but are profusely connected throughout their bases with one another and with larger and larger lymph vessels.

The lymph vessels are also distinguished as '**Precollector**' and '**Collector**' vessels which drain the fluid content called lymph into larger lymph trunks such as right and left thoracoabdominal lymph ducts (**Truncus thoracoabdominalis**). These ducts are connected at intervals through anastomoses and in *Gallus*, these ducts always terminate in the anterior venae cavae (Miyaki and Yasuda, 1977). Similar cervicothoracic ducts are also formed in the anterior part of the body. The medium-sized–as well as large-sized–lymph vessels are always provided with valves inside for controlling unidirectional flow of the lymph. The lymph valves occur in closer proximity and in greater number than the venous valves.

Histologically, the lymph vessels have been distinguished into two types: **Fibrotype** and **Myotype**. The walls of the larger vessels formed by the union of lymph capillaries con-

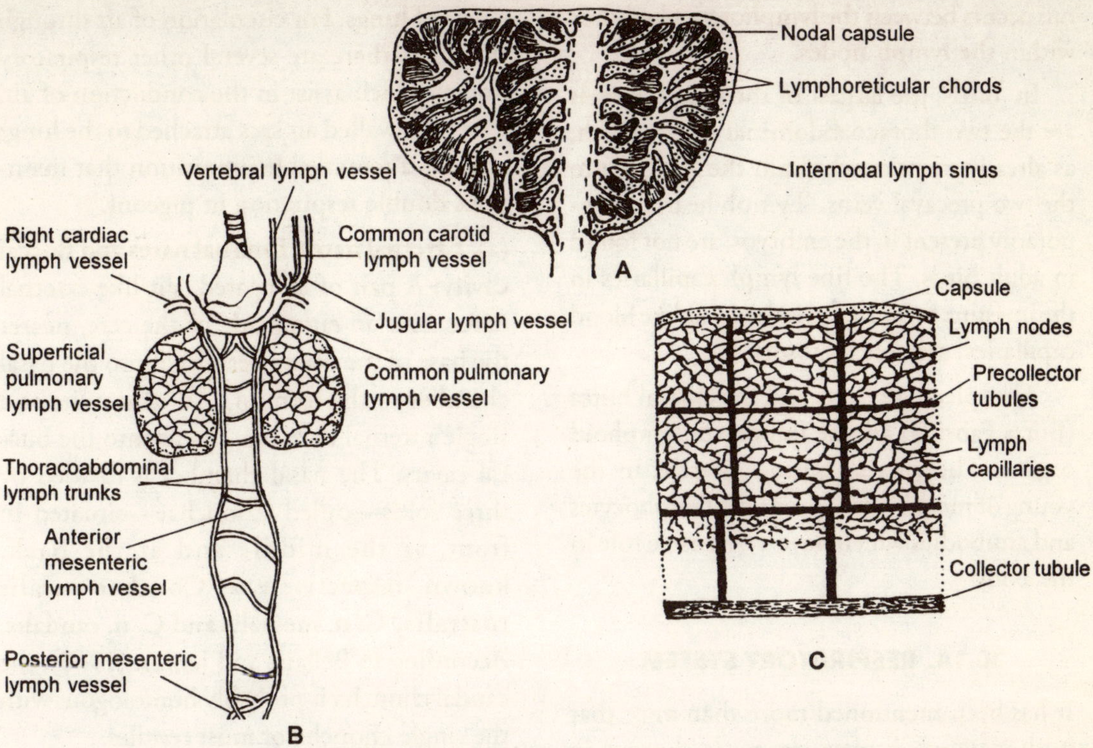

Fig. 30.27. Diagrammatic view of the lymplatic system in a pigeon. A. Structure of lymphoreticular chords and intranodal sinuses; B. Distribution of lymph capillaries and major lymph vessels in a pigeon; C. Schematic diagram showing lymph capillaries, precollector and collector vessels

tain elastic and collagenous fibres with some sparsely distributed smooth muscle cells, whereas the walls of still larger vessels are provided with three similar layers comparable to those of a fine artery. The 'Precollectors' and similar vessels fall under the first category while the large 'Collectors' and similar vessels fall under the second category.

Lymph nodes are very importent structures occurring at intervals along the circuitous course of the lymph vessels. The lymph nodes act as resistant boxes, checking flow of the lymph which has to filter out through these nodes. Avian lymph nodes occur as cervicothoracic and lumbar lymph nodes in many birds (e.g., in the order Anseriformes and in a number of other birds). The lymph nodes in birds consists of a system of lymph sinuses, lymphoreticular chords (formed of collagenous and recticular connective tissue fibres as well as some recticuloendothelial cells) and a thin capsule enclosing them. Lymphocytes, phagocytes and plasma cells are abundantly present within the capsule. These cells have important roles in the defence-mechanism of the body–in the destruction of germs and bacteria and in the production of antibodies.

It has been defined that the initial lymphatics always lie within organs and may be sinuses or precollectors. The initial lymphatics permit the entry of fluids from the interstitial space into the lumen of the initial lymph sinus. The precollectors are characterised by the presence of connective tissue trabeculae. An extensive system of lymph sinus occurs between the lymphoreticular hords within the lymph nodes.

In birds, the largest of the lymph vessels are the two thoracoabdominal ducts which, as already mentioned, drain the lymph into the two precaval veins. 'Lymph hearts', temporarily present in the embryos, are not found in adult birds. The fine lymph capillaries in the intestine form lacteals which, unlike blood capillaries, are able to absorb fat.

The spleen, thymus and the cloacal bursa (Bursa Fabricii) are the important lymphoid organs. The cloacal bursa—present in the young of most birds—produces lymphocytes and antibodies which have a defensive role in the body.

30.14. RESPIRATORY SYSTEM

It has been mentioned more than once that bird anatomy displays dramatic changes in all the systems owing to flight adaptation. The respiratory system in the pigeons, like in many other carinate birds, also shows remarkable modifications in its structures compatible with the complete aeration of blood and a high rate of respiration.

The lungbud arises as a ventral pharyngeal diverticulum from behind the last gill pouch. Soon this diverticulum divides into two lateral components which, being invested with mesoderm layer, extend caudally. The primary lung tube communicates with the ventral pharynx through an aperture called glottis. Further diversification of respiratory organs gradually takes place.

The respiratory organs in pigeons have been described here under two heads: **A. Respiratory organs,** and **B. Mechanism of respiration.**

A. Respiratory organs: In pigeons, the primary and the principal respiratory organs are a pair of lungs. For circulation of air through the lungs there are several other respiratory organs which assist in the conduction of air. The thin-walled air sacs attached to the lungs are also a remarkable innovation that maintains double respiration in pigeons.

(a). External nares, Internal nares and Buccal cavity: A pair of elongated, slit-like external nares, one on either side of the cere, nearer the base of the upper beak lead into the nasal chamber in the form of a 'V' and unite as a single internal nare that opens into the buccal cavity. The nasal chamber is divided by three folds—called Chonchae—situated in front, at the middle and at the back, known, respectively, as **Conchae nasalis rostralis, C. n. medialis and C. n. caudalis.** According to Bellairs and Jenkin (1960), the caudal choncha is probably homologous with the single choncha of most reptiles.

(b). Glottis and Larynx: The slit-like median longitudinal aperture lying behind the base of the tongue and on the dorsal aspect of the larynx is called the glottis. The aperture leads into the laryngeal cavity–guarded and controlled by the laryngeal cartilages and their muscles.

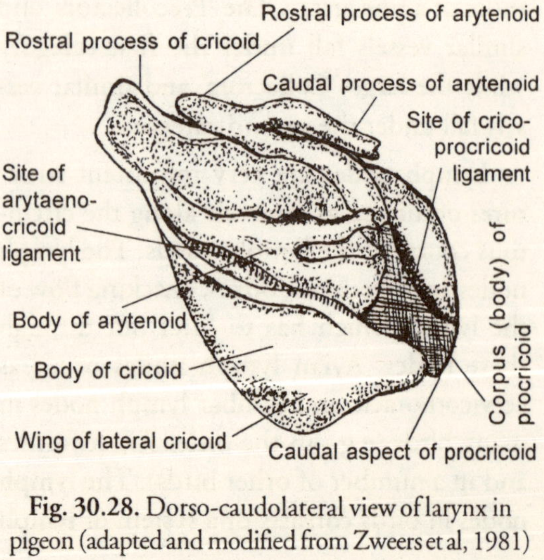

Fig. 30.28. Dorso-caudolateral view of larynx in pigeon (adapted and modified from Zweers et al, 1981)

The larynx in pigeon is formed by several cartilages into a dilated chamber at the rostral end of the trachea. The laryngeal cartilages are a pair of dorsally rod-shaped and bifurcated arytenoids, an expanded, plate-like ventrally placed single cricoid and a small median procricoid cartilage at the caudal end of the cricoid. According to Zweers et al (1981), it is comma-shaped with a bony body and a cartilagenous tail in *Gallus* and *Columba*. The procricoid forms joints with both the arytenoids and cricoid cartilages.

The larynx—being devoid of vocal cords in birds—does not function as the sound-producing organ as in other vertebrates. The sound is produced by a special organ—the **syrinx**—placed at the caudal end of the trachea.

(c) **Trachea:** The long, cylindrical wind-pipe or the trachea extends all along the neck region from the larynx in front to the syrinx behind; it lies ventral to the oesophagus. The trachea is well-guarded by tracheal cartilages which are 'complete rings, closely interlocked, with minimal intervals between them' (McLelland, 1965). The epithelial lining of the trachea (as also in the larynx, syrinx, primary bronchus, etc.) consists of pseudostratified ciliated columnar cells, goblet cells, basal cells and simple alveolar mucous glands.

In *Columba*, however, the tracheosyringeal cartilages along the dorsomedian line form incomplete rings and these are connected through narrow, thin bracelet-like cartilages (Mc Lelland, 1965). Especially in the family columbidae, these bracelets are covered by a thin tracheosyringeal membrane (King, 1989).

(d) **Syrinx:** At the site where the terminal part of the trachea, on entering the body cavity, divides into two primary bronchi, the last three or four tracheal rings of cartilages and the first half-ring of each bronchus become dilated to form a complex chamber called syrinx. This functions as the sound-producing organ in the pigeons, and in most other birds. From the standpoint of structural diversity and complexity, three main types of syrinxes have been distinguished: **tracheobronchial**, **tracheal**, and **bronchial**. Of these, the type of syrinx present in the family columbidae is called tracheobronchial syrinx. In this type of syrinx, the internal mucous membrane, coverings and cartilages of both the trachea and bronchus are specially modified. In tracheobronchial syrinx, both tracheal and bronchial elements of cartilages occur–the former as complete rings along midline before the point of bifurcation of the bronchi and as typically C-shaped half-rings in the paired primary bronchi.

Caudally, within the saccular cavity of the syrinx near the junction of the two bronchi, a small rod-shaped cartilagenous structure called '**pessulus**' occurs in the midline. Surrounding this pessulus lies a thin and delicate mucogenous membrane called **semilunar membrane** (Membrana semilunaris). In the syrinx of many passerine and non-passerine birds, however, both the pessulus and the semilunar membrane may be absent (Mayr, 1931; Ames, 1971; King, 1989). In pigeons, below the pessulus and the semilunar membrane, a pair of thin **tympaniform membranes** (Membrana tympaniformis medialis), being attached to the similar membrane above, act as the main vibrating membranes for producing sound vibrations. A similar smaller membrane, arising from the inner wall of the syrinx, is called **lateral** tympaniform membrane. The contraction and relaxation of the syrinx are controlled by a few muscles occuring on the external wall of the syrinx—M. **sternotrachealis**, M. **tracheolateralis** and a few smaller syringeal muscles. In pigeons, intrinsic syringeal muscles are lacking, but the extrinsic syringeal muscles—such as the paired M. tracheolateralis—are well-develpoed. This characteristic feature is quite prominent in the tracheobronchial type of syrinx.

(e) **Bronchus:** The syrinx divides caudally into a pair of lateral air-ducts (of lesser diameter than the trachea) called **primary bronchi** (sing. bronchus). In fact, these are extra-pulmonary parts of the primary bronchi lying between the syrinx and the lungs. The intrapulmonary primary bronchi lie within the lungs. The first cartilagenous ring of the primary bronchus is complete but the next few rings remain incomplete. The rings are absent in the intrapulmonary bronchi. Huxley (1882) used the terms '**vestibule**' for the swollen part where the primary bronchus enters into the lung and '**mesobronchus**' for the region of the primary bronchus lying caudal to the vestibule. The intrapulmonary primary bronchus (so-called mesobronchus) then extends caudally to terminate in the large abdominal sac (Saccus abdominalis). Here the terminal part of the primary bronchus dialates into a large funnel-shaped **saccobronchus**. The saccobronchus is connected to a number of **parabronchi** and to the abdominal sac. However, Akester (1960) mentions that the vestibulum as described by Huxley (1882) is not present in *Columba.*

In recent nomenclature, the primary bronchi give rise to four types of secondary bronchi: **medioventral, mediodorsal, lateroventral,** and **laterodorsal.** The large number of delicate branching air ducts that arise from the secondary bronchi are called **parabronchi** (Huxley, 1882; also known as **tertiary bronchi**).

(f) **Lungs:** The lungs in pigeons are smaller compared to the body size, elongated, less elastic, and externally divided into 5-6 lobes. Both the lungs are rigidly anchored to the dorsal bodywall and the investing membrane of the lung–the pleura is lacking here. However, a special pleural membrane, called '**pulmonary aponeurosis**' occurs horizontally covering the free ventral surface of the lungs. This layer is formed of a compact layer of fibrous tissues and the costopulmonary muscles present between the ribs and the lungs. Within the lungs numerous bundles of air-chambers, called **alveoli**, surround the parabronchi. The alveoli are supplied with large number of minute tubules called **air capillaries** which form loops and anastomosing networks opening into the parabronchi. The actual respiratory surface for the exchange of O_2-CO_2 is formed by the inner lining of the air capillaries which come in contact with the network of blood vessels.

The **atria** (sing. atrium) are the polygonal chambers which lead to the air capillaries from the parabronchi and may form many secondary bronchi.

(g) **Air-sacs (Sacci pneumatici):** The enormous demand for oxygen and, consequently, a high rate of respiration in birds are in conformity with their flight adaptations. The air-sacs that arise in all birds from the lungs are a very distinctive feature in flying birds. The thin-walled air-sacs originate as dilatation of the mucous membrane of the intrapulmonary bronchi. As the walls of air-sacs are not supplied with blood capillaries they do not directly participate in respiration, but actively do so in double respiration by transferring stored oxygenated air to the lungs. The anatomy of the air-sacs have been re-viewed by King (1966) and McLelland (1989). The

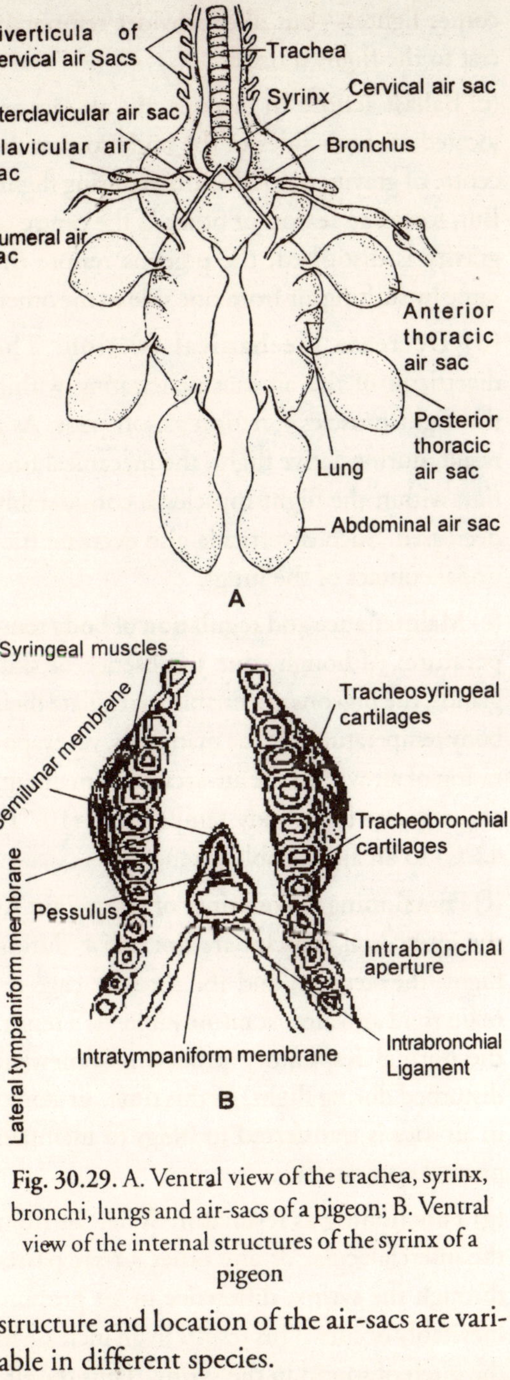

Fig. 30.29. A. Ventral view of the trachea, syrinx, bronchi, lungs and air-sacs of a pigeon; B. Ventral view of the internal structures of the syrinx of a pigeon

structure and location of the air-sacs are variable in different species.

In pigeons, nine major air sacs and a few other smaller and narrower air-sacs are present. Of the nine major air-sacs, four are paired and a single large median air-sac is present as an unpaired one.

On either side at the root of the neck and from the lungs laterally and caudally, the air-sacs arise as inflated balloons.

i. **Cervical air-sacs:** These small, spindle-shaped paired air-sacs arise from the rostral aspect of the lungs, one on either side, and extend forward, dorsal to the interclavicular air-sac, up to the base of the neck. From here, each sac proceeds anteriorly as a narrow tubular canal giving off 5-6 lateral diverticula on either side. These diverticula extend into the cervical vertebrae and the skull.

ii. **Interclavicular air-sacs:** Although this large, median triangular air-sac appears single, its origin lies in bilateral primordia from the lungs. This air-sac, also known as the clavicular air-sac, lies medially with its broad base covering the apical lungs. From either side of the broad base of the interclavicular air-sac arise two narrow diverticula that extend into the humerus and the clavicle in pigeons. These have been conventionally named as the **humeral** and **clavicular air-sacs**, although the recent nomenclature of these and of a few more diverticula in this region is variable. A narrow air-duct from either side of the narrower caudal end of the air-sac connects to the lung. From this region the interclavicular air-sac may give rise to diverticula which extend to the sternum, syrinx and scapula, as also muscles around the shoulder and region around the heart. This air-sac is connected to both the ventrobronchi and the dorsobronchi.

iii. **Anterior thoracic air-sacs:** Both the paired thoracic sacs (anterior and posterior) are intrathoracic in position. Each anterior thoracic sac arises as a balloon-like thin-walled air-sac from the front end of the lung and extends outwardly between the lung and the ribs. These air-sacs are connected to the ventrobronchi—the anterior thoracic sac is typically about one-third the volume of the posterior sac (NAA, 1993).

iv. **Posterior thoracic air-sacs:** These air-sacs, larger in size than the anterior ones, arise from the posterior aspect of the lungs, one on either side, and extend around the posteroir aspect of the thoracic cavity in front of the abdominal air-sacs. They remain rigidly anchored to the lateral thoracic wall. In pigeons, the right posterior thoracic air-sac is slightly smaller than the left.

v. **Abdominal air-sacs:** In pigeons, these are the largest in size and lie typically in the doro-caudal region of the coelom. Each abdominal air-sac arises as a pear-shaped elongated structure from nearly the terminal end of each lung and extends up to the end of the abdominal cavity. The right abdominal air sac is slightly larger in size than the left. From each abdominal air-sac arise two extremely narrow air-sacs—**pelvic** and **inguinal air-sacs.**

Functions of air-sacs: The role of air-sacs in carinate birds is undoubtedly very significant. They perform several functions which are briefly described:

(a) **Bellowing:** To maintain a steady current of air-flow through the lungs, the air-sacs perform a mechanical function by bellowing the air into the lungs.

(b) **Balloon action:** The air sacs–filled with hot air and inflated like balloons–considerably increase the buoyancy of the body, particularly during flight. In pigeons and in other birds, this increase in buoyancy not only decrease specific gravity of the body–which becomes lighter—but also provides temporary rest to the flight muscles.

(c) **Ballast action:** In pigeons, the air-sacs are located on both sides of the body so that the centre of gravity is not disturbed during flight. But, for some reason or other, if the centre of gravity is disturbed, the pigeons restore the same by shifting air from one side to the other.

(d) **Decrease mechanical friction:** The diverticuli of the air-sacs penetrating within the flight muscles function as soft pads. As a result, during active flight, the mecanical friction within the flight muscles is considerably decreased. Such diverticula also decrease frictional contact of the lungs.

(e) **Maintenance and regulation of body temperature:** Although, due to absence of skin glands, the pigeons are unable to regulate their body temperature like the mammals, yet evaporation of air within the air-sacs helps maintain and regulate their body temperature (107°F/42°C) to an appreciable extent.

(f) **Functioning as container of reserve air:** As the intercostal muscles are not active during flight, the sternum and the throacic cage remain rigid and almost immovable. As a result, the normal respiratory process is somewhat disturbed during flight. At this time, air stored in air-sacs is transferred to lungs to maintain normal respiration.

(g) **Functioning as resonator:** When air from the interclavicular or any other air-sac passes through the syrinx, difference in air pressure inevitably occurs. This results in an increase in the pitch of sound in the syrinx. Thus the air-sacs act as resonator of sound.

(h) **Regulation of moisture:** The evaporation of moisture through thin walls of the air-sacs partly regulates total water content of the body.

B. Mechanism of respiration in pigeon: Owing to the complex branching system of the bronchi within the lungs and the significant role played by the air-sacs in pigeons, the air flow within the lungs has been rendered rather tortuous. It has also to be remembered in this connection that in pigeons, as also in other flying birds, the mechanism of inspiration and expiration at rest and in flight, as also the rates of respiration, are different. However, the description and sketch of a section of the parabronchial wall of *Gallus* given here are almost the same as in *Columba.*

During inspiration, first air rich in oxygen reaches the lungs through the primary bronchus, and then, from the lungs, into the air sacs through the secondary bronchi. It has already been mentioned that the four types of secondary bronchi (medioventral, lateroventral, mediodorsal and laterodorsal) are distributed along the ventral, lateral and dorsal aspects of the lungs. These, in turn, at the tertiary level, are divided into numerous (about 1,800) **parabronchi** which have their walls invested with smooth muscles. Numerous small multicelluar cavities or atria occur in these walls. The atria are connected to innumerable air capillaries through infundibulum appearing tape-like or like the tube of a funnel.

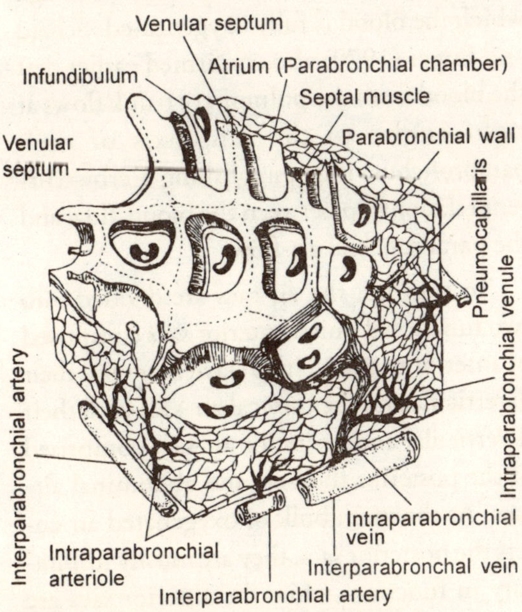

Fig. 30.30. Diagrammatic representation of a part of the parabronchial wall in *Gallus* showing atria, septa, infundibulum, blood vessels and pneumocapillaries (Based on and modified from NAA, 1993)

According to Gerisch (1971), the air capillaries, called **'Rami respiratorii'**, arise from the infundibula. He also suggests the presence of another type of tubes smaller than the infundibula. The air capillaries have also been termed variously–**Ductulus respiratorius** (Gerish, 1971); **Respiratory labyrinth** (Smith et al, 1986); **Pneumocapillaries** (King et al, 1993), etc.

The diameter of the parabronchi is about 0.5 mm, while that of the delicate tubules of air capillaries ranges between 3μm and 10μm. Within the lungs, the air capillaries–being richly supplied with blood capillaries–are the usual sites for O_2-CO_2 exchange. No tidal air remains in the lungs of pigeons. And, after complete expulsion of unoxygenated air from the lungs during expiration, no residual air remains in the lungs too. At this time, the oxygenated tidal air from the air-sacs enters into the lungs in the second phase. Thus, in pigeons and in other birds, the lungs are supplied twice with oxygenated air and, hence, the process is known as **'double respiration'**.

(i) Respiration at rest

While a pigeon is at rest or standing, the inspiration brings about an extension of the thoracic cage caused by the action of associated muscles. Due to contraction of the **external intercostal muscles,**[1] the sternum, ribs and the coracoid are moved forward and downward. As a result, the thorax expands, the areas within the cavities of air-sacs increase, and even the air-spaces within the lungs considerably increase. The caudal aspect of the thorax shows comparatively greater expansion (**see Figs. 23.2. A, B**).

During expiration at rest, the contraction of a few abdominal muscles exerts pressure on the lungs and air-sacs. Hence, during expiration, just as air is expelled from the lungs on one hand, the oxygenated air from the thoracic and abdominal air-sacs enters into the lungs for the second phase of respiration. In fact, most of the inspiratory air (about 90-95 percent) enters into the abdominal air-sacs. The rest of the air (about 5-10 percent) only enters into the thoracic air-sacs. During respiration at rest, only a small amount of air enters into the cervical and interclavicular air-sac. A pigeon completes the respiratory cycle, that is, inspiration and expiration about 30 times per minute (as against 14-20 times in human beings).

Although there are several opinions as to the flow of air current through the lungs, yet it may be stated in a general way that, during inspiration, most of the air passing through **primary bronchus (mesobronchus)**, **secondary bronchus, ostium** and **saccobronchus** first fill in the abdominal sacs and the posterior thoracic air-sacs (Bretz and Schmidt-Neilsen, 1972). The air from parabronchi enters into the air capillaries where the O_2-CO_2 exchange takes place through diffusion. The air, rich in CO_2, enters into anterior thoracic sacs wherefrom the air is expelled outside the body through primary bronchus, trachea and external nares. Because of the appreciable amount of CO_2 in the air-sacs, some authors believe that—in both inspiration and expiration—the flow of air current is unidirectional. But, according to Salt and Zeuthen (1960), air flows from the dorsal parabronchi to the ventral parabronchi during inspiration and the direction is reverse during expiration. J. Z. Young (1981) states that the posterior sacs contain only 4 percent of CO_2 but 17 percent of O_2 as against 7 percent CO_2 and 14 percent O_2 in the anterior sacs. As the air more rich in O_2 passes forward, it meets the blood of the lungs. Hence, according to Young (1981), it is not a system of **counter-currents** but that of **cross-currents** through which the blood is fully oxygeneated. Scheid and Piiper (1970) also confirmed earlier that the blood from the pulmonary trunk flows at right angles to the long axis of each parabronchus, thus establishing a cross-current relationship between the blood-flow and the parabronchial air-flow.

To sum up, the air-sacs are divided into two functional units: anterior sacs comprised of anterior thoracic, interclavicular with their diverticula and the cervical air-sacs with their diverticula; and posterior air-sacs comprised of the posterior thoracic and abdominal air-sacs. As the main bulk of oxygenated air enters the posterior sacs, they are mostly inspiratory in function, while the anterior sacs are

1. The external intercostal muscles occur more prominently within the anterior intercostal spaces where each fleshy belly of the muscle is attached to the ventral half of the vertebral rib, ventral to the uncinate process. The internal intercostal muscles, on the other hand, are more prominent in the posterior intercostral spaces. Fedde (1987) states that the external intercostal muscles are usually inspiratory in function, whereas the internal intercostals are expiratory.

mostly expiratory. But at inspiration, although not directly, some air enters the anterior sacs through posterior secondary and tertiary bronchi. At expiration, air from the posterior secondary and tertiary bronchi enter the anterior sacs and trachea to be expelled out.

(ii) Respiration during flight

It has been experimentally observed that a pigeon during flight has its metabolic rate 27 times more than that of the bird at rest. Hence, during flight, the respiratory rate of the bird also increases considerably. While a pigeon flies at 44 miles/70 km per hour, 76 times more air passes through its lungs than that entering the lungs in a bird at rest. This has been estimated to be 185 cc of air in a pigeon at rest, while the amount is 14 litre in its flight, that is, at this time, the respiratory rate increases by about 76 times. The increased activity of the pectoral muscles during flight not only causes the wing-beats but it also brings about greater movement of the sternum and ribs, and so, more rapid air flow through the lungs than in the bird at rest. It has also been evidenced that the interclavicular and cervical air-sacs having minor role in the respiration of a bird at rest actively participate during flight.

30.15. EXCRETORY SYSTEM

In pigeons, the excretory system consists of a pair of **metanephric kidneys** that are large and elongated. Each is divided into three lobes. The middle lobe is slightly larger than the other two. Each kidney is located at the back of the pelvis on either side of the middorsal line and is rigidly anchored to the dorsal wall of the pelvis. Histologically, the two layers of the kidney are quite distinct: an outer **cortex** and an inner **medulla**.

The actual excretory unit of a kidney is the **nephron**. Large number of nephrons are present in both the cortex and medulla, but their structures and functions are different. The cortical nephrons have similarities with those of reptiles and the medullary or the **juxtamedullary nephrons** with those of mammals. Wideman et al (1981) used two general terms for the avian cortical nephron: **Tubulus proximalis** and **Tubulus distalis**–corresponding to proximal convoluted tubule and distal convoluted tubule of mammals. Morild et al (1985)—who worked extensibly on the glomerular, extraglomerular and juxtamedullary cell components—report that the '**Macula densa**' is always present in birds, in both cortex and medulla. Both the cortex and medulla have convoluted and straight limbs and each has several bends.

The medullary tubule has a prominent **Loop of Henle** (also called **Ansa nephrica**) formed by both the proximal and distal tubules with a narrow intermediate connecting limb in the medullary region. The glomeruli of the cortical nephrons are smaller than those of the juxtamedullary region. The loop of Henle is either absent or ill-defined in the cortical nephrons of birds in general.

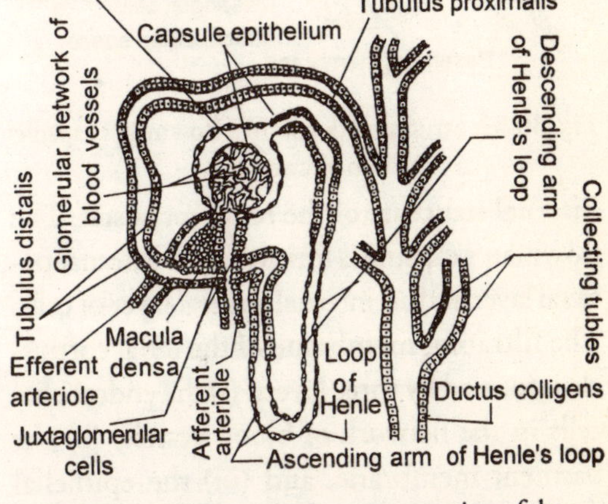

Fig. 30.31: Diagrammatic representation of the structure of a nephron with Bowman's Capsule and tubules

30.15.1. Bowman's Capsule, Glomerulus and Malpighian Corpuscle

In the cortical region of the kidney, the closed end of the uriniferous tubule becomes inflated like a balloon and its epithelial cells form a double-walled cup known as the **Bowman's capsule**. This encloses in its cavity a knot of blood vessels formed by the affarent and efferent arterioles which divide into a finer branching network. This tuft of knot of blood vessels is called **glomerulus**. The Bowman's capsule and the glomerulus, together, are known as the **Malpighian corpuscle**. At the beginning of the formation of glomerulus, the afferent arteriole forms a small number of branching loops which reunite and come out of the glomerulus as an efferent arteriole. The afferent arteriole, being wider and shorter, has an increased blood pressure which facilitates glomerular filtration of the fluid.

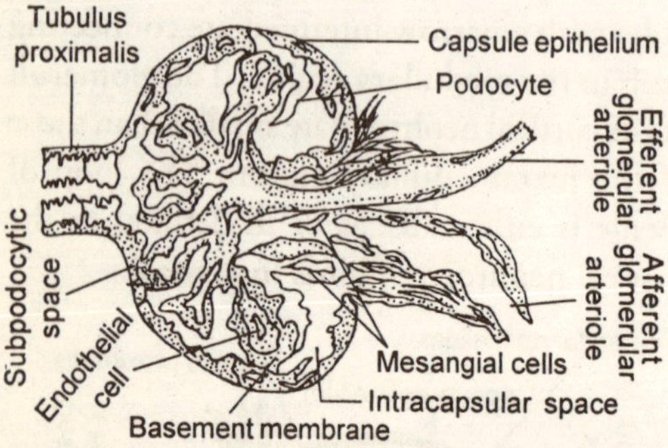

Fig. 30.32: Structural details of a Bowman's capsule with glomerulus

Internal structure of the renal corpuscle: The Bowman's capsule is formed of an external parietal layer and an internal visceral layer of cells. The **filtrating membrane** of the renal corpuscle is formed by three layers: (i) the endothelial cells in the network of blood vessels, (ii) the basment membrane, and (iii) the epithelial cells in the visceral layers of the Bowman's capsule. The visceral layer is formed of cells called **podocytes**. The podocytes of the visceral layer remain attached to the **basal lamina** through a number of tentacle-like cytoplasmic projections called '**pedicels**'. The **subpodocytic space** lies between the podocyte and the basement membrane. The narrow gaps between the adjacent foot-processes are known as the **filtration slits.** The endothelial cells of the glomerular network are flat, with round pores. Another kind of cells, called '**mesangial cells**', occur in the space between the endothelial cells and the basal lamina.

From the mediolateral aspect of the middle lobe of each kidney arises a narrow urinary duct or ureter on either side and it descends straight downward to open on either side of the urodaeum. The urinary bladder is absent in pigeons and in other birds (except in Ostrich). In avian kidneys, adequate uric acid is secreted. The urine, rich in uric acid, is collected by the collecting tubules of the medulla and transported to the ureter where absorption of water takes place. As a result, the urine becomes highly cocentrated and then, mixed with excreta, it is expelled through the cloaca. Some urine passes up the coprodeaum to the large intestine where further reabsorption of water takes place, the urine becomes more concentrated into a semi-solid '**guano**' which is expelled through the cloaca.

30.15.2. Functions of the Kidney

Of several functions of the kidneys, two main functions are specially noteworthy: (i) urine formation, and, through urine, expulsion of excess water, nitrogen, metabolites of sulphur and proteins from the body; and (iii) selective reabsorption of certain essential materials for the body from the glomerular filtrate.

(a) **Urine formation:** Three active principles are operative in the formation of urine:

1. Glomerular filtration: From the plasma fluid of blood, proportionate amount of water and crystalloids like sodium, potassium, chlorine, glucose, urea, etc. are filtered out by the glomerular network of the blood vessels. These accumulate within the Bowman's capsule. Only the blood corpuscles and the larger protein molecules cannot pass through the semi-permeable glomerular network.

Hence the glomerular filtration fully depends on the molecular size of the materials to be filtrated.

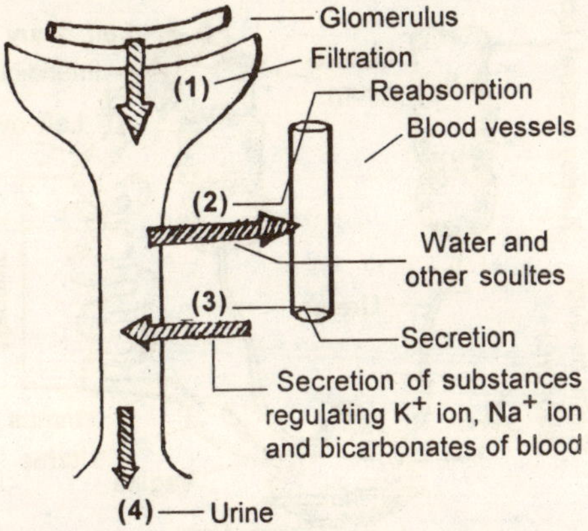

Fig. 30.33.: Three phases in the production of urine by a nephron. (1) glomerular filtration; (2) tubular reabsorption of water and many salts; (3) tubular secretion of certain substances regulating concentration of ions and bicarbonates in blood; (4) Production of urine at the end of phase-3

2. Tubular reabsorpotion: About 99% of water and most of the salts of various elements are reabsorbed through this process. The tubular reabsorption regulates the concentration of the body fluid and electrolytes. In reptiles and birds, the increased concentration of urea and uric acid is a notable feature. Hence, in most cases, the process acts as **selective reabsorption.**

3. Tubular secretion: Usually the selective reabsoption controls concentration of potassium ions (K+), hydrogen ions (H+) and bicarbonates. This process also helps expulsion of medicinal wastes. In most cases, the tubular secretion takes place through active transport.

Glomerular filtration is the first step in the formation of urine. The urine becomes largely modified after tubular reabsorption and tubular secretion.

(b) Other functions of kidneys

i. The kidneys produce a few new substances such as **ammonia, hippuric acid, inorganic phosphates.**

ii. The kidneys **regulate osmotic pressure** in blood and tissues.

iii. In an emergency, the enzyme renin acts on hypertensinogen of liver and forms **'Angiotensin I'**. The latter is acted upon by another enzyme to produce **'Angiotensin II'** which brings about a contraction of the branching arterioles and thus facilitates increase in blood pressure.

iv. **Erythropoietin,** another constituent of kidney, stimulates production of the red blood corpuscles.

v. The kidneys maintain a balance between the **acidity** and **alkalinity** of blood plasma.

vi. **The kidneys maintain** water balance in the body and expel metabolic wastes as **urea and uric acid** in a concentrated form.

30.16. REPRODUCTIVE SYSTEM

In birds, reproduction is usually seasonal, the gonads becoming more active in a particular season of the year. At this time, not in pigeons but in many other species of birds, sexual dimorphism is observed either in size or in the colour of feathers and feet or in the presence or absence in the crown or crest, etc. The weight of the testes enormously increases during breeding season.

The male and female reproductive systems of a pigeon are described:

30.16.1. Male Reproductive System

The male reproductive system in pigeon consists of a pair of oval whitish **testes** and a pair of narrowly coiled and opaque **vasa deferentia** each of which forms proximally another coiled **epididymis**.

(a) **Testis:** At the anterior end and ventral to each kidney lies the testis, attached firmly by a special segment of the peritoneal membrane called the **mesorchium**. Each testis is invested by a very thin fibrous layer called **tunica albuginea**. Numerous filamentous coiled **seminiferous tubules** producing different stages of the male gonad—**spermatogonia**, **spermatocytes** and **spermatozoa**—occur within the **seminiferous tubules**. The different stages of sperm-producing cells are actually formed by cell divisions of the epithelial cells of the seminiferous tubules. Amidst the epithelial cells, at intervals, are found some hormone-secreting interstitial cells called the **cells of Leydig**.

(b) **Vas deferens:** A narrowly coiled and opaque vas deferens, arising from the inner marginal wall of each testis, descends downward parallel to the ureter and opens into the urodaeum. Near the urodaeum each vas deferens swells to a small extent to form a sac-like **seminal vesicle** (Glomus seminale; NAA, 1993). Each vas deferens at the testis end forms a narrow coiled **epididymis** which is connected to the seminiferous tubules through a large number of narrow delicate tubules called **vasa efferentia**.

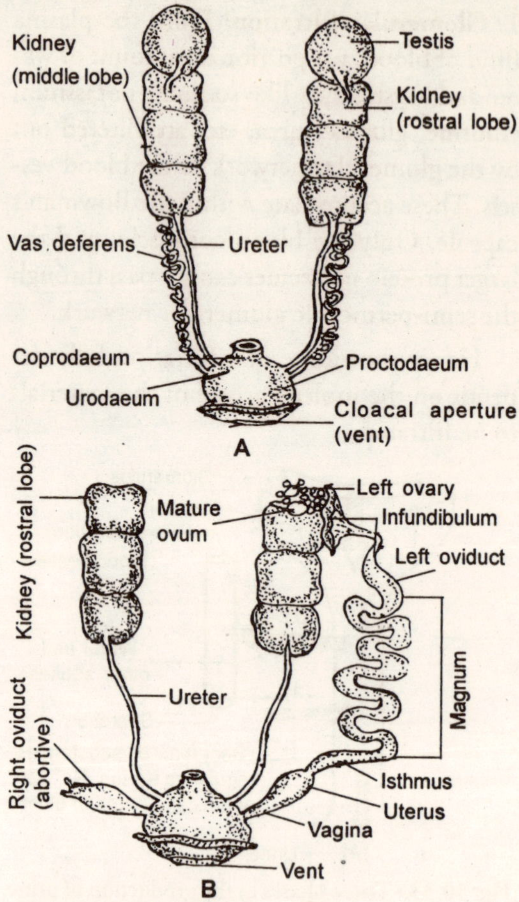

Fig. 30.34: Urinogenital systems of pigeon. A. Male urinogenital system; B. Female urinogenital system. Note that in recent nomenclature, all the five parts from infundibulum to vagina are subdivisions of the avian oviduct

30.16.2. Female Reproductive System

The female reproductive system in the adult consists of a well-developed ovary and oviduct on the left side only, the ovary on the right side is atrophied and only a rudiment of the right oviduct persists. In embryonic condition, however, both the ovaries and oviducts are bilaterally present. But, as development proceeds, the ovary and the oviduct of the right side gradually degenerate and, in the adult bird, these organs on the left side only persist and become functional.

(a) Ovary: The left ovary is quite large in size; it is located at the rostroventral aspect of the left kidney, being firmly attached to the dorsal body wall by a fold of the coelomic peritoneum called '**mesovarium**'. The enlargement in the size of the ovary is maximum during breeding season. Numerous small shallow sac-like follicles, called 'ovarian follicles', develop over the surface of the ovary, giving it a studded appearance. Within each follicle, an ovum representing a particular developmental stage in oogenesis is present. Each ovum is filled with sufficient yolk. The ovarian follicle supplies the oocyte with yolk collected from the liver. Different layers and zones of the follicle have been named—zona radiata, stratum granulosum, lamina perivitellina (vitelline membrane), lamina basalis folliculi, theca interna and theca externa).

(b) Oviduct: The left oviduct (Oviductus sinister) is a broad and elongated tubular structure that is suspended from the dorsal bodywall by the dorsal oviductal ligament which continues ventrally as the fan-shaped ventral oviductal ligament. The free edge of this ligament is supported by smooth muscles and, caudally, this ligament is expanded into a muscular cord called '**Funiculus musculosus**'. The oviduct is divided into five parts: **Infundibulum**, **magnum**, **isthmus**, **uterus**, and **vagina**. Although these subdivisions are almost universally accepted in birds, Schwarz (1969) suggested simply two subdivisions–cranial and caudal—of the oviduct. The infundibulum comprises a wide oviductal funnel which opens into the body cavity by an elongated slit-like aperture and a narrow tubular part following the funnel. The tubular part is also known as chalaziferous region as the tubular glands secrete chalazae of the egg shell. The aperture of the funnel is encircled by wavy lips called **infundibular fimbriae**.

The magnum is the longest part of the oviduct. The magnum mainly secretes albumen.

(c) Isthmus: A short narrower region immediately following the magnum is called the isthmus. The isthmus glands add some protein and water to the albumen, and, possibly, the first deposit of calcium also occurs in the isthmus.

(d) Uterus: It is a comparatively broader part that immediately follows the isthmus. Some authors have preferred the term 'Shell gland' for uterus which bears no homology with that of a mammal. The uterus has been divided into three zones: (i) the apical region—the initial short tubular portion where the first deposit of calcium occurs; (ii) the major uterine region—a pouch-like portion holding the egg for most of the period of shell formation, and (iii) the uterine recess—the funnel-shaped caudal region containing sufficient ester lipids in the ciliated cells.

The longitudinal primary folds of the uterus are intersected by transverse or oblique furrows forming leaf-like lamellae (King, 1979).

(e) Vagina: The uterus caudally leads into a short, muscular S-shaped tube—the vagina. It opens in the urodaeum of the cloaca. Possibly, the contraction of some sort of a sphincter muscle of the vagina helps expulsion of the egg.

30.16.3. Breeding, Fertilization and Development

During **breeding season**, the mates attract each other by their characteristic flight in which the wings are flapped dorasally and struck against one another. In pigeons, the breeding season is not very clearly defined, although each year there are 2 to 3 clutches, each with two eggs.

Internal fertilization takes place as a rule. During breeding season, the adult male and female pigeons attach themselves by their proctodaeum (there being no erectile copulatory organ in the pigeons) and this type of mating in birds is also known as **'cloacal kiss'**. At this time, the male transfers its spermatozoa into the cloaca (urodaeum) of the female. The spermatozoa then travel upward and are temporarily stored within the sperm-storage tubules (**Tubuli spermatici**) of the oviduct. From here the spermatozoa are released and each spermatozoon fertilises an ovum.

The **eggs** in pigeons are quite large, white, elliptical. Each shelled egg is filled with yolk material. The ovum is the female gamete that arises solely from the ovary after the second maturation division. In the ovum, the thin layer of cytoplasm forms the **blastoderm** or the **germinal disc** (**Discus germinalis**), as the case may be. The germinal disc is a greyish area about 1/8 in (3 mm) in diameter just beneath the cytolemma. According to Gilbert (1971 b), it is called **blastoderm** if fertilized **and blastodisc** if unfertilized. While the egg descends down the oviduct, the albumen layer is formed around the yolk circle and the germinal disc.

The albumen layer is also formed of three components: (i) **Chalaza**—the thin innermost layer of the dense albumen, (ii) the **liquid—thin albumen layer**, and (iii) the **thick, white dense albumen layer**. This layer is connected to the **internal shell membrane** at each end of the egg on either side. The shell membrane formed around the albumen on all sides has an internal and an external layer. The external layer forms a hard shell that protects the egg in all respect.

An **embryo** is formed within the shelled egg and further development proceeds. To activate development of the egg, a temperature ranging between 97°F/36°C and 104°F/40°C has to be maintained. To achieve this in the pigeons, both the parents sit on the eggs alternately, the process being known as **incubation**. Although incubation is done by both the parents in most species, there are exceptions where incubation is done by the females (e.g., Golden Eagle, Hornbill, Tailor bird, Baya Weaver bird), and in the ratite birds (except in Ostrich where both the parents sit), the males usually attend to incubation.

The incubation period, though slightly variable in different species, lasts in the pigeons for about 14-19 days. Then the hatchlings break open the shell and come out. The nidicolous (altricial) hatchlings are fed by both the parents with the highly proteinous milky secretion of the crop glands called **crop milk** or **pigeon's milk**. The young pigeons fly after about 30-35 days and become used to their usual diet.

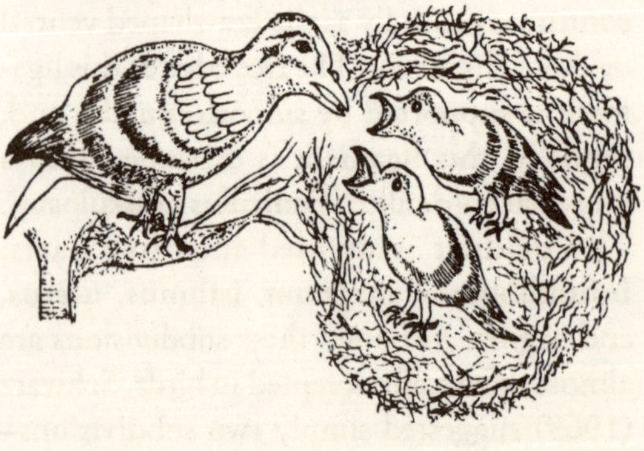

Fig. 30.35. An adult pigeon feeding the hatchlings

30.17. NERVOUS SYSTEM

The body cells in animals respond to stimuli originating from the changes in both external and internal environment. The modes of such responses are different in different cells and tissues—the endocrine glands secrete, the muscle cells contract, the nerve cells conduct impulses and the lymph cells engulf the microorganisms. The hormones secreted by the endocrine glands act in coordination with the nervous system in the integration of all bodily functions. If the function of the nervous system may be said to be very fast on electrochemical basis, that of the endocrine system is rather slower as it acts on chemical basis. But the two systems are so closely related and influenced by the other that these two systems are designated as **neuroendocrine system.**

Almost all the parts of the nervous system are derived from ectoderm and are formed of specialised types of nerve cells and tissues. During embryonic development the nerve cells arise from the neural tube and the neural crest cells, both derived from the ectoderm. The unit of nerve tissue is a **neuron** which consists of a nucleated cell body and its projections.

30.17.1. Structure of a Neuron

A neuron consists of a multi-cornered or a smooth-surfaced **cell body** or **cyton** containing a central nucleus and cytoplasm around it. Within the cytoplasm are found scattered numerous basophilic granules called **'Nissl's granules'**. These granules are formed of ribonucleic acid and ribosomes. These take part in the production of new cytoplasm and protein synthesis of neuron. The concentrations of nerve cells within the brain and spinal cord are known as **nuclei** and such concentrations outside the central nervous system are called **ganglia.**

The other part of a neuron constitutes projections of the cell bodies. These projections are mainly of two types: (i) **dendrites** which arise as shorter, narrower and delicate branching projections from the external wall of the nerve cell and establish connections with the dendrites of adjacent neurous; (ii) **axon** or **axis cylinder** which arises as a considerably long (much longer than the

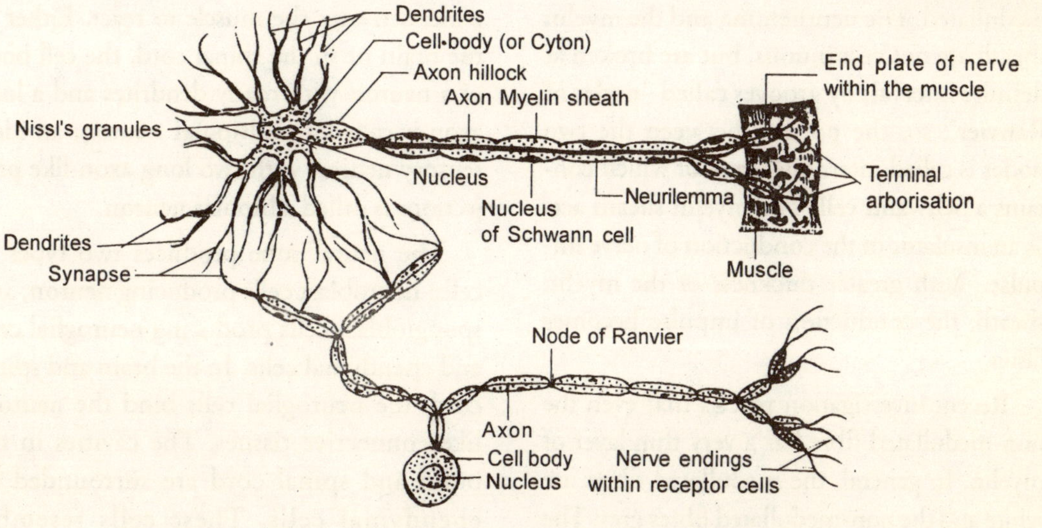

Fig. 30.36. Different parts of a sensory and a motor neuron with their synapses

dendrite's) projection from a blunt corner of the nerve cell and forms a bundle of terminal branches called '**terminal arborisation**'. From the long axon, one or more collateral branches may or may not develop.

The axon and its external sheath are together known as **nerve fibre**. The nerve fibre outside the central nervous system is ensheathed by a thin, continuous covering called **neurilemma**—formed of a special neuroglia of the central nervous system—and is then distributed in the peripheral nervous system.

It is to be noted here that the brain and the spinal cord arise from the neural tube, while the peripheral and most of the autonomic nervous system arise from the neural crest cells. **The neural tube, the neural crest cells and placodes of all the sensory organs are ectodermal.**

In the fibre, surrounding the axon and between the axon and the neurilemma, there is a shiny fatty layer called myelin sheath or medullary sheath. Thus, based on the presence or absence of the myelin sheath, the nerve fibre is designated as medullated or non-medullated. The neurilemma and the myelin sheath are not continuous, but are broken at definite intervals by grooves called '**nodes of Ranvier**'; so, the portion between the two nodes is called internode, each of which contains a **Schwann cell**. The myelin sheath acts as an insulator in the conduction of nerve impulse. With greater thickness of the myelin sheath, the conduction of impulse becomes faster.

Recent investigation reveals that even the non-medullated fibre has a very thin layer of myelin. In general, the medullated fibres are white and the non-medullated fibres grey. The bundles of nerve fibres within the central nervous system are called '**fibre tracts**', the nerve fibres aligned in parallel bundles and held together by the connective tissue outside the central nervous system are called '**nerve**'. Electron microscopic study reveals that the distal end of an axon divides into two or more branches, each of which forms a terminal swelling called '**terminal button**'—where adequate quantity of mitochondria are present. Most of the peripheral nerves are formed of medullated fibres and the autonomic nerves appearing grey have a thin layer of myelin.

The conduction of nerve impulse is a complex phenomenon. In the cranial and spinal nerves of the peripheral nervous system, two types of neurons are concerned in the conduction of stimuli and response such as (i) **sensory afferent neurons,** and (ii) **motor efferent neurons**. From the receptor cells and organs the sensory neuron conducts the stimulus first to the ganglia of the cranial and spinal nerves and, thence, to the brain or the spinal cord through the synapses formed by the dendrites of these two types of neurons. The effector organ then responds to the stimulus through motor nerve which, on reaching the motor end plate, activates the muscle to react. Either in the brain or in the spinal cord, the cell body of a neuron with many dendrites and a long axon is called a **multipolar neuron**; an ideal sensory neuron with two long axon-like projections is called a **bipolar neuron.**

The neural tube produces two types of cells: neuroblast cells producing neuron, and spongioblast cells producing neuroglial cells and ependymal cells. In the brain and spinal cord, the neuroglial cells bind the neurons like connective tissues. The cavities in the brain and spinal cord are surrounded by ependymal cells. These cells resemble epithelial cells and are sometimes ciliated. In the brain and spinal cord, the cell bodies, dendrites, neuroglial cells and the non-medullated axons together form the **'grey matter'**; the medullated white bundles of nerve fibres are called **'white matter'**. In general, the white matter is concerned with the conduction of the impulse and the grey matter in coordinating impulses and effecting response.

Based on polarisation, a stimulus may be conducted on both sides of a nerve cell. It has already been mentioned that the dendrites (and also the axon-ends) form synapses with those of the other neuron for the conduction of stimulus, but the transmission of stimulus is not direct through the synapse; instead, usually the axon-end of the neuron secretes a complex bio-chemical substance called acetylcholine which activates dendritic ends of the other neuron to receive the stimulus which is thus conducted through chains of neurons. The post-ganglionic fibres of the sympathetic nerve secretes a hormone called norepinephrine which also produces impulse. Norepinephrin is a catecholamine group of hormone secreted by the adrenal medulla. Another hormone—epinephrine—of this group also performs similar function.

30.17.2. Functional Divisions of the Nervous System

The functions of the nervous system may be precisely subdivided into two parts: **somatic** and **visceral**. The somatic domain is restricted to the functions of the voluntary muscles of the body, bones, skin and derivatives of the skin. The visceral functions include stomach and other organs of the alimentary tract, heart and other organs of the circulatory system, respiratory organs, urinogenital organs, and endocrine glands.

Both somatic and visceral parts are each constituted of two antagonistic nervous elements—**afferent** or **sensory**, and **efferent** or **motor**. Their precise functions are enumerated:

i. **Somatic sensory** nerve fibres carry impulses from the skin and its derivatives, sense organs, voluntary muscles, skeletal joints, etc. to the central nervous system.
ii. **Somatic motor** nerve fibres carry responses from the central nervous system to the effector organ where the response is translated into action.
iii. **Visceral sensory** fibres carry impulses from the viscera, taste buds and other internal organs to the central nervous system.
iv. **Visceral motor** fibres carry responses from the internal organs supplied with involuntary muscles, heart and glands—both exocrine and endocrine.

It may be specially mentioned here that in the medulla oblongata of the brain and in the spinal cord, the above-mentioned four types of nerve fibres with different properties are serially arranged from the dorsal to the ventral aspect. From this arrangement of the location of nerve fibres it may be presumed that the dorsal aspect of the central nervous system is sensory and the ventral aspect motor.

A. Brain

The brain in pigeon has a structural correspondence with that of other vertebrates, and, more specifically, with that of amniota. As an integrating system of the body, the brain in pigeon is very highly developed largely due to its volant adaptation.

The nervous system in pigeons, as in other vertebrates, is chiefly subdivided into two parts: the **central nervous system**—comprising the brain and the spinal cord, and the **peripheral nervous system** such as the cranial nerves and the spinal nerves that arise from the central nervous system and are distributed outwardly, that is, towards periphery of the body.

Besides the above two subdivisions of the nervous system, a few authors have described the **autonomic nervous system** as an independent system but most authors (Weichert and Presch, 1977; Young, 1981; Romer and Parson, 1986) consider the autonomic system as a part of the peripheral nervous system and prefer to call it **Peripheral autonomic system.**

(i) Development and Differentiation of the Brain: A Brief History

The condensation and infolding of the ectoderm cells in the mid-dorsal region of the embryo form the neural plate which, in the form of two neural folds, join from either side to form the neural tube. At this time, the thick and swollen anterior part of the neural tube, due to unequal growth and two transverse fissures, is subdivided into three primary brain vesicles. From the front end backwards, these primary vesicles are known as the **'Forebrain'** or **'Prosencephalon, 'Midbrain'** or **Mesencephalon** and **'Hindbrain'** or **'Rhombeneephalon'**.

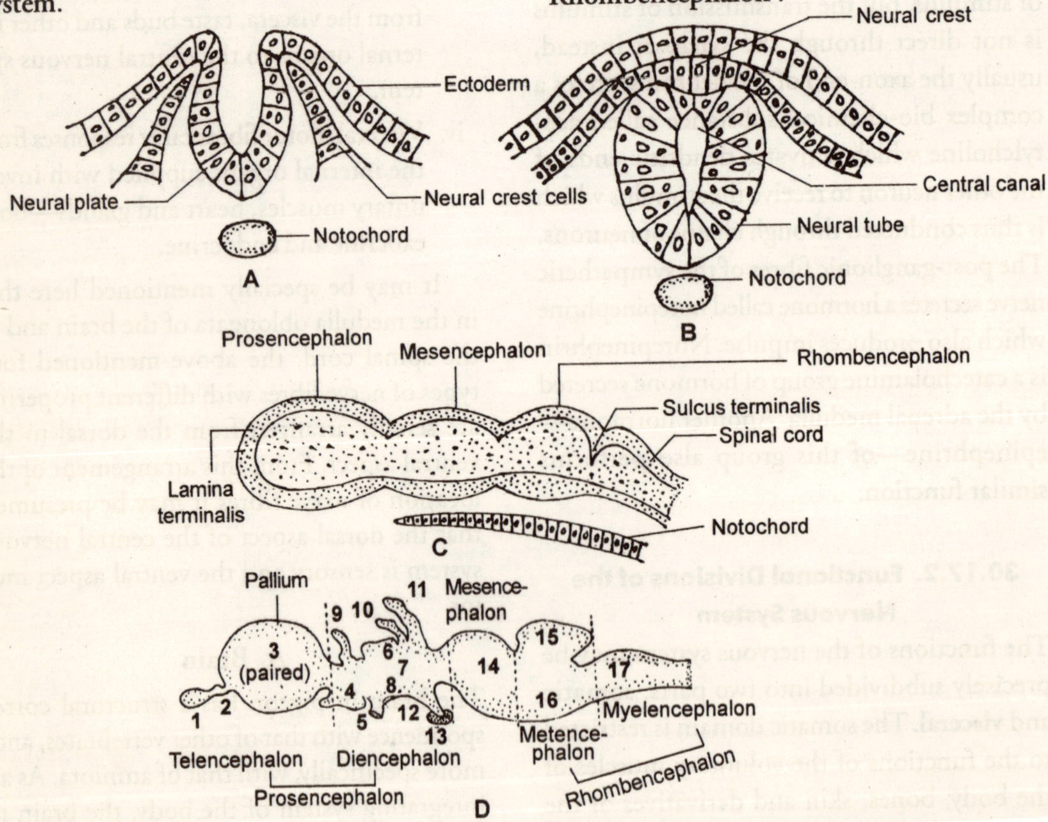

Fig. 30.37. Different stages in the development and differention of the brain in pigeon. A. Formation of neural plate and neural crest cells from the ectoderm; B. Formation of neural tube, central canal and the neural crest; C. Development of primary brain vesicles; D. Later differentiation and subdivisions of the brain: 1. Olfactory lobe; 2. Olfactory stalk; 3. Cerebral hemispheres (paired); 4. Lamina terminalis; 5. Optic stalk; 6. Epithalamus; 7. Thalamus; 8. Hypothalamus; 9. Paraphysis; 10. Parietal organ; 11. Pineal body; 12. Infundibulum; 13. Pituitary gland; 14. Tectum (Midbrain); 15. Cerebellum; 16. Pons; 17. Medulla oblongata

The rate of growth of the posterior part of the neural tube is much less compared to that of the anterior part forming the brain, and, hence, the posterior part of the brain continues behind as the spinal cord. Gradually the prosencephalon is subdivided into **telencephalon** and **diencephalon**, and the rhombencephalon into **metenecephalon** and **myelencephalon**. The mesencephalon or the midbrain remains undivided, except that, in the snakes and mammals, the two optic lobes are divided into four by a transverse fissure through the middle.

The rostral wall of the telencephalon is called **'Lamina terminalis'** which marks the anteriormost boundary of the brain. Telencephalon further differentiates anteriorly ahead of lamina terminalis to form a pair of cerebral hemispheres whose thickened dorsal wall forms the pallium. The original pallium or the archipallium is transformed in different vertebrates into palaeopallium and neopallium. The lateral and ventral walls of the cerebral hemispheres form thickened areas called **'Corpora striata'**. Rostroventrally, the two cerebral hemispheres give rise to two small olfactory lobes. This rostral part of the telencephalon is also known as 'Rhinencephalon'.

The diencephalan or the third part of the forebrain is partially covered by the cerebral hemispheres in front and the optic lobes behind. The diencephalon is formed dorsally by the **epithalamus**, medially by the **thalamus** and ventrally by the **hypothalamus**. From the front end of the epithalamus, a paraphysis and, from the cadual end, a club-shaped **pineal organ** or, at its front, a shorter and narrower parapineal organ—or both—may develop as projections in different vertebrates. The thin dorsal ependymal layer of the diencephalon joins the richly vascularised **pia mater** to form the **'Tela choroidea'**. The infoldings of the tela choroidea, rich in blood capillaries, form the **anterior choroid plexus** over the third ventricle as its roof. At the ventral surface of the diencephalon, the hypothalamus grows downward in the form of a funnel-shaped infundibulum which, at its tip, bears the **pituitary organ**—the master endocrine gland. Just in front of the infundibulum and ventral to the diencephalon the two optic nerves form the **optic chiasma**.

The thick dorsal and lateral walls of the midbrain or the mesencephalon form the tectum which develops into two distinct optic lobes. At the ventral surface of the mesencephalon, in the region of the basal plate, numerous motor nuclei and the ascending and descending fibre tracts are present. This thick area is called **'tegmentum'**. Bundles of longitudinal fibres forming the thick, pillar-like **crura cerebri** ventrally connect the diencephalon with the rhombencephalon.

In the hindbrain or rhombencephalon, the dorsal outgrowth of the metncephalon forms the **cerebellum** which assumes a distinctly lobulated structure in flying birds. This part is concerned with the maintenance of equilibrium and speed in flight and in coordinating various reflex activities. The outer cerebral cortex is composed of grey matter; the inner medulla has a layer of white matter. The last part of the hindbrain constitues the myelencephalon forming the **medulla oblongata**. Behind the cerebellum this part assumes a tringular shape which becomes cylindrical further caudally. The medulla joins the spinal cord at the level just in front of the first pair of spinal nerves. The medulla oblongata is regarded by many authors as the oldest part of the brain because of its uniform structure in almost all the vertebrates. Although the lateral and ventral walls of the medulla are adequately thick, its dorsal wall retains thin epithelial characteristic. The roof of the medulla is formed by the richly vascularised pia matter in conjunction with the tela choroidea known here as the **posterior choroid plexus.**

In the adult amniota and also in a few reptiles, the brain stem consisting of the metencephalon and the myelencephalon remains almost unchanged, but in a few reptiles and in birds and mammals, marked changes occur in the ventral surface of the brain stem due to concentration of a large number of nuclei and fibre tracts. These fibre tracts become specially developed and swollen in the mammalian brain to form the **Pons Varolii** which transversely connects the cerebrum with the cerebellum. Pons is also moderately developed in birds.

(ii) Ventricles of the Brain

The longitudinal cavities of the two cerebral hemispheres are called lateral ventricles. Further ahead, the narrow cavities of the olfactory lobes are known as **rhinocoel.** On either side, the narrow rhinocoel is connected to the lateral ventricle of that side. Within the diencephalon the internal cavity of the thalamus is called the third ventricle. The two lateral ventricles comminucate with the third ventricle through an aperture called **'Foramen of Monro'**. In the midbrain the cavity of the two optic lobes on either side is called **optocoel.** The fourth ventricle within the medulla oblongata is quite large. The third ventricle communicates with the fourth ventricle through a narrow canal known as **'iter'** or the **'Aqueduct of Sylvius'**. A narrow canal from the fourth ventricle also communicates with the narrower slit-like metacoel of the cerebellum in pigeons.

(iii) Flexures of the Brain

Only in *Branchiostoma*, the brain and the spinal cord are aligned in a straight line. In others, particularly during development in the embryo of a vertebrate, unequal growth in different parts of the brain within limited space gives rise to several flexures. During embryonic conditions, three such flexures like the **cephalic flexure** occurring dorsal to the diencephalon, the **pontine flexure** lying ventral to the cerebellum, and the **cervical flexure** lying dorsal to the medulla oblongata change the original position of the brain—either temporarily or permanently.

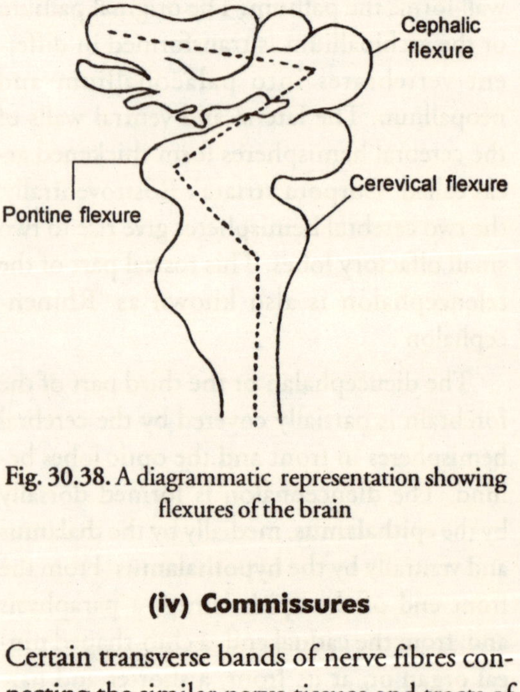

Fig. 30.38. A diagrammatic representation showing flexures of the brain

(iv) Commissures

Certain transverse bands of nerve fibres connecting the similar nerve tissues and tracts of both sides within the brain and spinal cord are known as commissures. Of several commissures in the brain, two commissures are specially noteworthy: (i) **Anterior Commissure**—which arises from the lamina terminalis and lies in front of the third ventricle, connecting the corpora striata of both sides; (ii) **Posterior Commissure**—which lies dorasally at the junction of the diencephalon and mesencephalon.

Likewise, in the brain, certain commissures also develop within the grey and white matters of the spinal cord, such as, the **Dorsal grey commissure** and **Ventral grey commissure**—lying dorsal and ventral to the central canal, respectively, within the grey matter and a **White commissure** within the white matter.

(v) Meninges of the Brain

Both the brain and spinal cord are enclosed by one or more protective coverings formed of stiff connective tissues. These membranes are called meninges (sing. meninx). In the process of evolution of higher vertebrates, these membranes become more diversified.

A vascularised stiff membrane covering the bone is called periosteum which encloses the skull cavity and neural canal of the vertebrae.

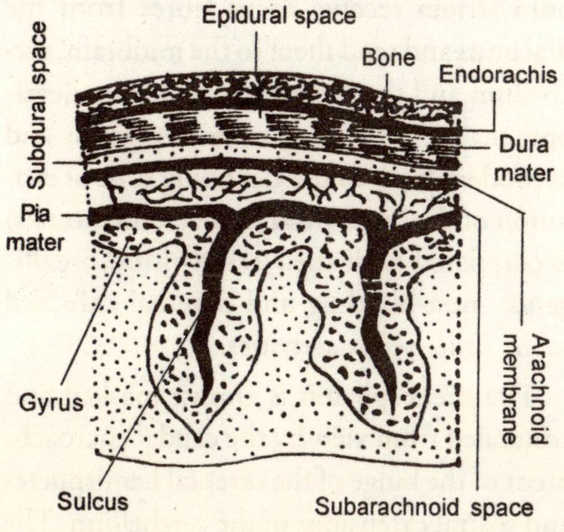

Fig. 30.39. A sectional view showing meninges of the brain

This membrane is known as endorachis. This is not a meninx of the brain or spinal cord, but is closely related to the underlying meninges.

In amphibia, reptiles and birds, two meninges appear in the brain: the external '**dura mater**' formed of connective tissues, and the internal '**pia-arachnoid membrane**' which is partly derived from the neural crest cells. The space between the pia-arachnoid and the dura mater is called the **subdural space** and that between the dura mater and endorachis is known as **epidural space**. The epidural space is mainly filled with alveolar tissues, fats and venous capillaries.

Although earlier workers identified only two meninges in birds, the dura mater (which also consists of two closely adherent layers) and the pia mater (as against three distinctly separable meninges in mammals), Bohme (1970) and Jones and Dolman (1978) have confirmed the presence of separate Pia and Arachnoid layers and subarachnoid space in *Columba* and *Gallus*. The subarachnoid space is filled with cerebrospinal fluid.

The cerebrospinal fluid slowly flows through the ventricles of the brain, central canal of the spinal cord, subarachnoid space, etc. and it exits through three apertures present at the dorsal surface of medulla oblongata to reach the space between the meninges. As a result, the entire central nervous system is bathed in the **cerebrospinal fluid**. The medial aperture is known as '**Foramen of Magendie**' and the lateral apertures as '**Foramina of Luschka**'.

In pigeons, as also in the adult vertebrates in general, the main divisions, subdivisions and parts of the brain are enumerated below (summarised from the schemes adopted by Kent, 1983, and Romer and Parason, 1986).

(vi) Divisions of the Brain

Table 30.2: Primary Divisons, Subdivisons and Parts of the Brain in a Pigeon

Divisions	Subdivisions	Parts
1. Forebrian (Prosencephalon)	Telencephalon	Olfactory lobes/Olfactory bulbs (also called Rhinencephalon); Cerebral hemispheres (Cerebrum); Basal nuclei (Corpus striatum); Cerebral cortex (Pallium); Lateral ventricles
	Diencephalan	Epithalamus; Thalamus; Hypothalamus; Pineal organ and Pituitary organ; Third Ventricle and associated structures
2. Midbrain (Mesencephalon)	Tectum mesencephali; Tegmentum mesencephali	Two optic lobes (Corpora bigemina); Tegmentum (Crura cerebri); Aqueduct of Sylivius and other associated nuclei, commissura and fasciculi
3. Hindbrain (Rhombencephalon)	Metencephalon	Cerebellum; Part of Medulla oblongata; Tegmentum; Part of Fourth Ventricle, and associated structures
	Mycelencephalon	Most of Medulla oblongata and Fourth Ventricle

(vii) Brain in Pigeon

(a) **Structure:** Compared to body size, the brain in pigeon is large, slightly elongated, and laterally broad.

Because of a poor sense of smell, the olfactory lobes are smaller. However, compatible with a high degree of intelligence and dynamic activities of life, the cerebral hemispheres are quite large and caudolaterally broad and extended to overlap a part of the cerebellum. The pallium is thin and small (represented mainly by the archipallium and palaeopallium as in reptiles), but the Corpora striata within the cerebrum are thick and extensive and divided into three layers: **Epistriatum, Mesostriatum and Hypostriatum.**

In pigeons, as in other birds, hypostriatum assumes a remarkable development. The corpora striata receive nerve fibres from the thalamus and send them to the midbrain, cerebellum and the medulla. The greater development of the hemispheres in pigeons and in modern birds in the ascending scale of evolution of the brain (next to that in mammals) is correlated with their acute vision, intelligence, nest-building and parental care and many other reflex activities.

The diencephalon is almost masked and concealed from view by the caudal encroachment of the bulge of the cerebral hemispheres and rostral extensions of the cerebellum. The pineal organ is small and third ventricle is slit-like narrow. The thalami region is very

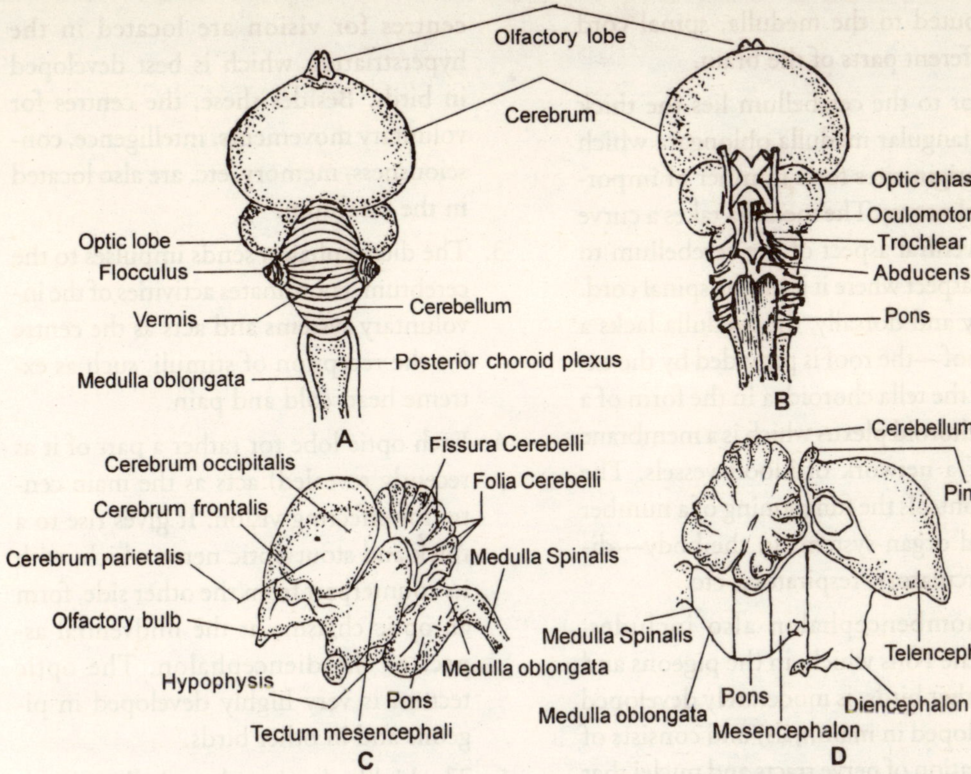

Fig. 30.40. Diagrammatic representation of the brain in pigeon: A. Dorsal view; B. Ventral view; C. Lateral view; D. Mediolateral view showing relationship of the different parts of the brain

thick and well-developed and are connected with the corpora striata as well as nerve fibres of the spinal cord. Ventrally, the pituitary gland is well-developed and with it is connected the hypophysial portal system.

In the midbrain, the two optic lobes are very large and rounded, and they are displaced somewhat laterlly due to the caudal extension of the cerebrum. The aqueduct of the diencephalon communicates with the narrow slit-like cavities of the optic lobes called **optocoel** (present in anamniota; not present in birds). The tectum of the midbrain, giving rise to the optic lobes, is very thick and well-developed. Likewise, the tegmentum—comprising the motor nuceli and fibre-tract regions—is also thick and well-developed in pigeons. The two optic nerves, each arising from the optic lobe on either side, form a distinct optic chiasma at the midventral aspect of the two optic lobes. While partial exchange of fibres takes place at the chiasma, the rest of the nerve of one side enters the opposite eye.

The cerebellum in pigeons is large, rostrally and caudally elongated and each has a central lobe called Vermis—formed of a large number of folds and grooves and a lateral lobe on either side containing a flocculus. Within the cerebellum, the bundles of nerve cells forming nuclei are stratified into two distinct layers: an outer layer of grey mater and an inner layer of white mater. These layers with their branches and subbranches are arranged like petals of a flower and form the characteristic **'Arbor vitae'** within the cerebellum. Quite a few nerve tracts from the cerebellum are distributed to the medulla, spinal cord and to different parts of the brain.

Posterior to the cerebellum lies the thick flat and triangular medulla oblongata which provides origin-sites for a number of important cranial nerves. The medulla takes a curve from the ventral aspect of the cerebellum to the dorsal aspect where it joins the spinal cord. Proximally and dorsally, the medulla lacks a nervous roof—the roof is provided by the extension of the tella choroidea in the form of a **posterior choroid plexus** which is a membrane formed of a network of blood vessels. The medulla controls the functioning of a number of essential organ-systems in the body—digestive, circulatory, respiratory, etc.

The rhombencephalon also includes, ventrally, the Pons which, in the pigeons and in many other birds, is moderately developed (best developed in mammals) and consists of a concentration of nerve tracts and nuclei that transversely connect the cerebrum with the cerebellum.

(b) Functions of the Brain

1. The olfactory lobes being very small and rudimentary, the sense of olfaction is poorly developed in pigeons.
2. The large cerebral hemispheres—mainly due to enormous development of the Corpora striata—control various important functions in the pigeons. The corpora striata arises as thickenings of the lateroventral walls of the cerebrum and are arranged in three layers around the lateral ventricles. The corpora striata control various reflex activities of the body and transmit impulses to different parts of the brain.

   In pigeons, the centre for hearing is located in the broad neostriatum and the centres for vision are located in the hyperstriatum which is best developed in birds. Besides these, the centres for voluntary movements, intelligence, consciousness, memory, etc. are also located in the cerebrum.
3. The diencephalon sends impulses to the cerebrum, coordinates activities of the involuntary systems and acts as the centre for the reception of stimuli, such as extreme heat, cold and pain.
4. Each optic lobe (or rather a part of it as recently revealed) acts as the main centre for effecting vision. It gives rise to a thick and stout optic nerve which, with its counterpart from the other side, form an optic chiasma at the midventral aspect of the diencephalon. The optic tectum is very highly developed in pigeons and in other birds.
5. The highly developed cerebellum in pigeons acts as the main organ for voluntary movments, migratory movements and for balancing and steering the body in flight.
6. The medulla oblongata, providing origin-sites for a number of cranial nerves, controls the involuntary functions of various systems in the body.

B. Spinal Cord

Caudally, the medulla oblongata becomes narrower and cylindrical, and the site at which it enters the neural canal of the first cervical vertebra (Atlas) to join the spinal cord marks the commencement of the spinal cord. It then runs through the neural canal of the vertebral column. Alternatively, it may be said that the spinal cord begins at the site where the distribution of the outer grey matter and inner white matter of the brain is just reversed.

The meninges of the brain also extend as enclosures of the spinal cord. Two longitudinal grooves running along the mid-dorsal and mid-ventral lines of the spinal cord are known, respectively, as the **dorsal fissure** and the **ventral fissure.**

The spinal cord becomes swollen at two places to form two complex systems of nervous network. These are **'Brachial plexus'** which sends nerves to the wings, and the **'Lumbosacral plexus'** (also known as **synsacral plexus** in birds) which sends nerves to the two hindlimbs and adjacent parts.

The dorsal and ventral fissures incompletely subdivide the spinal cord into two equal halves. These two halves are connected with one another through commissures of nerve tissues that bridge them along the mid-region.

During its development, the rate of growth of the spinal cord is much slower than that of the vertebral column, and, hence, the length of the spinal cord is much less than that of the vertebral column. Caudally, the spinal cord becomes gradually narrower, appearing like a spearhead and is known as **'Conus terminalis'**. Laterally, in this region, below the level of the plexus synsacralis, Baumel (1979, 1988) has vividly described in *Columba livia* the presence of **'Plexus pudendus'**–comprised of the ventral roots of spinal nerves numbering 27 to 31, and the **'Plexus caudalis'** comprised of the ventral roots of spinal nerves numbering 32 to 37 or 38.

In a cross-section of the spinal cord, inner to the meninges, the outer layer of white matter formed of numerous medullated nerve fibres practically surrounds the inner layer of grey matter formed of non-medullated nerve fibres. In the amniotes, the grey matter assumes the shape of H- or a butterfly around the central canal of the spinal cord. The comparatively narrower arms on the dorsal side of the 'H' are called **'dorsal columns'** and the wider ventral arms are called **'ventral columns'**. In the connecting arm of the 'H', the portions dorsal and ventral to the central canal are known, respectively as the **'dorsal and ventral gray commisssures'**.

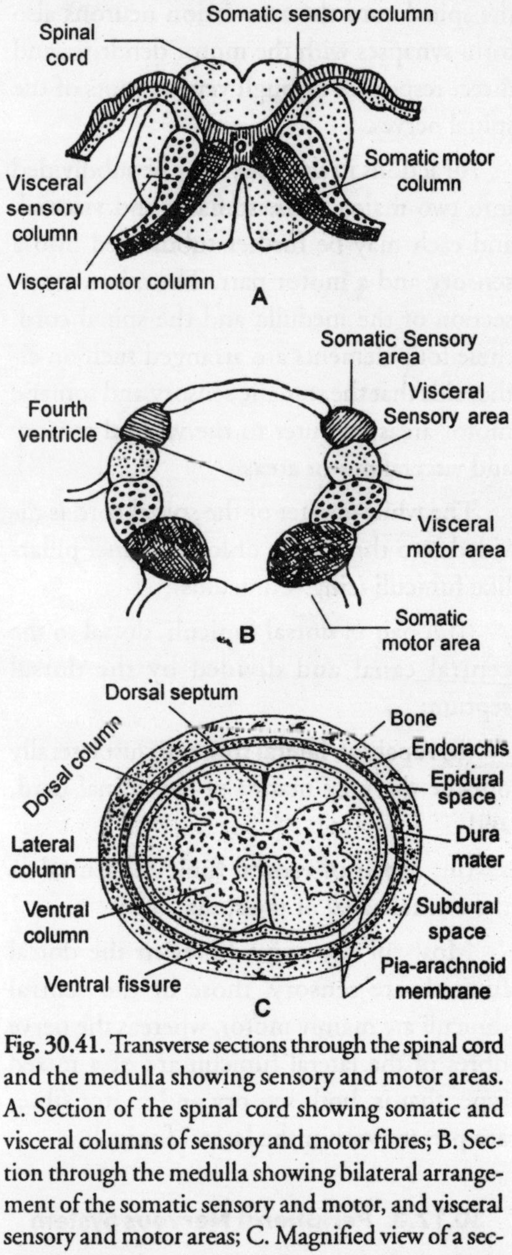

Fig. 30.41. Transverse sections through the spinal cord and the medulla showing sensory and motor areas. A. Section of the spinal cord showing somatic and visceral columns of sensory and motor fibres; B. Section through the medulla showing bilateral arrangement of the somatic sensory and motor, and visceral sensory and motor areas; C. Magnified view of a section through the spinal cord showing columns of grey and white matter in the spinal cord

The smaller bundles of nerve cells in the dorsal columns are mostly **association neurons**. Their dendrites form synapses with the axons of sensory or afferent nerve fibres which transmit the impulse through the dorsal roots of the spinal nerves to the spinal cord. Within the spinal cord the association neurons also form synapses with the motor dendrites and direct responses through ventral roots of the spinal nerves.

All actions in the body may be subdivided into two major parts: **somatic** and **visceral**, and each may be further subdivided into a **sensory** and a **motor** part. Thus, in a cross-section of the medulla and the spinal cord, these four elements are arranged such on either side that the somatic sensory and somatic motor areas lie outer to the visceral sensory and visceral motor areas.

The white-matter of the spinal cord is divided into three pairs of longitudinal pillar-like **funiculi** (sing. Funiculus):

(i) A pair of **dorsal funiculi**, dorsal to the central canal and divided by the dorsal septum;

(ii) A pair of **lateral funiculi**, just laterally outside the grey matter of the spinal cord, and

(iii) A pair of **ventral funiculi**, ventral to the central canal of the spinal cord.

Most of the nerve fibres in the dorsal funiculi are sensory, those of the ventral funiculi are mainly motor, whereas the nerve fibres of the lateral funiculi are of a mixed type, that is, both sensory and motor fibres are present within the lateral funiculi.

30.17.3. Peripheral Nervous System

The nerves and ganglia that arise from the central nervous system and are distributed to the various organs or periphery of the body are called the peripheral nervous system. The autonomic nervous system controlling activities of organs supplied with involuntary muscles in the body is also a special division of the peripheral nervous system.

The peripheral nervous system comprises the cranial and the spinal nerves, arising in pairs from the brain and spinal cord, respectively. Although the spinal nerves arise metamerically from the spinal cord, the cranial nerves (either 10 pairs or 12 pairs) do not arise metamerically and, in many cases, their dorsal and ventral roots do not connect with one another. In some cases, the dorsal root and the gangalia are not discernible. For this reason, from the point of view of origin and distribution, the cranial nerves are not considered homologous.

(A) Cranial Nerves

In birds and mammals, 12 pairs of cranial nerves arise from different parts of the brain, and, as already mentioned, these are either **sensory**, bringing in stimuli from the sensory organs to the brain, or **motor**, that conduct responses from the brain or motor organs concerned, or **mixed** nerves, which are formed of both sensory and motor fibres.

Based on the anatomy of human body and that of the mammals in general, the names of 10 or 12 pairs of cranial nerves were already fixed some three or four centuries ago. And so, after the discovery by Pincus (1894) of a small nerve arising from the rostral part of the cerebrum (actually from the lamina terminalis), and extended up to the Jacobson's organ, it could only be designated as '**Nervous terminalis**' or the '**Zero nerve**'. However, this nerve is not present in birds.

In the amniote vertebrates, in general, 12 pairs of cranial nerves are present (10 pairs in the anamniota). Their names, nature (i.e., sensory or motor or mixed), origin from the brain, distribution (ie., the organs they supply) and precise functions have been summarised below in a tabulated form:

Table 30.3: Cranial Nerves in Pigeon

No.	Nerve	Nature of the nerve	Origin	Distribution	Function
1.	**Olfactory**	Sensory	Unlike a typical sensory nerve, this nerve is formed of sparsely distribiuted medullated fibres of the olfactory placode and is joined to the rostral end of the small olfactory lobe.	Mucous membrane of the nasal cavities	This nerve helps in the perception of smell which is poor in birds
2.	**Optic**	Sensory	The thick optic tectum formed by the lateroventral wall of the optic lobes on either side of origin of the optic nerve.	The two optic nerves form the optic chiasma on the ventral aspect of the diencephalon and from the chiasma each nerve enters the eye of the opposite side. Many authors consider the optic nerves as fibre tracts, rather than cranial nerves. The cell bodies of nerve fibres are located in the retina of the eye.	The optic nerves effect vision through the eye
3.	**Oculo-motor**	Motor	Most of the nerve is formed of somatic motor fibres. The paired nerves arise—one on	Branches from this nerve are distributed to four of the six eye muscles. These muscles are: **superior**	The movements of the eyeball within the orbit, and of the upper eyelid, lens and iris are all per-

Contd.

No.	Nerve	Nature of the nerve	Origin	Distribution	Function
			either side of the midventral line of the mesencephalon	rectus, inferior rectus, internal (or anterior) rectus and inferior obliquus. Some fibres supply the muscle elevating the upper eyelid. Some autonomic fibres accompanying the somatic motor fibres supply the ciliary apparatus concerned with the accommodation of the eye and also supply sphincter muscle of the iris, which regulate the size of the pupil	formed by the innervation of this nerve. This nerve also receives a few proprioreceptive somatic sensory fibres
4.	Trochlear	Motor	This paired, narrower somatic motor nerves arise from the dorsocaudal surface, one on either side of the dorsomedian line of the midbrain	This nerve supplies the superior oblique eye muscle	Helps in movement of the eyeball. The nerve also contains a few proprioreceptive somatic sensory fibres
5.	Trigeminal	Mixed	This large fifth cranial nerve arises from the anterior medulla oblongata rostrolaterally. The fifth caranial nerve is	Soon after the origin, the nerve forms a distinct 'Gasserian ganglion' which sends nerve fibres to the so matic sensory	This nerve sends sensory fibres to different parts of the eyes and motor fibres to the adjoining muscles that

Contd.

No.	Nerve	Nature of the nerve	Origin	Distribution	Function
			considered as the nerve of the mandibular arch	columns of grey matter. From the ganglion this nerve divides into three main branches: (i) **Ophthalmic** (smallest of the three branches) which sends sensory fibres to the conjunctiva, cornea, lachrymal gland, part of the mucous layer of the nose and skin of eyelids; (ii) **maxillary** which sends somatic sensory fibres to the upper jaw; (iii) **mandibular** which, unlike the first two branches, is a mixed nerve ie, it contains both somatic sensory and visceral motor fibres. This branch supplies different parts of the lower jaw, its visceral motor fibres are directly concerned with the contraction of the muscles of the lower jaw. Its somatic sensory fibres supply the lower lip. Visceral	move the upper lip or the jaw, lower eyelid, tongue, etc. and also activate the muscles of the buccal cavity and pharyngeal wall

Contd.

No.	Nerve	Nature of the nerve	Origin	Distribution	Function
				sensory and somatic motor fibres are absent in the mandibular branch of the nerve. The author (Bhattacharya, 1980) while working on the jaw muscles of *Columba livia* observes that the outer mandibular branch of the Vth cranial nerve passes through the gap formed by the biforked insertion-aponeurosis of the M. pseudotemporalis superficialis which is one of the mandibular adductors. Thus, in pigeons, the bifurcation of the insertion-aponeurosis of this muscle and passage of the nerve through the aponeurosis (unlike the passage of the nerve over the aponeurosis as in other birds) are rather unusual. This was later confirmed by van Gennip (1986).	

Contd.

No.	Nerve	Nature of the nerve	Origin	Distribution	Function
6.	**Abducens**	Motor	It is a pair of small somatic motor nerves that arise—one on either side of the mid-ventral surface of the medulla, slightly behind the level of the trigeminal nerve.	This nerve supplies the external rectus muscle of the eye and a fine branch from this nerve also supplies the nictitating membrane.	Helps in movement of the eye-ball and the nictitating membrane
7.	**Facial**	Mixed	This nerve, formed of visceral sensory and visceral motor fibres, arises laterally from the medulla oblongata, caudal to the origin of the trigeminal nerve. Just after the origin, the nerve divides into two branches: the **palatine** and the **hyomandibular** of which the hyomandibular is the main branch. A geniculate ganglion is formed at the point where the hyomandibular branch leaves the medulla. Both the nerves are divided into several smaller branches.	The palatine branch is distributed to the muscles of the palate, buccal cavity and pharynx, nasal mucosa, etc. together with their associated glands. The palatine branch (N. palatinus) has two distinct branches, the Ramus dorsalis and the Ramus ventralis. The hyomandibular branch supplies the hyoid apparatus, the depressor mandibulae muscle and the muscles of the neck region.	The innervation of the palatine branch causes movements of the buccal cavity and palate, and that of the hyomandibular branch is concerned with the movements of the hyoid apparatus, the lower jaw and the neck. In fact, after giving off branches to the depressor mandibulae muscle, the main hyomandibular nerve divides into two branches, Ramus hyoideus and Ramus cervicalis to bring about movements of the hyoid apparatus, tongue, buccal floor and neck regions.

Contd.

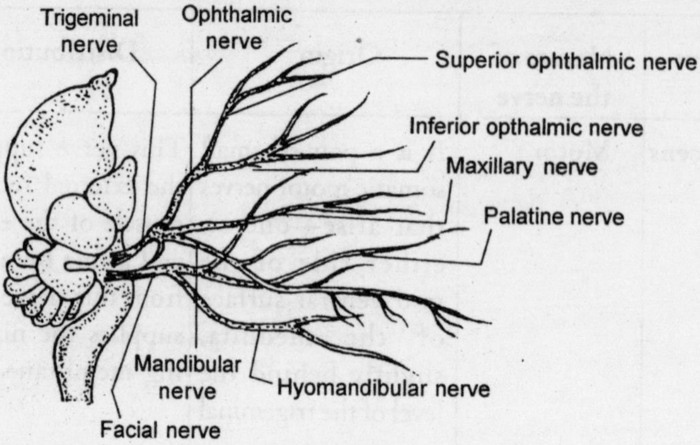

Fig. 30.42. Origin and distribution of the Vth and VIIth cranial nerves of pigeon

No.	Nerve	Nature of the nerve	Origin	Distribution	Function
8.	Auditory or Acoustic	Sensory	It is a purely somatic sensory nerve which arises one from either side of the medulla oblongata, just behind the origin of the facial nerve. Near origin an acoustic ganglion is formed and in devepolment this ganglion is closely associated with the geniculate ganglion of the facial nerve.	In higher forms, the auditory nerve is divided into two main branches: a vestibular branch supplying vestibule of the inner ear and a Cochlear branch supplying cochlea of the inner ear. The fibre tracts for hearing and equilibrium pass on to the cerebellum where the centre for equilibrium is located and to the cerebral cortex where the centre for hearing is located.	This nerve functions both for hearing and maintenance of equilibrium.

Contd.

No.	Nerve	Nature of the nerve	Origin	Distribution	Function
9.	**Glossopharyngeal** This nerve is also known as **'Vestibulocochlear nerve'**.	Mixed	This ninth cranial nerve arises, one from either side of the medulla oblongata, behind the auditory nerve. This nerve at its base forms two distinct afferent ganglia, the proximal **superior ganglion** and the distal **petrosal ganglion.** In recent nomenclature, both these ganglia have been termed as 'Ganglion proximale and 'Ganglion distale'. The proximal ganglion of the ninth joins the large proximal ganglion of the Vagus nerve (tenth cranial nerve) and both the ganglia lie in a fossa at the base of the skull. The distal ganglion of the ninth nerve lies away from the proximal ganglion and very close to the superior cervical ganglion of the autonomic nervous system.	This nerve has several branches each supplying the muscles and mucous membrane of the organs such as larynx, pharynx, tongue, palate, trachea, salivary glands, taste buds, etc. The ninth nerve has prominent connections with the tenth as well as twelvth cranial nerves.	This nerve has both sensory and motor components which are either concerned with taste perception or movements of the pharynx and tongue. Some visceral motor fibres pass directly to the muscles of the pharynx.

Contd.

No.	Nerve	Nature of the nerve	Origin	Distribution	Function
10.	Vagus	Mixed	Each vagus nerve arises broadly from either side of the medulla oblongata, just behind the ninth nerve and, as already mentioned, the proximal ganglia of the ninth and tenth nerves are fused to each other. However, the proximal ganglion of the tenth nerve is much larger in size than that of the ninth. The distal ganglion of the vagus nerve lies at some distance from the proximal ganglion at the root of the thorax.	The vagus nerve, during its long course, gives off a number of branches to larynx, oesophagus, crop, proventriculus, gizzard, heart and lungs. It has several connections, such as vagoglosopharyngealis, vagohypoglossalis and vagohypoglosso-cervicalis, etc.	The main part of the vagus nerve is formed of preganglionic visceral motor fibres. The vagus nerve controls the movements of several vital organs — airflow through the glottis, peristalsis of the digestive canal, heart beat, expansion and contraction of the lungs, functioning of the crop, etc.

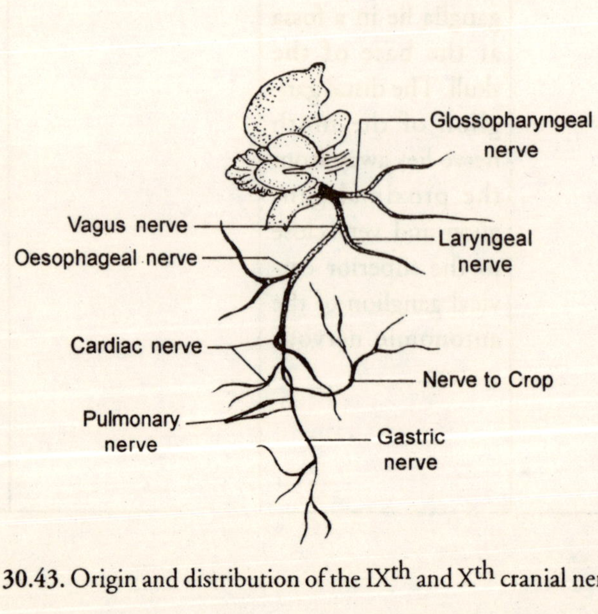

Fig. 30.43. Origin and distribution of the IXth and Xth cranial nerves of pigeon

No.	Nerve	Nature of the nerve	Origin	Distribution	Function
11.	Spinal acessory	Motor	The eleventh and twelveth cranial nerves are present only among the amniotes. The eleventh nerve arises from the caudo-lateral aspect of the medulla oblongata, just behind the origin of the vagus nerve with which it is intimately related. The eleventh nerve has a cranial portion derived from some posterior visceral motor fibres of the vagus and a spinal portion formed of somatic motor fibres of some of the spino-occipital nerves.	The cranial portion of the nerve is distributed to the muscles of the pharynx and larynx; the spinal portion is distributed to the muscles of the neck, sternum, clavicle, jaws, etc. The spinal accessory nerve assists the vagus nerve in its overall functions.	The rotational movements of the neck, and movements of the pharynx, larynx and various other visceral organs are attributed to the functions of this nerve.
12.	Hypoglossal	Motor	The nerve arises from the caudolateral medulla oblongata, behind the origin of the eleventh nerve. This nerve is also found only in the amniotes and it is a purely somatic motor nerve. At a distance from the origin the hypoglossal nerve has connec.	The hypoglossal nerve is distributed to the intrinsic muscles of the tongue and to a few muscles of the lower jaw and neck.	The movements of the tongue almost solely and those of the lower jaw and neck partly are accomplished by the supply of this nerve.

Contd.

No.	Nerve	Nature of the nerve	Origin	Distribution	Function
			tions with the ninth and tenth cranial nerves, as well as with the first spinal nerve.		

(B) Spinal Nerves

It has already been mentioned that in pigeons, 38 pairs of spinal nerves are present, each pair arising from the spinal cord, one on either side, more or less in correspondence with the vertebral column, but not strictly due to differential growth of the two. Each spinal nerve arises from the spinal cord by two roots, the **dorsal root** and the **ventral root.** The dorsal root, just outside the spinal cord, forms a prominent ganglion called dorsal root ganglion which contains the sensory neurous of the dorsal root. The axons of these neurons enter into the spinal cord and join the dorsal column of grey matter. As the dorsal root transmits sensory impulses toward the spinal cord, it is called **afferent sensory,** while the ventral root is called **efferent motor.**

Sensory fibres may be either **somatic** or **visceral.** The somatic sensory fibres which carry impulses from the skin, voluntary muscles and skeletal parts form synapses with the neurons that join the somatic sensory columns of grey matter. On the other hand, the visceral sensory fibres coming from visceral structures end in the visceral sensory columns of grey matter. The ventral or the efferent motor roots are located in the ventral column of the grey matter.

In pigeons, a spinal nerve is formed by the union of the dorsal and ventral roots and the spinal nerve thus formed finds its exit through the **intervertebral foramen** to supply the muscles of the body, wings and legs.

The spinal nerves have been named according to the regional subdivisions of the vertebral column such as, 12 pairs of cervical spinal nerves, 8 pairs of thoracic, 12 pairs of lumbosacral nerves and 6 pairs of caudal. The last three pairs of spinal cervical and the first

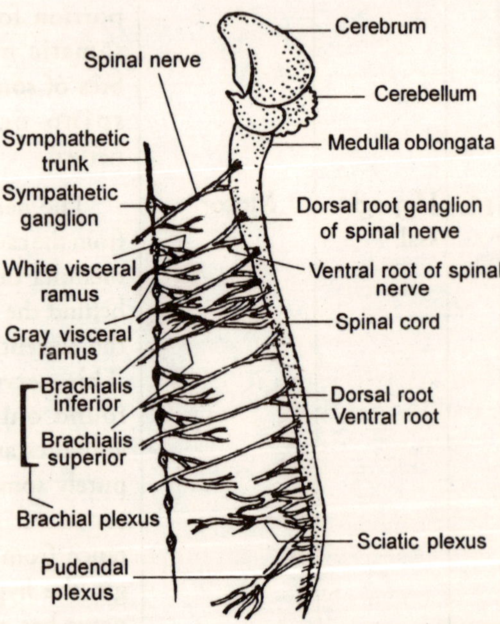

Fig. 30.44. Diagram showing spinal cord, symphatheic trunk and spinal nerves with plexuses in a pigeon

two pairs of thoracic nerves unite to form the **brachial plexus** (or the **cervico-brachial plexus**, as it is called now) and the second, third, fourth and fifth lumbosacral nerves unite to form the **sciatic plexus** (also called **lumbosacral plexus**). The lumbosacral plexus may be divided into separate lumbar, sacral and pudendal plexus, supplying, respectively, the thigh, lower legs, and the cloaca and tail regions.

(C) Autonomic Nerves

The autonomic division of the peripheral nervous system (also known as **sympathetic** and **parasympathetic** nervous system) comprises of delicate and complex nerve trunks with chain ganglia, as well as preganglionic and postganglionic fibres which, being intimately connected with either the cranial or the spinal nerves (and, hence, with the central nervous system), control the movements of organs supplied with involuntary muscles. The autonomic system usually depends for its function on that of the brain and spinal cord.

In the autonomic system, outgoing impulses to the effector organs must travel through two types of neurons—the **preganglionic** and **postganglionic** fibres and with them the ganglia that serve as centres for relaying the impulses. The cell bodies of the preganglionic fibres lie in certain ganglia which occur at a distance from the central nervous system and within which the synapses between the pre-and postgaglionic fibres take place. The axons of the postganglionic fibres then supply the involuntary muscles, cardiac muscles, glands and various other internal organs in the body.

The autonomic nervous system is divided into sympathetic and parasympathetic nervous systems, both of which are provided with preganglionic and postganglionic fibres. Usually, the actions of these two systems are antagonistic.

In pigeons, a pair of sympathetic nerve trunks extend longitudinally, one on either side of the ventral aspect of the vertebral column. Each trunk is provided with a series of ganglia at intervals forming the '**chain ganglia**'. A group of such ganglia at the anterior end of the sympathetic trunk are fused to form three large cervical ganglia of which the superior cervical ganglion is the largest. The preganglionic fibres—which partly contribute to the formation of the white visceral rami of all the thoracic—and the first three lumbar spinal nerves connect to the corresponding chain ganglia.

Some other preganglionic fibres may pass through the sympathetic trunk ganglia (without forming synapses) to connect to the three large prevertebral ganglia in the abdominal region. These are also sympathetic ganglia of the thorax and abdomen, but are different from those of the sympathetic trunk. The postganglionic fibres arising from the prevertebral ganglia are distributed to the smooth muscles and various internal organs in the body under involuntary control.

The postganglionic fibres forming the grey visceral ramus are distributed via spinal nerves to the skin, smooth muscles in the wall of the blood vessels, arrector pili muscles, skin glands, etc. Other postganglionic fibres arising from the anterior ganglia supply the intrinsic muscles of the eye, lachrymal gland, mucous membrane of the palate, mouth, etc. The postganglionic fibres arising from the coeliac and superior mesenteric ganglia supply the oesophagus, stomach, small intestine, proximal part of the large intestine, liver, pancreas, etc.

As the sympathetic nervous system is connected to the medulla oblongata only in the thoracic and abdominal regions, it is known as **'thoraco-lumbar outflow'**. On the other hand, the parasympathetic nervous system is designated as **'cranio-sacral outflow'**, as preganglionic motor fibres of only a few cranial and sacral spinal nerves are connected to this system.

Parts of the third, seventh, ninth and tenth cranial nerves are formed of preganglionic parasympathetic fibres. The ganglia in which these fibres terminate are situated at a longer distance, close to the organs supplied by the parasympathetic system. Hence, the preganglionic visceral motor fibres of this system are very long and the postganglionic fibres are much shorter.

Of the various organs supplied by the preganglionic parasympathetic fibres, the intrinsic muscles of the eye, submaxillary gland, sublingual gland, parotid gland, heart, larynx, trachea, bronchi and lungs, digestive canal and digestive glands, kidney and genital organs are the main. The parasympathetic system connected to the spinal nerves is formed of preganglionic efferent fibres of the second, third and fourth sacral spinal nerves. Here too the postganglionic fibres are shorter.

(D) Functions of the Sympathetic and Parasympathetic Nervous Systems

Although the structures and functions of the sympathetic and parasympathetic nervous systems are more elaborate in mammals, in birds too these are reasonably well-developed.

On stimulation the sympathetic system brings about reactions, such as contraction of arrector pili muscles, constriction of cutaneous blood vessels, dilatation of the pupil, increasing rate of heart-beat, retardation of peristalsis in the digestive tracts, increase of blood pressure and in the amount of blood sugar, etc. The sympathetic system assists in the defence mechanism of the animal at times it is threatened with danger or adverse conditions.

The posganglionic sympathetic fibres secrete a hormone of catecholamine group called norepinephrine which—in somewhat different concentration—is also secreted by the adrenal medulla along with epinephrine. The preganglionic fibres of both the sympathetic and parasympathetic systems, however, secrete a chemical called acetylcholine at the synapses during conduction of impulse.

As already mentioned, the functions of the parasympathetic system in general are opposite to those of the sympathetic system, such as constriction of the pupil, dilatation of the cutaneous blood vessels, contraction in the walls of the alimentary canal causing peristalsis, increase in the secretion of the salivary glands and gastric glands, etc.

30.18. SENSE ORGANS

Different specialised groups of cells, derived from the ectoderm and termed receptor cells, perceive different kinds of stimuli due to changes in the environment—external or internal to the animal body. These stimuli-producing impulses are then transmitted through nerve fibres to the central nervous system which translate the specific stimuli into appropriate responses or sensations.

In vertebrates, the receptor cells are integrated as specialised sense organs such as olfactory organs for the peception of the sense of smell, visual organ or the eye for the perception of sight, acoustic or auditory organ for the perception of the sense of hearing, etc.

The receptor cells have been named differently—based on the nature of the stimuli they perceive—and are divided into two main categories such as **'exteroceptors'**—receiving stimuli from the external environment, and **'interoceptors**—receiving stimuli from the internal environment. The common receptor organs in the first category are **photoreceptors, phonoreceptors, olfactoreceptors, tangoreceptors, gustatoreceptors, thermoreceptors,** etc., and those of the second category are **proprioreceptors.** The sensory nerve fibres involved in the first category are somatic sensory, and those of the second category localised in the tendons, skeletal joints and muscles; the muscles of the digestive tract and the smooth muscles of the heart are visceral sensory.

The different kinds of sensory organs present in pigeons are:

1. **Organs of touch:** In pigeons the cutaneous receptors of epidermis function as the organs of touch. The Merkel's corpuscles present in the body skin, buccal cavity and tongue form the cup-shaped nerve-net at the ends of the main nerves. **Grandry's corpuscles** present in the skin of the eye perceive the sense of touch and the **Herbst's corpuscles** present in the dermis transmit the sense of cold, heat, vibration, etc. to the brain.

2. **Organs of smell:** A pair of olfactory organs, located caudally in the upper beak, are connected to the pair of olfactory nerves. Of the three conchae situated in the lateral walls of the nasal cavity, only superior conchae are provided with the olfactory epithelium. Except some sea-birds—birds, in general, possess a poor sense of smell.

3. **Taste papillae:** The receptor cells for the perception of taste are called the **'gustatoreceptor cells'** or **taste buds.** These are formed of two types of cells–taste cells and supporting cells. Although the taste buds are not present in the keratin investment of tongue in pigeons, the epithelial covering of the buccal cavity and pharynx–as well as the basal region of the tongue–possess ample taste buds.

4. **Organs of sight: the eye:** In pigeons, compared to body size, a pair of large eyes effecting vision are present within the large orbits, one on either side of the cranium. The two orbits are separated by a thin membranous interorbital septum. In all vertebrates, in general, the eyeball(**Bulbus oculi**), eye muscles and nerves conform to a uniform pattern. The pigeon's eyeball is somewhat biconvex in shape.

In early development, different parts of the eye develop from three sources: the retina and optic nerve arise from the diencephalon, the lens arises from the superficial ectoderm of the head region and the other parts are derived from the adjacent mesenchyme.

The beginning of eye formation is marked by a pair of laterally expanded **optic vesicles.** Each optic vesicle grows larger and its distal end touches the superficial ectoderm, whereas the proximal end remains narrower as the **optic stalk.** The optic vesicle then forms a double-walled optic cup whose outer layer contains cells along with **pigment granules.** The granules gradually form the **pigment layer of the retina.** The inner layer (which later fuses with the outer layer) is transformed into the **sensory or nervous portion of the retina.** Within the optic cup, a **lens placode** is formed from the superficial ectoderm.

The axons from the sensory neurons of the retina are channelised in a bundle through the optic stalk to form the optic nerve. The cornea is formed from the mesenchyme tissue which extends between the superficial ectoderm and the optic cup. The **choroid fissure** is formed as a groove at the ventral side of the optic cup and the optic stalk. Most of the retina is quite thick, it forms the sensory area; only the inner layer of the retina at the rim of the optic cup is thin—it fuses here with the pigment layer to form the ciliary body and iris. The mesoderm tissue contributes to the formation of the iris, muscles of the iris and lens, vascularised choroid layer, inner layer of cornea and covering layer of the thick sclera.

In pigeons, the eyeball is protected externally by the upper and lower eyelids, the lower being more mobile than the upper. A third eyelid called the **nictitating membrane** lying at the corner of the upper eyelid lies inner to the **upper** and lower eyelids. This semi-transparent membrane can be moved a little obliquely across the eye very fast. The third eyelid keeps the eyes clean and provides special protection to the eyes. The **lachrymal gland** and the **Harderian gland** are connected to the eyes.

The walls of the eyeball are formed mainly of three layers: **sclera, uvea,** and **retina.**

i. Sclera: In the eyeball the sclera forms the thick, tough and white outermost coat of connective fibres. At the open front of the eye the sclera forms a transparent membrane called the cornea, the rest of the opaque and white sclera is called the 'white of the eyeball'. In most birds (including pigeons) and in many other vertebrates, a ring of overlapping bony ossicles—called **scleral plates,** numbering about 14 or 15—lie within the sclera and around the cornea. These provide shape, rigidity and protection to the eyeball. The cornea has five layers of which the external epithelium of the cornea fuses with a thin and delicate membrane along the inner sides of the eyelids to form the **conjunctiva.**

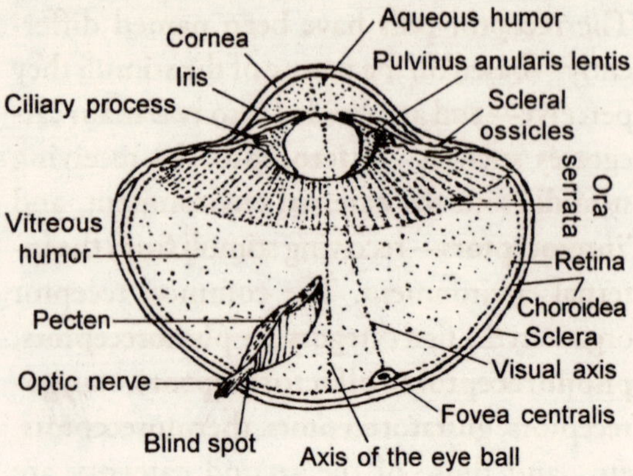

Fig. 30.45. Internal structures of the eye of a pigeon

ii. Uvea: The **'Tunica vasculosa bulbi'** (NAA, 1993), the vascular layer of the eyeball inner to the sclera, is called the uvea which is formed of three main parts: **choroid, ciliary body** and **iris.**

The choroid consists of numerous delicate blood vessels and pigment granules within the meshwork of connective tissues. The choroid maintains the main supply of blood vessels to the retina and it consists of five layers of which the outermost suprachoroid layer and the innermost basal layer are loosely attached to the sclera and the retina, respectively. From the site where the uvea turns inward from the fibrous sclera, the choroid extends anteriorly to form the circular coloured iris which is richly supplied with striated muscles.

The iris is heavily pigmented and has an aperture at its centre—the **'pupil'**—which controls entry of light into the eye like a camera and regulates the brightness and quality of the retinal image. The iris is thin at its attachment with the ciliary body, thicker in the middle and then again becomes thin near the border of the pupil. Of several muscles of the iris which control the size and movement of the aperture of the pupil, the sphincter and a dilator muscle are main. Both these muscles are striated and coloured. The anterior surface of the iris is lined by a single layer of flat epithelial cells, while the posterior surface is lined by darkly pigmented cells. The surface of the iris of columbiform birds (e.g., pigeons, doves) possesses reflective cells or **iridocytes** forming the '**Tapetum lucidum iridis**'.

Another important structure, the **ciliary body**, is located behind the iris and at the junction of the sclera and cornea. The ciliary body—which is mainly formed of radially disposed striated muscle fibres—remains attached to and suspends the biconvex lens from both sides with the help of a sort of tentacular ligament. The ciliary body is, in fact, an extension of the choroid layer. In birds, the ciliary processes are numerous and irregular and they are usually formed of non-nervous parts of the retina. The muscles of the ciliary body are usually two—one placed anteriorly, the anterior ciliary muscle or the **Crampton's muscle**, and the other placed posteriorly—called the posterior ciliary muscle or the **Brucke's muscle**. Both these muscles are concerned with the accommodation of the lens and cornea (Troilo and Wallman, 1987; Martin, 1987).

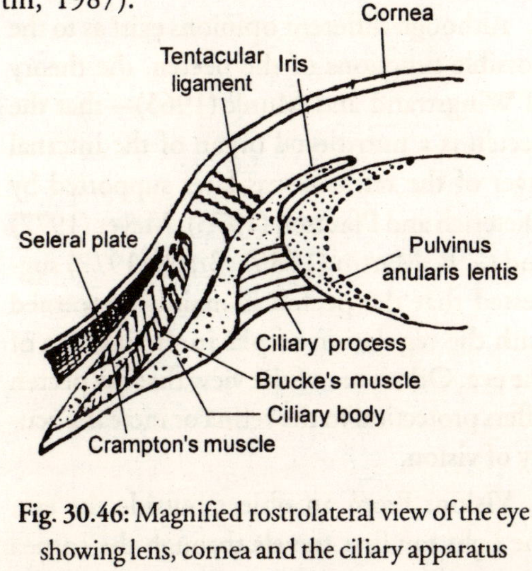

Fig. 30.46: Magnified rostrolateral view of the eye showing lens, cornea and the ciliary apparatus

The iris, ciliary body and lens divide the cavity of the eyeball into an anterior chamber and a posterior chamber, the latter being more spacious than the former. The anterior chamber is filled with a watery fluid, the **aqueous humor** and the posterior chamber is filled, with a jelly-like fluid called '**vitreous humor**'.

Due to action of the Brucke's muscle, the lens is drawn towards the anterior chamber and the curvature of the lens is increased as the shape of the eye remains unchanged due to the presence of the sclerotic plates. This results in a suitable accommodation of the lens for near vision of the bird. This function is aided by the contraction of the circular sphincter muscle of the iris. The diameter of the cornea decreases as the Crampton's muscle draws the cornea from both sides. Thus this muscle, too, assists in the function of the Brucke's muscle in the near vision of an object.

iii. Accommodation of Lens and Cornea: The acuity and speed with which a bird sees an object placed near or at a distance result from the changes in the position of the lens and cornea. This is called accommodation of the eye. In speedy walking or flight of the pigeons, the accommodation of the lens and cornea is a remarkable feature (Martin, 1987). The mechanisms of accommodation (though not very well-understood in all cases) in birds and mammals are different.

In pigeons and other birds, a ringlike band or pad occurs on either side of the eye. This is known as '**Pulvinus anularis lentis**' which is unique in birds. This ringlike pad surrounds inwardly a sac-like hydrostatic chamber called '**Vesicula lentis**' which, in turn, surrounds the central core of the lens. As a result, the refraction of light rays in the eye is suitably changed. According to Martin

(1985, 1987), the pulvinus and the vesicula lentis transfer pressure from the ciliary muscles to the central core of the lens. Thus accommodation of the lens becomes possible.

iv. Retina: Retina is the innermost layer of the eyeball containing photoreceptive cells. The entire retinal surface or the deepest region in the posterior half of the eyeball is called **fundus** of the eye. The boundary of the fundus or of the sensory retina is marked by a thin terminal region called '**Ora serrata**'.

Within the fundus, the retina mainly consists of two layers: the outer pigment layer which remains attached to the choroid layer and the inner nervous layer of retina. The latter is further subdivided into four regions: (i) the neuroepithelial layer, (ii) the bipolar layer, (iii) the ganglion cell layer, and the (iv) optic fibre layer. The optic fibre layer forms the optic nerve (**Stratum neurofibrarum**). The neuroepithelial layer of the inner retina contains the visual photoreceptor cells, the **rods** and **cones**.

The imaginary median longitudinal line which passes through the cornea, pupil and the lens to reach the posterior pole of the eyeball is called the **optical axis**. But the actual visual axis lies somewhat angulary with the optical axis. The site at which the optic nerve leaves the eyeball is devoid of rods and cones. This site on the retina is called '**blind spot**' where no image of any object is registered. On one side of this spot, where the visual axis meets the fundus, an area of shallow depression on the sensory retina is called '**fovea centralis**' which is marked by longer, thinner and tightly packed cells with increased density and regular arrangments, and it also contains specific types of ganglion cells (Hayes et al, 1991). This is the site of most acute vision in the pigeon and, in other birds, capable of high spatial resolution, dectection of movements, etc.

The rods and cones are the cells of different shapes. The rod cells are narrow, thread-like, whereas the cone cells are thick and short. These cells are associated with the underlying bipolar nerve cells and the latter, in turn, are connected with the axons and dendrites of larger ganglion cells lying below. The rod cells can discriminate between light and darkness, and are able to control movements and flight routes. The cone cells can determine the acuity of vision and colour of the the object.

v. Pecten: The site of exit of the optic nerve from the sensory retina is marked by linear arrangement of the optic fibre layer. Pecten is a toothed comb-like vascular but non-nervous and non-muscular rectangular plate-like structure attached to the internal aspect of the above-mentioned linear band and is covered by a layer of the non-sensory retina. It stands erect within the vitreous body. The shape and size of pecten vary in different species of birds. At least three types of pecten are observed in different birds. In pigeons and other flying birds, the type of pecten is called '**Pecten oculi pliactus**'—its comb-like teeth are arranged on a basal membrane called '**Pons pectinus**'. Plenty of pigment granules and glial cells are present in the pecten.

Although different opinions exist as to the possible functions of the pecten, the theory of Wingstrand and Munk (1965)—that the pecten is a nutritional organ of the internal layer of the retina—was later supported by Dieterich and Pfautsch (1973), Meyer (1977) and G. R. Martin (1985). Brach (1977) suggested that the pecten is mainly concerned with the regulation of pH in the interior of the eye. Others are of the view that the pecten offers protection to the retina or increases acuity of vision.

5. Vision: From an object outside the eye, the light ray first travels through the cornea and pupil to reach the lens. The parallel light rays, while passing through the lens, are refracted and are then reflected on the fovea of the retina. The photoreceptive cells of the retina produce inverted image of the object and the image is transmitted through the optic nerve to the brain. The visual centre of brain through a complex process converts the inverted image of the object into a real one which is seen by the animal.

6. **Organs of hearing: the ear:** In pigeons, as well as in all higher vertebrates, the organs of hearing perform dual functions—hearing and maintenance of equilibrium of the body. In pigeons, the ear is well-developed and divided into three parts: the **external ear**, the **middle ear**, and the **internal ear**. The epithelium of the internal ear is derived from the ectoderm. Each ear on either side of the head first arises as a placode of the superficial ectoderm which invaginates as an auditory pit. The pit is closed to form the **auditory** or **otic vesicle** which loses its connection with the ectoderm. The otic vesicle lies close to the brain and is enclosed by mesenchymal tissue.

i. **External ear:** At the caudolateral aspect on either side of the head a circular patch of skin called **tympanum** lies covered by feathers. From the tympanum a short canal called the **external auditory meatus** extends up to the middle ear.

ii. **Middle ear:** The middle ear appears as a hollow tubular chamber within which lies a rod-shaped bony structure called **columella (columella auris)**. The columella transmits the sound waves to the internal ear. The two projections, one lying at either end of the columella—such as, a trifurcate extra-columellar cartilage at the tympanum end and a broad foot-plate at the vestibular end—assist in the transmission of sound waves. The movements of the columella are regulated by the muscles attached to it. The eustachian canal on either side connects the buccal cavity with the middle ear.

iii. **Internal ear:** The internal ear is known as the **'membranous labyrinth'** (=**Labyrinthus vestibularis**). The vestibular organ in birds is contained within and enclosed by the auditory capsule. In pigeons and in other birds, the vestibular organ is mainly formed of three structures: the **utriculus**, the **sacculus** and the **ductus semicirculares**. The semicircular ducts arise from the utriculus. Smith (1985) states that though the utriculus, sacculus and semicirculur canals are separate structures, the sensory receptors within them have possibly the same function. The vestibular labyrinth as a whole maintains equilibrium of the body, reacting to angular, linear and other positional changes.

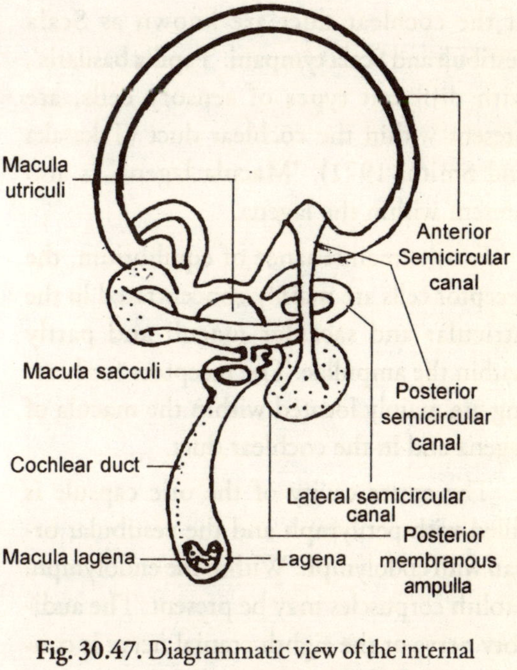

Fig. 30.47. Diagrammatic view of the internal structures of the inner ear of a pigeon

The utriculus is a chamber of the membranous labyrinth and it gives rise to three thin-walled semicircular canals of which two are vertical and one horizontal. The vertical canals are known as **anterior** and **posterior semicircular canals (Ductus semicircularis anterior et posterior)** while the horizontal canal is called **lateral semicircular canal (Ductus semicircularis lateralis)**. Each semicircular duct becomes swollen at its end forming a membranous ampulla. The sensory terminal parts of both the utriculus and sacculus each form a swollen structure known, respectively, as the **'Macula utriculi'** and **'Macula sacculi'**. Both these structures have hair cells, supporting cells, basal cells, etc. which are sensitive to linear acceleration and gravity.

The sacculus is filled with a fluid called **endolymph.** The narrow ventral outgrowth of the sacculus is called **'Ductus cochlearis'** (also known as **Scala media**) which ends blindly as a swollen structure called the **'Lagena'**. Perilymphatic spaces on either side of the cochlear duct are known as **Scala vestibuli** and **Scala tympani**. **'Papilla basilaris'**, with different types of sensory cells, are present within the cochlear duct (Takasaka and Smith, 1971). **'Macula lagena'** is also present within the lagena.

For the maintenance of equilibrium, the receptor cells are mainly concentrated in the utricular and saccular maculi and partly within the ampullae. The receptors for hearing are mainly located within the macula of lagena and in the cochlear duct.

The entire cavity of the otic capsule is filled with perilymph and the vestibular organ with endolymph. Within the endolymph, otolith corpuscles may be present. The auditory nerve or the eighth cranial nerve is connected with the internal ear.

7. Hearing: The structure of the internal ear of a pigeon has many similarities with that of a mammal. The hair cells or the receptors for hearing, occurring on the basilar membrane of the cochlea, are placed on the tectorial membrane. Based on the intensity of sound vibrations, part of the tectorial membrane containing hair cells undergo gentle vibrations. When this stimulus reaches the hearing centre of the brain through the auditory nerve, the pigeon becomes capable of hearing.

8. Ear as organ of equilibration: The maintenance of equilibrium is an essential function of the vestibular organ (Membranous labyrinth). A pigeon's equilibrium of the body may be disturbed either at rest or in flight. When the body tilts to one side and the equilibrium is disturbed, greater quantity of endolymph flows to that side and the otoliths lying within the endolymph come in contact with the receptors for the maintenance of equilibrium and stimulate the receptors. These acquaint the bird about its changed position relative to the gravitational force and the bird's normal position—either in flight or at rest—is restored.

30.19. ENDOCRINE GLANDS

A general description of the endocrine glands in birds has been provided in some details in Chapter 13.

Quite a number of ductless glands are located at different sites in the body. Their secretions, called hormones, get directly mixed with blood and—through the general circulation of blood—are conveyed to organs at distant sites to bring about profound changes in the body. The hormones are usually either proteinous or steroid by nature and their chemical composition is either simple or complex.

The pigeons have a well-developed endocrine system. The endocrine glands are: **pituitary, thyroid, parathyroid, adrenal, pancreas, pineal body** and **gonads.**

1. Pituitary gland (Hypophysis cerebri): This small and somewhat spherical gland is located on the ventral aspect of the diencephalon of the brain and connected to it by an infundibular stalk. The pituitary gland, due to its multiple functions, is also referred to as the '**master endocrine gland**'. It is situated in a depression of the basisphenoid called the '**sella turcica**'. The pituitary gland has a dual embryonic origin which is purely ectodermal.

From the point of view of origin, the gland is mainly divided into two parts: **adenohypophysis** and **neurohypophysis**. Based on location the gland is also divided into three main parts or lobes: **anterior, intermediate** and **posterior** known, respectively, as **Pars distalis, Pars intermedia** and **Pars nervosa**. Pars tuberalis of the anterior lobe is a modification of the Pars distalis. Adenohypophysis or **Rathke's pouch** is a dorsal evagination derived from the ectodermal epithelium of the primitive mouth cavity. The neurohypophysis arises from the infundibulum as a ventral evagination of the embryonic diencephalon.

In pigeons, the intermediate lobe is absent, yet the hormone '**intermedin**' secreted by this lobe is traceable. The neurohypophysis forms the posterior lobe. Although the anterior pituitary is formed of three types of secretory cells, such as basophil, acidophil and chromophobe, this part secretes at least seven important types of hormones — **Somatotrophic** or **Growth hormone** (STH or Gh), **Thyroid-stimulating hormone** (TSH), **Follicle stimulating hormone** (FSH), **Adrenocorticotrophic hormone** (ACTH), **Luteinising hormone** (LH) or **Luteotrophic hormone** or **Prolactin** (LTH), and **Chromatophorotrophic hormone** (MSH).

The posterior lobe of the pituitary representing neurohypophysis is not strictly a part of the endocrine gland, as the hormones are secreted elsewhere by the neurosecretory cells which pass down the secrections to the posterior lobe. The two hormones are called **Oxytocin** and **Vasopressin.** While the former helps in contracting myoepithelial cells and smooth muscles of the uterus, the latter has an antidiuretic function,—reabsorption of water from the urine.

2. Thyroid gland: In pigeon, at the junction of the neck and trunk where the trachea divides into two bronchi, a pair of thyroid glands, one on either side, occur as the second important endocrine gland, next to pituitary. **Thyroxin**, the hormone secreted by this gland, contains about 65% of iodine and has a profound role in regulating metabolism of the body. The thyroid gland consists of a large number of thyroid follicles, each lined by a single layer of epithelial cells, secreting inside the cavity of the follicle a viscid fluid called **colloid.** The colloid is mainly formed of an iodised glycoprotein called **thyroglobulin** which, on hydrolysis, forms **triiodothyronine.** Both thyroxine and triiodothyronine are amino acids which pass from the gland to the bloodstream.

The thyroid hormones, besides controlling rate of metabolism of all the tissues in the body, also assist in the development of the nervous system, in the process of growth of the body, and in reproduction.

3. Parathyroid gland: Usually at the caudolateral aspect (sometimes caudodorsal) of each thyroid gland, paired parathyroid glands occur as small rounded bodies of which one is superior and one inferior in position. Histologically, this gland is formed of two types of epithelial cells: (i) Chief cells that are much larger in number than the other type of cells, and (ii) the oxyphil cells. The chief cells secrete the **parathyroid hormone** (PTH) or **Parathormone**, and the oxyphill cells possibly help building the chief cells. The parathyroid hormone regulates the amount of calcium, phosphate and magnesium ions in blood. It also helps in the activation of Vitamin D.

4. Adrenal gland: This paired, yellowish and somewhat tringular gland occurs one on either side at the cranio-medial or apical position of the kidney. Each gland is enclosed by a thin connective tissue membrane.

In pigeons the adrenal gland is quite well-developed and has two distinct parts–the outer **cortex** and the inner **medulla**. Both the parts secrete certain essential hormones for general metabolism and reproduction. The adrenal cortex has a mesodermal origin, while the medulla is derived from the neural crest cells of the ectoderm.

The **adrenal cortex** secretes three types of complex steroids—**glucocorticoids (Adrenocorticoids)**, **mineralocortricoids,** and **sex steroids**. The glucocorticoids play an important role in carbohydrate and protein metabolism. The mineralocorticoids and the sex steroids have important functions in regulating the electrolytes and water balance in the body, as well as assisting in the functions of gonadal hormones like oestrogen, androgen, progesterone, etc.

The two types of cells of the **adrenal medulla** secrete hormones called **epinephrine** or **adrenalin** and **norepinephrine** which are comprised of closely related substances under the catecholamine group. The medullary portion of the gland is supplied by preganglionic fibres of the sympathetic nervous system.

The histophysislogy of adrenal medulla and its relation with the cortex has been extensibly studied by Ghosh and his associates (2001, 2002).

5. Pancreas: The pancreas is the second largest digestive gland in the body and serves both as an exocrine and an endocrine gland. The endocrine part of the gland consists of certain groups of cells called **'Islets of Langerhans'** which remain scattered within the pancreatic tissue. The islets mainly consist of **beta cells(β-cells)** which secrete the hormone **insulin** that decreases the sugar level in blood. When its secretion at a desired level is inhibited, the blood sugar level rises.

On the other hand, another kind of cells called **alpha cells (α-cells)**—slightly larger in size than the beta cells but fewer in numbers—secrete the other hormone called **glucagon** which increases the blood sugar level. Hence, a balanced secretion of these two hormones play significant role in carbohydrate metabolism.

6. Pineal body (Syn: Epiphysis cerebri): The avian pineal body is an endocrine organ containing photoreceptive cells that produce the hormone **melatonin.**

7. Gonads: Both testis and ovary, besides producing spermatozoa and ova, secrete certain important sex hormones. The functioning of the accessory sex organs and appearance of secondary sex characteristics in birds—change in the colour of the beaks and seasonal change of colour of plumage, sexual dimorphism, seasonal photoperiodic changes—are largely attributed to the secrection of **testosterone**, a hormone secreted by the interstitial cells of the testis.

The ovaries too are partly endocrine in nature. They secrete the hormones **oestrogen**, **progesterone** and **androgen**. These are steroid hormones; their synthetic forms have also been produced. The development of the secondary sex characteristics in female are mainly due to the influence of oestrogen. Whereas progesterone, possibly secrected by the ovarian follicle, may have a role in the ovulation of the bird. Androgen, however, is secreted under certain conditions. If the left ovary is removed in birds, the rudimentary right ovary may be modified into a testis-like organ which may secrete androgen responsible for the expression of male characteristics.

In the mammals in general—and in the placental mammals in particular—the hormones attain greater functional divergence. With the beginning of and during intrauterine development of the embryo, the role of progesterone, a steroidal hormone secreted by the corpus luteum of the ovary, and an additional hormone called **relaxin**–a non-steroidal hormone possibly secreted at some site of the ovary—have great functional significance in the development of the mammals, as well as of a few other vertebrates.

Epilogue

As already mentioned in the 'Preface', this book has been a prime endeavour of the author to impart an elementary knowledge in Ornithology to the students and researchers interested in the study of birds. Ornithology now-a-days encompasses manifold activities of avian life, including intricate behavioural patterns in birds. Only a few selected chapters have been included in this book which is intended to be used more as a guide line to the study of birds than an in-depth study of all their activities. Each aspect of bird life is now treated as a specialised branch with ample scope of further exploration.

Therefore, within the limited scope of this book, chapters have been selected to highlight the structures and functions in birds in the context of current trends of researches in avian biology. I owe an apology to the readers for my inability to include and discuss all the aspects of bird life such as, mate choice, mate guarding, nest mortality, adaptations of bird chicks, island birds, predation on bird population, etc. Some of the chapters have been discussed very briefly, some in more details. In all the topics appropriate references have been cited for further studies of students, teachers and researchers.

The author sincerely hopes that anyone interested in the study of birds will find the book useful. On the other hand, due consideration has been given by the author in the planning of the book in such a way, that the students of both pass and Honours courses in zoology at the undergraduate level and those at the postgraduate level are largely benefited.

Some of the chapters have been discussed very briefly without going into much technicality for the purpose of just introducing the subject to the readers who may explore deeper to gather detailed information through exhaustive references cited in the text.

Quite a long time has elasped between the beginning of writing the manusrippt and its publication, and because in each year many new facts are being added, the readers are reminded of some minor changes in the data, hypotheses and analyses that may have occurred within this period.

It would be highly encouraging for the author to welcome valuable suggestions from the readers towards improvement of the text in future.

References

Able, K. P., and V. P. Bingman.,1988. The development of migratory orientation. In: Acta XIX Internatl. Ornithol. Cong. (H. Ouellet ed.). Natl. Mus. of Natural Sciences, Ottawa Univ. Press, Canada. Vol. II. pp. 1932-1940.

Able, K. P., and M. A. Able. 1990. Calibration of the magnetic compass by a migratory bird's celestial rotation. *Nature*, 347: 378-380.

Adret-Hausberger, M. 1988. Species specificity and dialects in Starlings's whistles. In: Acta XIX Internatl. Ornithol Cong. (H. Ouellet ed.). Natl. Mus. of Natural Sciences, Ottawa Univ. Press, Canada. Vol. II. pp. 1585-1597.

Akester, A. R. 1960. The comparative anatomy of the respiratory pathways in the domestic fowl (*Gallus domesticus*), pigeon (*Columba livia*), and domestic duck (*Anas platyrhynchos*). J. Anat. 94: 487-505.

Ali, S. 1979. The Book of Indian Birds. 11th ed. (Revised and enlarged). Bombay Nat. Hist. Soc., Bombay.

Ali, S. and S. D. Ripley. 1987. Compact Handbook of the Birds of India and Pakistan, together with those of Bangladesh, Nepal, Bhutan and Sri Lanka. 2nd ed., Oxford Univ. Press, Oxford; New York, Toronto, etc.

Ames, P. L. 1971. The morphology of the syrinx in passerine birds. Bull. Peabody Mus. Nat. Hist. (Yale Univ.) 37: 1-194.

Arora, M. P. 1992. Animal behaviour. 1st ed., Himalaya Publishing House, Mumbai.

Assenmacher, I. 1973. The peripheral endocrine glands. In: *Avian Biology* (D. S. Farner and J. R. King, eds.) Vol. 3. Academic Press, London.

Aymar, G. C. 1935. *Bird flight.* Bodley Head, London.

Bailey, R. E. 1953. Accessory reproductive organs of male fringillid birds: seasonal variations and response to various sex hormones. Anat. Rec. 115: 1-19.

Baker, E. C. S. (1922-1927). Fauna of British India: Birds, 4 Vols. Taylor and Francis, London.

Baker, M. C. 1975. Song dialects and genetic differences in white-crowned sparrows (*Zonotrichia leucophrys*). Evolution, 29 : 226-241.

Baker, M. C, 1982a. Vocal dialect recognition and population genetic consequences. Am. Zool. 22: 561-569.

Baker, M. C., 1982b. Genetic population structure and vocal dialects in *Zonotrichia* (Emberezidae). In: Acoustic communication in birds. Vol. 2. D. E. Kroodsma and E. H. Miller (eds.). Academic Press, New York. pp. 209-235.

Baker, M. C. 1983. The behavioural response of female Nuttall's white-crowned sparrows to male song of Natal and alien dialects. Behav. Ecol. Sociobiol 12: 309-315.

Baker, M. C., and M. A. Cunningham. 1985. The biology of bird-song dialects. Behav. Brain Sci. 8: 85-100.

Ballmann, P. 1969a. Die Vogel aus der altburdigalen Spalten füllung von Wintershof (West) bei Eichstätt in Bayern. Zitteliana (München) 1: 5-60.

Baptista, L. F., 1975. Song dialects and demes in sedentary populations of the white-crowned sparrow (*Zonotrichia leucophrys nuttalli*). Univ. Calif. Berkeley Publ. Zool. 105: 1-52.

Baptista, L. F. and L. Petrinovich. 1986. Song development in the white-crowned sparrow: social factors and sex differences. Anim. Behav. 34: 1359-1371.

Baumel, J. J. 1975c. Heart and blood vessels. Aves. In: Sisson and Grossman's; *The Anatomy of the domestic animals* (R. Getty. ed.) 5th ed. Vol. 2. pp. 1968-2009. Saunders, Philadelphia.

Baumel, J. J. 1979. Osteologia/Arthrologia/ Systema cardiovasculare. In: Nomina Anatomica Avium (NAA) (J. J. Baumel, A. S. King, A. S. Lucas, A. M. Breazile, and H. E. Evans, eds.). pp. 53-121, 123-173, 347-407. Academic Press, London.

Baumel, J. J., A. F. Dalley and T. H. Quinn. 1983. The collar plexus of subcutaneous thermoregulatory veins in the pigeon *Columba livia*: its association with oesophageal pulsation and gular flutter. Zoomorphology 102: 215-239.

Baumel, J. J. 1988. Functional morphology of the tail apparatus of the pigeon *Columba livia*. Adv. Anat. Embryol. Cell Biol. no. 110: 115 pp. Springer Verlag, Heidelberg.

Baumel, J. J. 1993. Osteologia/ Arthrologia/ Systema Cardiovasculare. In: NAA (J. J. Baumel, L. M. Witmer, R. J. Raikow eds.). pp. 45-132, 133-188, 407-476. Nuttal Ornithological Club, C/o Museum of Comparative Zoology, Harvard University, U. S. A.

Beck, W., and W. Wiltschko. 1988. Magnetic factors control the migratory direction of Pied Flycatchers (*Ficedula hypoleuca* Pallas). Acta XIX Int. Ornithol. Congr., Ottawa, Canada, 1986. Publ. Natl. Mus. Nat. Sci., Ottawa University Press, pp. 1955-1962.

Beddard, F. E. 1898. The structure and classification of birds. Longmans, Green and Co., London. pp. xx+ 584.

Beddard, F. E. 1911. On the alimentary tract of certain birds and on the mesenteric relations of the intestinal loops. Proc. Zool. Soc. London, pp. 225-231.

Beibach, H., W. Friedrich., and G. Heine. 1986. Interaction of body mass, fat, Foraging and stopover period in trans-Sahara migrating passerine birds. Oecologia (Berlin), 69: 370-379.

Bellairs, A. d'A., and C. R. Jenkin, 1960. The skeleton of birds. In: **Biology and Comparative Physiology of birds** (A. J. Marshal. ed.) Vol.1, pp. 241-300. Academic Press, London.

Benowitz, L. I., and H. J. Karten,1976a. The tractus infundibuli and other afferents to the parahippocampal region of the pigeon. Brain Res. 102: 174-180.

Benton, M. J. 1985. Classification and phylogeny of the diapsid reptiles. *Zoological Journal of the Linnean Society*, 84: 97-164.

Benton, M. J. 1990a. Origin and interrelationships of dinosaurs. In: *The Dinosauria* (D. B. Weishampel, H. Osmólska and P. Dodson eds.) University of California Press, Berkeley. pp. 11-30.

Benton, M. J. 1997. Vertebrate Palaeontology. 2nd ed. Chapman and Hall. London/ New York.

Berger, A. J. 1960a. Some anatomical characters of the Cuculidae and the Musophagidae. Wilson Bull. 72: 60-104.

Berkhoudt, H. 1977. Taste buds in the bill of the Mallard (*Anas platyrhynchos*): their morphology, distribution and functional significance. Netherl. J. Zool. 27: 310-331.

Berkhoudt, H. 1985. Structure and function of avian taste receptors. In: Form and Function in birds, Vol. 3., Academic Press, London.

Berthold, O. 1984. The endogenous control of bird migration: a survey of experimental evidence. Bird Study, 31: 19-27.

Berthold, P. 2003. Genetic basis and evolutionary aspects of bird migration. *Advances in Study of Behabiour*. 33: 175-229.

Bhadury, J. L., B. Biswas., and S. K. Das. 1957. The arterial system of the domestic pigeon, *Columba livia* Gmelin. Ant. Anz. 104: 1-14.

Bhatt, D., and V. K. Sethi. 2006. In: 0-14, Communication: Abstract, 24th Internatl. Ornithol. Cong., Hamburg, Germany.

Bhattacharyya, B. N. 1980. The morphology of the jaw and tongue musculature of the common pigeon, *Columba livia*, in relation to its feeding habit. Proc. Zool. Soc. (Calcutta) 31: 35-127.

Bhattacharyya, B. N. 1994. Diversity of feeding adaptations in certain columbid birds: a functional morphological approach. J. Biosci., Vol. 19, No. 4: 415-427 (Publ. of the Indian Academy of Sciences).

Bhattacharyya, B. N. 1997. Food and feeding adaptations in certain columbid and passerine birds in relation to agriculture in India. Proc. Zool. Soc. (Calcutta) 50 (2): 166-170.

Bhattacharyya, B. N. 1998. Functional morphology of the feeding apparatus of the Common Green Pigeon, *Treron phoenicoptera* (Latham). Proc. Zool. Soc. (Calcutta) 51 (1): 1-44.

Bingman, V. P., and K. P. Able. 1979. The sun as a cue in the orientation of the White-throated Sparrow, a nocturnal migrant. Anim. Behav. 27: 621-622.

Bingman, V. P., and K. P. Able. 1983. Magnetic field orientation of migratory Savannah Sparrows with different first summer experience. Behaviour, 87: 43-53.

Blanford, W. T. 1895-1897. Fauna of British India: Birds, Vol. 3-4; London.

Bock, W. J. 1960. Secondary articulation of the avian mandible. Auk 77: 19-55.

Bock, W. J. 1964. Kinetics of the avian skull J. Morph. 114: 1-52.

Bock, W. J. 1966. An approach to functional analysis of bill shape. Auk: 10-51

Bock, W. J. 1972. Morphology of the tongue apparatus of *Ciridops anna* (Drepanididae). Ibis 114: 71-78.

Bock, W. J. 1974. The avian skeleto-muscular system. In: Avian Biology (D. S. Farner and J. R. King, eds.). Vol. 4: 120-250. Academic Press, London/New York.

Bock, W. J., J. Morony., and J. Farrand. 1980. *Reference list of Birds of the World*. Amer. Mus. Nat. Hist. Publ.

Bohme, G. 1970. Eine organartige Bildung im IV. Ventrikel beim Huhn. Verh. Anat. Ges. 64: 245-250.

Bologna, G. 1995. The Macdonald Encyclopedia of Birds of the World. pp. 1-511. Little Brown and Company, London.

Brach, V. 1977. The functional significance of the avian pecten: a review. Condor 79: 321-327.

Brodkorb, P. 1971. Origin and evolution of birds. In: *Avian Biology*. Vol. 1 (D. S. Farner and J. R. King, eds.). pp. 19-55. Academic Press, New York.

Brown, N. L., and B. K. Follett, 1977. General Comparative Endocrinology; 33, 267.

Bruderer, B., and L. Jenni. 1988. Strategies of Bird Migration in the area of the Alps. Acta XIX Int. Ornithol. Congr., Ottawa University Press, pp. 2150-2161.

Burton, P. J. K. 1974. Jaw and tongue features in Psittaciformes and other orders with special reference to the anatomy of the Tooth-billed pigeon (*Didunculus strigirostris*). J. Zool. London. 174: 255-276.

Bühler, P. 1981. Functional anatomy of the avian jaw apparatus. In: *Form and Function in Birds* (A. S. King and J. McLelland, eds.). Vol. 2. Academic Press, London/New York. pp. 439-468.

Bühler, P. 1985. On the morphology of the skull of *Archaeopteryx*. In: *The Beginnings of Birds* (Hecht, M. K., Ostrom, J. H., Viohl, G; Wellnhofer, P, eds.). Proc. Internatl. Archaeopteryx Conference, Eichstätt, pp. 135-140.

Bühler, P., L. D. Martin, and L. M. Witmer, 1988. Cranial kinesis in the late Cretaceous birds, *Hesperornis* and *Parahesperornis*. Auk 105: 111-122.

Buttendiek, E. 1980. Die Benennung des skeletts beim Truthuhn (*Meleagris galapavo*) unter Berücksichtigung der Nomina Anatomica Avium, 1979. Inaug. Diss. Tierärtzl, Hochschule, Hannover.

Cameron, A. C. 1938. Birds drinking in the dry interior. Emu 38: 336-337.

Catchpole, C. K. 1980. Sexual selection and the evolution of song in European warblers of the genus *Acrocephalus*. Behaviour 74: 149-166.

Chatterjee, S. 1991. Cranial anatomy and relationships of a new Triassic bird from Texas. Phil. Trans. R. Soc., London, B 332: 277-342.

Chatterjee, S. 1995. The Triassic bird *Protoavis*. Archaeopteryx 13: 15-31.

Chiappe, L. M. 1993. Enantiornithine (Aves) tarsometatarsi from the Cretaceous Lecho Formation of northwestern Argentina. *American Museum Novitates*, 3083, 1-27.

Chiappe, L. M. 1995a. The first 85 million years of avian evolution. Nature 378: 349-355.

Clench, M. H. 1970. Variability in body pterylosis, with special reference to the genus *Passer*. Auk 87: 650-691.

Cochran, W. W., H, Mouritsen, and M. Wikelski. 2004. Migratory songbirds recalibrate the magnetic compass daily from twilight cues. Science, 304: 405-408.

Cooper, A., and D. Penny, 1997. Mass survival of birds across the Cretaceous-Tertiary boundary: Molecular evidence. Science 275: 1109-1113.

Cracraft, J. 1968. The lacrimal-ectethmoid complex in birds: a single character analysis. Amer. Midl. Nat. 80: 316-359.

Cracraft, J. 1986. The origin and early diversification of birds. Paleobiology 12: 383-399.

Cracraft, J. 1988. The major clades of birds. In: Benton, M. J. (ed.). The phylogeny and classification of the tetrapods. Vol. 1: Amphibians, Reptiles, Birds. Syst. Assoc. Special Vol 35A: 339-361. Clarendon Press, Oxford.

Cranfield, I. ed., 2001. In: Dinosaurs and other prehistoric creatures; Greenwich Edition, London.

Currie, P. J. 1987. Bird-like characteristics of the jaws and teeth of troodontid theropods (Dinosauria, Saurischia). J. Vert. Paleontol. 7: 72-81.

Darwin, C. 1859. On the origin of species by means of Natural Selection, London.

de Beer, G. R. 1954a. *Archaeopteryx* and Evolution. *The advancement of Science. London. 42: 160-170.*

de Beer, G. R. 1956. The evolution of ratites. Bull. Brit. Mus. (Nat. Hist.), Zoology. Vol. 4, No. 2: 57-70.

Delius, J. D., and J. Emmerton. 1978. Sensory mechanisms related to homing in pigeons. In: Animal orientation, navigation, and homing. K. Schmidt-Koenig and W. T. Keeton (eds.) Springer-Verlag, Berlin. pp. 35-41.

Dhindsa, M. S., and H. S. Toor. 1980. Extent of bird damage to rice nurseries in the Punjab and its control. Indian J. Agric. Sci. 50: 715-719.

Dhindsa, M. S. 1986. Ecology of Indian weaverbirds. Indian Rev. Life Sci. 6: 101-140.

Dhindsa, M. S., P. S. Sandhu., H.K. Saini, and H. S. Toor. 1991. House Crow damage to sprouting sunflower. Trop. Pest. Manage. 37: 179-181.

Dhindsa, M. S., and H. K. Saini. 1994. Agricultural Ornithology: an Indian perspective. J. Biosc. Vol. 19., no. 4: 391-402.

Dieterich, C. E., and M. Pfautsch. 1973. Fine structure observations of the pecten oculi capillaries of the chicken. Freeze etching, scanning, and transmission electron microscopic investigations. Z. Zellforsch. Milkrosk. Anat. 146: 473-489.

Dolbeer, R. A. 1988. Current status and potential of lethal means of reducing bird damage in Agriculture. Acta XIX Int. Ornithol. Congr., Ottawa, Canada., Vol. I. (H. E. Ouellet ed.). Publ. Natl. Mus. Nat. Sci., Ottawa University Press; pp. 474-483.

Dorst, J. 1962. The migration of birds. Heineman, London.

Dyke, G. J., and J. Thorley. 1998. Reduced cladistic consensus and the interrelationships of *Protoavis, Avimimus* and Mesozoic birds. Orcheopteryse (acc).

Emlen, S. T. 1967. Migratory orientation in the Indigo Bunting, *Passerina cyanea*. Part II. Mechanism of celestial orientation. Auk 84: 463-489.

Emlen, S. T. 1970. Celestial rotation: its importance in the development of migratory orientation. Science 170: 1198-1201.

Emlen, S. T. 1975. Migration: orientation and navigation. In: *Avian biology*, Vol. V (D. S. Farner and J. R. King ed.). pp. 129-219. Academic Press, New York.

Epple, A., and M. H. Stetson (eds.). 1980. Avian Endocrinology. Academic Press, New York.

Falconer, I. R. 1971. In: "Physiology and Biochemistry of the domestic fowl" (D. J. Bell and B. M. Freeman eds.). p. 459, Academic Press, London.

Falkmer, S., and G. Patent. 1972. In: Handbook of Physiology, Sect. 7, Vol. I, *Endocrine Pancreas*, (D. Steiner and N. Frienkel, eds.), Chap 1, Waverley Press, Baltimore.

Falls, J. B. 1982. Individual recognition by sound in birds. In: Acoustic communication in birds. Vol. 2. (D. E. Kroodsma and E. H. Miller, eds.). Academic Press, New York, pp. 237-278.

Farmer, S. W., H. Papkoff., and T. Hayashida. 1974. Endocrinology 95, 1560.

Farner, D. S., and J. R. King. 1973. Avian Biology, Vol. III, Academic Press, New York, London.

Feduccia, A., and H. B. Tordoff. 1979. Feathers of *Archaeopteryx*: asymmetrical vanes indicate aerodynamic function. Science, 203, 1021-1022.

Feduccia, A. 1995. Explosive evolution in Tertiary birds and mammals. Science 267: 637-638.

Feduccia, A. 1996. The Origin and Evolution of Birds. Yale Univ. Press, New Haven and London.

Fisher, H. I. 1955. Some aspects of the kinetics in the jaws of birds. Wilson Bull. 67: 175-188.

Fisher, J., and R. T. Peterson. 1964. *The World of Birds*. MacDonald, London.

French, E. I., and R. D. Hodges. 1977. Cell Tissue Research, 178; p. 397.

Friant, M. 1945. I. Les carpiens du Nandou (Rhea). II. Development et interpretation de la ceinture scapulaire du Nandou (Rhea). C. R. Acad. Sci., Paris, 221: 641-643; 711-713.

Fujita, H. 1963. Electron microscope studies on the thyroid gland of domestic fowl with special reference to the mode of secretion and the occurrence of a central flagellum in the follicle cell. Z. Zellforsch, Mikrosk. Anant. 60: 615-632.

Fürbringer, M. 1888. Untersuchungen zur Morphologie und Systematik der Vögel. Amsterdam, 2 vols. xiix, 1751 pp.

Furr, B. J. A. 1969b. Progesterone concentrations in the blood and ovaries of laying, moulting and broody hens. Gen. Comp. Endocrinol. 13, Abstr. 57.

Gadow, H. 1889. On the taxonomic value of the intestinal convolutions in birds. Proc. Zool. Soc., London.

Gadow, H., and E. Selenka. 1891. Vögel: I. Anatomischer Theil. Vol. 6. P:t. 4. In: *Bronn's Klassen und Ordnungen des Thier-Reichs*. C. F. Winter, Leipzig, pp. 1-1008.

Galton, P. M. 1970a. The posture of hadrosaurian dinosaurs. Journal of Paleontology, 44: 464-473.

Gauthier, J. 1986. Saurischian monophyly and the origin of birds. In: K. Padian (ed.), The origin of birds and the evolution of flight. Mem. Calif. Acad. Sci B: 1-55.

Gauthreaux, S. A. Jr. 1980. The influences of long-term and short-term climatic changes on the dispersal and migration of organisms. In: Animal migration, orientation, and navigation. S. A. Gauthreaux. Jr. (ed.). Academic Press, New York. pp. 104-174.

George, J. C., and A. J. Berger, 1966. Avian Myology. Academic Press, New York.

Gerisch, D. 1971. Die Bronchi atriales in der Lunge des Haushuhnes (*Gallus gallus domesticus* L.). Ein beitrage zur Morphologie und Nomenklatur.Inaug. Diss. Hanover.

Gerritsen, A. F. C., and J. Sevenster. 1985. Foraging behaviour and bill anatomy in Sandpipers. Fortschr Zool 30: 237-240.

Gilbert, A. B. 1971b. The egg: its physical and chemical aspect. In: Thé physiology and biochemistry of the Domestic Fowl (D. J. Bell and B. M. Freeman eds.). vol. 3., Academic Press, London.

Ghosh, A. 1962. In: 'Progress in Comparative Endocrinology' (K. Takewaki, ed.). p. 70., Academic Press, New York.

Ghosh, A. 1973. Histophysiology of the adrenal medulla. Presidential address, Indian Science Congress; Diamond Jubliee Session, Chandigarh; Pp. 1-24.

Ghosh, A., S. W. Carmichael, and M. Mukherjee. 2001. Avian adrenal medulla: cytomorphology and function. Acta Biologica Szegediensis. Vol. 45 (1-4): 1-11.

Ghosh, A., M. Mukherjee and S. Ghosh. 2002. Stress and avian adrenal function. Biomedical Research 13 (2/3): 69-74.

Glick, B. 1959. The experimental production of the stress picture with cortisone and the effect of penicillin in the young chickens. Ohio. J. Sci 59: 81-87.

Glick, B. 1980. The Thymus and Bursa of Fabricius: Endocrine organs? In: *Avian Endocrinology* (A. Epple and M. H. Stetson eds.). Academic Press, New York.

Goldmith, A. R., and B. K. Follett. 1980. *Anterior Pituitary Hormones*. In: Avian Endocrinology (A. Epple and M. H. Stetson eds.). Academic Press, New York.

Goodman, D. C., and H. I. Fisher. 1962. Functional anatomy of the feeding apparatus in waterfowl (*Aves: Anatidae*). Southern Illinois University Press, Carbondale. pp. 1-193.

Gregory, W. K. 1935. Remarks on the origins of the Ratites and Penguins; with discussion by Robert Cushman Murphy. Proc. Linn. Soc. New York, Nos. 45-46: 1-18

Gwinner, E., and W. Wiltschko. 1978. Endogenously Controlled changes in migratory direction of the Garden Warbler, *Sylvia dorin*. J. Comp. Physiol. 125: 267-273.

Gwinner, E., 1986. Circannual rhythms in the control of avian migrations. Adv. Study-Behav. 16: 191-228.

Hall, B. K., and H. P. Hughes. 1970. Z. Zellforsch. 10c. 1. In: Avian adrenal medulla: structure and function (A. Ghosh); Avian Endocrinology. (A-Epple and M. H. Stetson eds.). Academic Press, New York.

Harvey, S., and C. G. Scanes. 1977. J. Endocrinol 73, 321.

Hayes, B., G. R. Martin., and M. Brooke. 1991. Novel area serving binocular vision in the retinae of procellariiform sea birds. Brain Behav. Evol. 37: 79-84.

Heidweiller, J., and G. A. Zweers. 1992. Development of the drinking mechanisms in the chicken (*Gallus gallus* L.). Zoomorphol 111: 217-228.

Heidweiller, J., J. A. van Loon., and G. A. Zweers. 1992b. Flexibility of the drinking mechanism in adult chickens (*Gallus gallus* L.) (Aves). Zoomorphol 111: 141-159.

Heilmann, G. 1926. The Origin of Birds. Appleton, Dover; Reprint 1972.

Higuchi, H., and P. Berthold. 2002. Migration and conservation of cranes and stocks. Abstract Vol., xxiii Internatl. Ornithol. Congr., Beijing, China, P. 42.

Hofer, H. 1945. Untersuchungen über den Bau des Vogelschädels, besonders über den der Spechte und Steisshühner. Zool. Jb. Abt. Anat. Ontogenic Tiere 69: 1-158.

Hofer, H. 1955. Neuere Untersuchungen zur Kopfmorphologie der Vögel. Acta XI Internatl. Ornithol. Congr. (A. Portmann, and I. Sutter. eds.). pp: 104-137.

Holmes, W. N., and J. Cronshaw. 1980. Adrenal cortex: structure and function. In: Avian Endocrinology (A. Epple and M. H. Stetson, eds.). Academic Press, New York.

Homberger, D. G. 1980. Funktional-morphologische Untersuchengen Zur Radiation der Ernährungsund Trinkmethoden der Papageien (Psittaci). Bonn Zool Monogr 13: 1-192.

Homberger, D. G. 1986. The lingual apparatus of the African Grey Parrot, *Psittacus erithacus* Linnè (Aves: Psittacidae): Description and theoretical mechanical analysis. Ornithological Monogr: 1-233.

Homberger, D. G., and R. A. Mayers. 1989. Morphology of the lingual apparatus of the Domestic Chicken, *Gallus gallus*, with special attention to the structure of the Fasciae. The American Journal of Anatomy 186: 217-257.

Hou, L., Z. Zhou., L. D. Martin., and A. Feduccia. 1995. A beaked bird from the Jurassic of China. Nature 377: 616-618.

Huxley, T. H. 1867. On the classification of birds and the taxonomic value of the modification of certain of the cranial bones observable in that class. Proc. Zool. Soc., London: 415-472.

Huxley, T. H. 1882. Respiratory organs of *Apteryx*. Proc. Zool. Soc., London., pp: 560-569.

Islam, M. Z., and A. R. Rahmani (eds.). 2002. Threatened Birds of India, '**Buceros**', Vol. 7. No. 1 & 2., Publ. Bombay Natural History Society (BNHS).

Jenkins., P. F. 1978. Cultural transmission of song patterns and dialect development in a free-living bird population. Anim. Behav. 25: 50-78.

Johnson, R. 1984. The cranial and cervical osteology of the European Oystercatcher *Haematopus ostralegus* L. J. Morph 182: 227-244.

Kardong, K. V. 1995, 1998. Vertebrates: Comparative anatomy, function, evolution. Wm Brown, Dubuque, Iowa.

Katz, Y., V. Liepa., and J. Viksne. 1988. Orientation Research in the Latvian SSR in 1982-1985. In: Acta XIX Internatl, Ornithol. Congr., Ottawa, Canada, 1986; Publ. Natl. Mus. Nat. Sci., Ottawa University Press, pp. 1919-1931.

Keast, A. 1969. Evolution of mammals on southern continents. Quart. Rev. Biol. 44: 121-167.

Kent, G. C. 1983. Comparative anatomy of the vertebrates. (5*th* ed.). The C. V. Mosby Company, St. Louis, Toronto, London.

Keeton, W. T. 1969. Orientation by pigeons: Is the sun necessary? Science 165: 922-928.

Keeton, W. T. 1971. Magnets interfere with pigeon homing. Proc. Natl. Acad. Sci., USA 68: 102-106.

Kiefer, S., and H. Hultsch. 2006. In: 0-14, Communication: Abstract, 24th Internatl. Ornithol. Cong., Hamburg, Germany.

King, A. P., M. J. West., and D. H. Eastzer. 1980. Song structure and song development as potential contribution to reproductive isolation in cowbirds (*Molothrus ater*) J. Comp. Physiol. Psychol. 94: 1028-1039.

King, A. P., M. J. West., and D. H. Eastzer. 1983b. Epigenesis of cowbird song–a joint endeavour of males and females. Nature (London), 305: 704-706.

King, A. S. 1966. Structural and functional aspects of the avian lungs and air sacs. In: *International Review of General and Experimental Zoology*. (W. J. L. Felts and R. J. Harrison, eds.). Vol. II, pp. 171-267. Academic Press, New York.

King A. S., and J. McLelland. 1980-1989. *Form and Function in Birds*. 4 vols. Academic Press, London.

King, A. S., and J. McLelland. 1984. *Birds. Their structure and function*. Bailliere Tindall, London.

Kooloos, J. G. M., and G. A. Zweers. 1989. Mechanics of drinking in the mallard (*Anas platyrhynchos*, Anatidae). J. Morph 199: 327-347.

Kooloos, J. G. M., and G. A. Zweers. 1991. Different integration of pecking, drinking and filter feeding mechanisms in water fowl. Acta Biotheor 39: 107-140.

Kramer, G. 1957. Experiments in bird orientation and their interpretation. Ibis 99: 196-227.

Kroodsma, D. E. 1974. Song learning, dialects, and dispersal in the Bewick's wren. Z. Tierpsychol. 35: 352-380.

Kroodsma, D. E. 1979. Vocal dueling among male marsh wrens: evidence for ritualized expressions of dominance/subordinance. Auk, 96: 506-515.

Kuenzel, W. J. 1983. Behavioural sequence of food and water intake: its significance for elucidating central neural mechanisms controlling feeding in birds. Bird Behav 5: 2-15.

Landolt, R. 1987. Vergleichend funktionelle Morphologie des Verdauungstraktes der Tauben (Columbidae) mit besonderer Berücksichtigung der adaptiven Radiation der Fruchttauben (Treronidae). Zool. Jahrb. Anat. 116: 169-215, 285-316.

Lehninger, A. L. 1970. Biochem. J. 119, P. 129.

Lehrman, D. S. 1963. On the initiation of incubation behaviour in doves. Anim. Behav. 11: 433-438.

Leipa, V. 1978. Orientation of European Robin in circular cages under artificial light cues. In: Orientatsija ptits (Bird orientation). H. Mihelsons, P. Blums, and J. Baumanis.

Leipa, V. 1982. Testing the significance of internal clock in the night orientation of the Robin (*Erithacus rubicula* L.) under natural sky. Proc. Int. Ornithol. Congr. 18: 235.

Lighthill, Sir James. 1975. *Mathematical biofluid dynamics.* Society for Industrial and Applied Mathematics. Philadelphia, Pennsylvania.

Lindsay, B. 1885. On the avian sternum. *Proc Zool. Soc., London*: 684-716.

Lofts, B., and R. K. Murton. 1973. On male reproduction. In: 'Avian Biology' (D. S. Farner and J. R. King, eds.), Vol. III. P. 1. Academic Press, New York.

Lorenz, K. Z. 1939. Vergleichende Verhaltens forschung. Zool. Anz. Suppl. 12. pp. 69-102.

Lowe, P. R. 1935. On the relationship of the Struthiones to the Dinosaurs and to the rest of the avian class, with special reference to the position of *Arachaeopteryx*. Ibis, London. (13) 5: 398-432.

Manger Cats-Kuenen, C. S. W. 1961. Casque and bill of *Rhinoplax vigil* (Forst) in connection with the architecture of the skull. Verh. Kon. Ned. Akad. Wet, Amsterdam. Afd. 2.53: 1-51.

Malinovský, L. 1962. Contribution to the anatomy of the vegetative nervous system in the neck and thorax of the domestic pigeon. Acta. Anat. 50: 326-347.

Marler, P. 1984. Song Learning: innate species differences in the learning process. In: The biology of learning. (P. Marler and H. S. Terrace, eds.). Dahlem Konferenzen. pp. 289-329.

Marler, P., and R. Pickert. 1984. Species-universal microstructure in the learned song of the swamp sparrow (*Melospiza georgiana*). Anim. Behav. 32: 673-689.

Martin, G. R. 1985. Eye. In: *Form and Function in Birds.* (A. S. King and J. McLelland, eds.). Vol. 3, pp. 311-373.

Martin, G. R. 1987. How do birds accommodate? Nature 238: 383.

Martin, L. D. 1987. The beginning of the modern avian radiation. Docum. Lab. Géol. Lyon., no. 99, pp. 9-19.

Martin, L. D. 1995. The Enantiornithes: Terrestrial birds of the Cretaceous. Cour Forschungsinst Senckenberg 181: 23-36 (In: G. A. Zweers and J. C. Vanden Berge, 1997/1998.

Mason, C. W., and H. Maxwell-Lefroy. 1912. The food of birds in India. Mem. Dept. Agri. India., Entom. Ser 3. (Pusa). pp. 1-371.

Matthews, G. V. T. 1963. The astronomical basis of "nonsense" orientation. Proc. Int. Ornithol. Congr. 13: 415-429.

Mayr, E. 1931. Die Syrinx einiger Singvögel aus Neu-Guinea. J. F. Ornith. 79: 333-337.

Mayr, E., and D. Amadon. 1951. A classification of Recent Birds. Amer. Mus. Nat. Hist. (AMNH) Publ. no. 1496., New York.

Mc. Atee, W. L. 1908. Food habits of the grosbeaks. Bull. Bur. Biol. Surv., U. S. Dep. Agricult (Washington) 32: 1-92.

McGowan, C. 1985. Tarsal development in birds: evidence for homology with the theropod condition. J. Zool. London, Ser. A., 206: 53-67.

Mckinney, F. 1988. Mate guarding and the confidence of paternity in Waterfowl. Acta

XIX Internatl. Ornithol. Congr., Ottawa, Canada (H. Ouellet, ed.). Publ. Natl. Mus. of Nat. Sci., University of Ottawa Press.

McLelland, J. 1968. The hyoid muscles of *Gallus gallus*. Acta Anat. 69: 81-86.

McLelland, J. 1979. Digestive system. In: *Form and Function in Birds.* (A. S. King and J. McLelland, eds.). Vol. 1, pp. 69-181. Academic Press, London, New York.

McLelland, J. 1989. Larynx and trachea. In: *Form and Function in Birds.* (A. S. King and J. McLelland, eds.). Vol. 4: 69-103, Academic Press, London.

Meddle, S. L., and J. C. Wingfield. 2002. Seasonal and behavioural induced changes in the neuroendocrine system of birds. Abs. Vol., XXIII Int. Ornithol. Congr., Beijing, China.

Meyer, D. B. 1977. The avian eye and its adaptations. In: Handbook of Sensory Physiology (F. Crescitelli, ed.). Vol. 7. no. 5, pp. 549-611. Springer, Berlin.

Midtgard, U. 1981. The Rete tibiotarsale and arterio-venous association in the hindlimb of birds: A comparative morphological study on counter-current heat exchange systems. Acta Zool. 62: 67-87.

Midtgard, U. 1984a. The blood vascular system in the head of the Herring Gull (*Larus argentatus*). J. Morph. 179: 135-152.

Millington, S. J., and T. D. Price. 1985. Song inheritance and mating patterns in Darwin's finches. Auk 102: 342-346.

Mindell, D. P. 1997. Avian Molecular Evolution and Systematics. Academic Press, London and New York.

Miyaki, T., and M. Yasuda. 1977. On the thoracic duct and the lumbar lymphatic vessel in the fowl. Jap. Vet. Sci. 39: 559-570.

Monroe, B. L., and C. G. Sibley. 1993. A World Checklist of birds. Yale University Press, New Haven and London.

Moore, F. R. 1978b. Sunset and the orientation of a nocturnally migrant bird. Nature (London), 274: 154-156.

Moore, F. R. 1988. Sunset and migratory orientation : observational and experimental evidence. Acta XIX Int. Ornithol. Congr., Ottawa, Canada. Natl. Mus. Nat. Sci., Ottawa Univ. Press. pp. 1941-1954.

Moreau, R. E. 1972. The Palaearctic-African bird migration systems. Academic Press, London.

Morild, I., R. Mowinckel., A. Bohle., and J. A. Christensen. 1985. The juxtaglomerular apparatus in the avian kidney. Cell Tissue Res. 240: 209-241.

Morony, J. J., W. J. Bock., and J. Farrand. 1975. *Reference list of birds of the world.* Amer. Mus. Nat. Hist., New York.

Müller, H. J. 1963. Die Morphologie und Entwicklung des Cranium von *Rhea americana* Linné. Z. Wiss. Zool. 168: 35-118.

Müller, G. B., and P. Alberch. 1990. Ontogeny of the limb skeleton in *Alligator mississippiensis*: developmental invariance and change in the evolution of archosaur limbs. J. Morph. 203: 151-164.

Nicholls, T. J., and G. P. Graham. 1972. In: Male reproduction; Avian Endocrinology (A. Epple and M. H. Stetson, eds.)., 1980. Biol. Reprod. 6, 179.

Noskov, G. A., N. P. Lovchenko, and T. A. Rymkevich. 2002. Principles of photoperiodic control of annual cycle of events in birds. Abs. Vol. XXIII Int. Ornithol. Congr., Beijing, China.

Novas, M. A., and P. F. Puerta. 1997. New evidence concerning avian origins from the Late Cretaceous of Patagonia. Nature, 387-392.

Oates, E. W. 1889-91. Fauna of British India: Birds. Vols. 1, 2, London.

Olson, S. 1985. The fossil record of birds. In: D. S. Farner., J. R. King., and K. A. Parkes (eds.). Avian Biology, Vol. 8: 79-256. Academic Press, Orlando.

Ostrom, J. H. 1973. The ancestry of birds. Nature 242: 136.

Ostrom, J. H. 1974. *Archaeopteryx* and the origin of flight. Quart. Rev. Biol., 49: 27-47.

Ostrom, J. H. 1976. *Archaeopteryx* and the origin of birds. Biol. J. Linn. Soc. 8: 9-182.

Ostrom, J. H. 1978. The osteology of *Compsognathus longipes* Wagner. Zitteliana 4: 73-118.

Ostrom, J. H. 1979. Bird flight: How did it begin? American Scientist 67: 46-56.

Ostrom, J. H. 1985. The meaning of *Archaeopteryx* . In: M. K. Hecht., J. H. Ostrom; G. Viohl., and P. Wellnhofer (eds.). The begining of Birds. Proc. Internat. *Archaeopteryx* Congr., Jura-Museum. Eichstatt. pp. 161-176.

Ostrom, J. H. 1991. The question of the origin of birds. In: H. P. Schultze., and L. Trueb (eds.). Origins of the higher groups of Tetrapods: Controversy and Consensus. Comstock Publ. Assoc., Ithaca, New York. pp. 467-484.

Owen, R. 1866. On the anatomy of Vertebrates. Vol. II., Birds and Mammals. Longmans Green, London, I-VIII, 586 pp.

Parker, T. J. 1892. Observations on the anatomy and development of *Apteryx*. Philos. Trans. B. 182: 25-134.

Parker, T. J., and W. A. Haswell. 1951, 1962. A Text book of Zoology. Vol. II, *6th* ed. (Reprinted), Macmillan, New York.

Payne, R. B. 1973. Behaviour, mimetic songs and song dialects and relationships of the parasitic indigo birds (*Vidua*) of Africa. Ornithol. Monogr. 11.

Payne, R. B. 1981a. Song learning and social interaction in indigo buntings. Anim. Behav. 29: 688-697.

Payne, R. B. 1981b. Population structure and social behaviour: models for testing the ecological significance of song dialects in birds. In: Natural selection and social behaviour. R. D. Alexander and D. W. Tinkle (eds.). Chiron Press, New York. pp. 55-90.

Payne, R. B. 1982. Ecologial consequences of song matching: breeding success and intra-specific song mimicry in indigo buntings. Ecology 63: 401-411.

Payne, R. B. 1983a. Bird songs, sexual selection, and female matching strategies. In: Social behaviour of female vertebrates. S. K. Wasser (ed.). Academic Press, New York. pp. 55-90.

Pearce, R. B., J. Cronshaw., and W. N. Holmes. 1977. In: Adrenal cortex: structure and function. Cell. Tissue Res. 183, 203.

Pennycuick, C. J. 1968. A wind-tunnel study of gliding flight in the pigeon *Columba livia. Journal of Experimental Biology* 49: 509-526.

Pennycuick, C. J. 1972. *Animal flight.* Arnold, London.

Pennycuick, C. J. 1975. Mechanics of flight. In: *Avian Biology,* Vol. V (D. S. Farner and J. R. King. eds.) pp. 1-75. Academic Press, New York.

Peters, J. L. 1931-70. Chech-list of birds of the world. I-XV. Harvard, Cambridge, Mass.

Pough, H. F., J. B. Heiser., and W. N. McFarland. 1990, 1991. Vertebrate life. Macmillan, New York.

Pulido, F., and J. H. Rappole. 2002. The evolution in avian migration. Abs. Vol. XXIII Int. Ornithol. Congr., Beijing, China.

Pumphrey, R. J. 1961. In: P. D. Sturkie: Avian Physiology (2nd ed.). 1965.

Pycraft, W. P. 1900. On the morphology and phylogeny of the Palaeognathae (Ratitae and Crypturi) and Neognathae (Carinatae). Trans. Zool. Soc., London 15: 149- 290.

Pycraft, W. P. 1901. Some of the points in the morphology of the palate of the Neognathae. Journ. Linn. Soc. London, Zool 28: 343-357.

Radesäter, T., and S. Jacobsson. 1988. Intra- and intersexual functions of song in the Willow Warbler (*Phylloscopus trochilus*). XIX Int. Ornithol. Congr., Ottawa, Canada, Natl. Mus. Nat. Sci., Ottawa Univ. Press, pp. 1382-1390.

Ramzan, M., and H.S. Toor. 1972. Studies in damage to guava fruit due to roseringed parakeet, *Psittacula krameri* (Scopoli), at Ludhiana (Pb.). *Punjab Hortic. J.* 12: 144-145.

Ramzan, M., and H.S. Toor. 1973. Damage to maize crop by roseringed parakeet, *Psittacula krameri* (Scopoli) in the Punjab. J. Bombay Nat. Hist. Soc. 70: 201-204.

Rana, B. D. 1991. Ecology of the Indian Ring Dove in an arid environment. Abs. no. 51, Int. Symp. on Environmental and Hormonal approaches to Ornithology. Srinagar, Garhwal University, U. P., India.

Ray, B., B. Basu., and S. Guha. 1984. Zoology, Vol. I (In Bengali version). The Central Book Agency, Kolkata 700 012.

Ray, B., B-N. Bhattacharyya., B. Basu., and P. Ray Basu. 2003. Zoology, Vol-I, Rev. 2nd. ed. (In Bengali version). Abhinaba Prokashan, Kolkata - 700 009.

Rich, P. V. 1985. Osteological Comparison of the Scrubbirds, *Atrichornis,* and lyrebirds, *Menura* (Passcriformes: Atrichornithidae and Menuridae). Records Australian Mus. 37: 1-173.

Richards, L. P., and W. J. Bock. 1973. Functional anatomy and adaptive evolution of the feeding apparatus in the Hawaiian Honeycreeper genus *Loxops* (Drepanididae). Amer. Ornith. Union, Ornith Monogram., no. 15. 173 pp.

Riddle, O. 1937. On carbohydrate metabolism in pigeons. *Cold Spring Harbour Symp.* Quant. Biol, 5: 362-374.

Ringer, R. K. 1976. Thyroid gland in birds: structure and function. In: 'Avian Physiology' (P. D. Sturkie ed.), 3rd edn., p. 348. Springer-Verlag, New York.

Romer, A. S. 1966. Vertebrate Palaeontology, 3rd ed., University of Chicago Press, Chicago.

Romer, A. S., and T. S. Parsons. 1986. The Vertebrate Body, 6th ed. W. B. Saunders, Philadelphia.

Romanoff, A. L. 1960. *The avian embryo*. MacMillan, New York.

Rowan, W. 1931. The riddle of migration. Williams and Williams, Baltimore, XIXV+ 151 pp.

Ruben, J. 1991. Reptilian physiology and the flight capacity of *Archaeopteryx*. Evolution 45: 1-17.

Saini, H. K., and H. S. Toor. 1991. Feeding ecology and damage potential of feral pigeons *Columba livia* in agricultural habitat. Gerfaut 81: 195-206.

Saini, M. S., M. S. Dhindsa., H. K. Saini., and H. S. Toor. 1994. Food of Jungle Babbler and Common Babbler: a comparative study. J. Bombay Nat. Hist. Soc.

Sakai, Y. 1960. Thyroid gland in birds: structure and function.

Sanz, J. L., L. M. Chiappe., and A. D. Buscalioni. 1995. The osteology of *Concornis lacustris* (Aves: Enantiornithes) from the lower Cretaceous of Spain and a reexamination of its phylogenetic relationships. Am. Mus. Nov 3133: 1-23.

Sauer, E. G. F., and S. T. Emlen. 1971. Celestial rotation and stellar orientation in migratory warblers. Science, 173: 459-461.

Scheid, P., and J. Piiper. 1970. Analysis of gas exchange in the avian lung: theory and experiments in the domestic fowl. Respir. physiol. 9: 246-262.

Sclater, P. L. 1858. On the general geographical distribution of the members of the Class Aves. J. Proc. Linnean Soc. (Zool.), 2, 130-145.

Sengupta, Sudhin. 1979 (Sep.); Jatinga Bird Mystery; Science Reporter. Pp: 587-590.

Sengupta, Sudhin. 1986 (Dec.); Bird phenomena of Jatinga: A magnetic cue; Science Reporter. Pp: 748-752.

Sereno, P. C., and C. Rao. 1992. Early evolution of avian flight and perching: new evidence from the Lower Cretaceous of China. Science, 255, 845-848.

Sibley, C. G., and J. E. Ahlquist. 1990. Phylogeny and classification of Birds: A study in Molecular Evolution. Yale Univ. Press, Yale.

Simonetta, A. M. 1960. On the mechanical implications of the avian skull and their bearing on the evolution and classification of birds. Quart. Rev. Biol., 35: 206-220.

Sinha, D., and A. Ghosh. 1961. Some aspects of adrenocortical cytochemistry in the domestic pigeon. Endokrinologie, vol. 40: 270-280.

Slater, P. J. B., S. A. Ince, and P. W. Colgan. 1980. Chaffinch song types: their frequencies in the population and distribution between repertoires of different individuals. Behaviour, 75: 207-218.

Slater, P. J. B., S. A. Ince., 1982. Song development in Chaffinches: what is learnt and when? Ibis, 124: 21-26.

Slater, P. J. B., F. A. Clements., and D. J. Goodfellow. 1984. Local and regional variations in Chaffinch song and the question of dialects. Behaviour, 88: 76-97.

Smith, C. A. 1985. Inner ear. In: *Form and Function in Birds* (A. S. King and J. McLelland eds.). Vol. 3, pp. 273-310. Academic Press, London.

Smith, J. H., J. L. Meier., C. Lamke., P. J. G. Neill., and E. D. Box. 1986. Microscopic and submicroscopic anatomy of the parabronchi, air sacs, and respiratory space of the budgerigar (*Melopsittacus undulatus*). Amer. J. Anat. 177: 221-242.

Sorjonen, J. 1983. Transmission of the two most characteristic phrases of the song of the thrush nightingale (*Luscinia luscinia*) in different environmental conditions. Ornis Scand 14: 278-288.

Sorjonen, J. 1986. Singing strategies in northern European passerines. Univ. Joensuu Publ. Sci. 9: 3-16.

Steiner, H. 1956. Die taxonomische und phylogeneltische Bedeutung der Diastataxie des Vögel flügels. J. Orn., Lpz., 97: 1-20.

Stettenheim, P. 1972. The integument of birds. In: *Avian Biology* (D. S. Farner and J. R. King eds.). Vol. 2. pp. 1-63. Academic Press, London.

Stresemann, E, 1927-1934. Sauropsida: Aves, etc., xi, 899 pp. In: Kuekenthal, *Handbook der Zoologie*, 7: Hälfte II.

Takasaka, T., and C. A. Smith. 1971. The structure and innervation of the pigeon's basilar papilla. J. Ultra. Res. 35: 20-65.

Terrill, S. B. 1988. The relative importance of ecological factors in Bird Migration. In: Acta XIX Int. Ornithol. Congr., Ottawa, Canada, 1986. Publ. Natl. Mus. Nat. Sci., Ottawa University Press, pp. 2180-2190.

Thomson, A. L. 1936. Bird migration H. F. G. & Witherby, London.

Thomson, A. L. (ed,). *'A New Dictionary of Birds'*. 1964. Nelson, London.

Tinbergen, J. M. 1995. The Macdonald Encyclopedia of birds of the world. Little Brown and Company, U. K. pp. 1-512.

Tortora, G. J., and N. P. Anagnostakos. 1990. Principles of Anatomy and Physiology (*6th* ed.).

Toor, H. S., H. Kaur, and M. S. Dhindsa. 1986. Community structure and feeding ecology of birds at a grain store in Punjab (India). Trop. Sci. 26: 233-247.

Turner, C. D., and J. T. Bagnara. 1976. General Endocrinology (*6th* ed.). W. B. Saunders Co. London.

Van Gennip, E. M. S. J. 1986. The osteology, arthrology, and myology of the jaw apparatus of the pigeon (*Columba livia* L.) Netherl. J. Zool. 36: 1-46.

Van Tyne, J., and A. J. Berger. 1976. *Fundamentals of Ornithology* (2nd ed.). John Wiley and Sons, New York, London, Sydney, Toronto.

Vander, A. J., J. H. Sherman., and D. S. Luciano. 1975. Human Physiology: The Mechanisms of Body Function. 2nd. ed., McGraw-Hill, New York.

Von Frisch, K. 1967. The dance language and orientation of bees. Belknap, Cambridge, Mass.

Voous, K. H. 1980. New developments in avian systematics: a summary of result. Proc. XVII Int: Ornithol. Congr. pp. 1232-1234.

Wallace, G. J. 1963. An introduction to Ornithology. The Macmillan Co., New York.

Wallraff, H. C. 1977. Selected aspects of migratory orientation in birds. Vogelwarte. 29: 64-76.

Watanabe, T. 1960. On the peripheral courses of the vagus nerve in the fowl. Jap. J. Vet. Sci. 22: 145-154.

Weller, M. W. 1968. Plumages and Wing Spurs of Torrent Ducks, *Merganetta armata*. Wildfowl 19: 33-40.

Wellnhofer, P. 1974. Das fünfte skelettexemplar vou *Archaeopteryx*. Paleontographica A 147: 169-216.

Wellnhofer, P. 1992. A new specimen of *Archaeopteryx* from the Solenhofen limestone. In: K. E. Campbell Jr. (ed.). Papers in Avian Palaeontology Honouring Pierce Brodkorb. Nat. Hist. Mus-. Los Angeles County Sci Ser 36: 3-23.

Wideman, R. F., E. J. Braun., and G. L. Anderson. 1981. Microanatomy of the renal cortex in the domestic fowl. J. Morp 168: 249-267.

Wiltschko, W. 1982. The migratory orientation of Garden Warblers, *Sylvia borin*. In: Avian navigation. F. Papi and H. G. Wallraff (eds.). Springer-Verlag, Berlin. pp. 50-58.

Wiltschko, W. 1983. Compasses used by birds. Comp. Biochem. Physiol. 76A: 709-717.

Wingstrand, K. G., and O. Munk. 1965. The pecten oculi of the pigeon with particular regard to its function. Biol. Skr. Danske Vid. Selsk. 14: 1-64.

Wolfson, A. 1952. The cloacal protruberance, a means for determining breeding condition in live male passerines. Bird Banding 23: 159-165.

Wright, A., J. G. Phillips., and D. P. Huang. 1966. In: Adrenal Cortex; W. N. Holmes and J. Cronshow (Avian Endocrinology: A. Epple and M. H. Stetson eds., 1980).

Witmer, L. 1990. The Craniofacial air sac system of Mesozoic birds (Aves). Zool. J. Linn. Soc. 100: 327-378.

Witschi, E. 1961. Sex and secondary sexual characters. In: Biology and Comparative Physiology of Birds (A. J. Marshall ed.). Vol. 2, pp. 115-168. Academic Press, New York.

Young, J. Z. 1981. The life of Vertebrates (3rd ed.). Butler and Tanner Ltd., Frome, Somerset, U. K.

Yudin, K. A. 1965. The Phylogeny and classification of the Charadriiformes. In: Fauna of U. S. S. R. Vol. 2: Birds. Part I, New Series No. 91, 262 pp. USSR. Acad. Sci. Zool. Institute [in Russian].

Zeigler, H. P., P. W. Levitt., and R. R. Levine. 1980. Eating in the pigeon (*Columba livia*). Movement patterns, stereotypy and stimulus control. J. Comp. Physiol. Psyohol. 74: 783-794.

Zinoviev, A. V. 2002. Stages in the evolution of the foot in jacanas (Jacanidae). In: Abstract Vol., IOC., Beijing, China.

Ziswiller V., and D. S. Farner. 1972. Digestion and digestive system. In: Avian Biology (D. S. Farner and J. R. King eds.). Vol. II, Academic Press, London. pp. 343-430.

Ziswiller, V. 1985. Palate. In: A Dictionary of Birds (B. Campbell and E. Lack, eds.). pp. 427-429. T and A. D. Poyser, Calton, U. K.

Zusi, R. L. 1962. Structural adaptations of the head and neck in the Black Skimmer, *Rynchops nigra* L. Publs. Nuttal Ornith. Club 98: 5-10.

Zusi, R. L. 1978. The interorbital septum in cardueline finches. Bull. Brit. Ornith. Club 98: 5-10.

Zusi, R. L. 1984. Functional and evolutionary analysis of rhynchokinesis in birds. Smithsonian Contrib. Zool 395: 1-40.

Zweers, G. A. 1974. Structure, movement and myography of the feeding apparatus of the mallard (*Anas platyrhynchos* L.). Neth. J. Zool 24: 323-487.

Zweers, G. A., H. C. Van Pelt., and A. Beckers. 1981. Morphology and mechanics of the larynx of the pigeon (*Columba livia* L.): a drill-chuck system (Aves). Zoomorph. 99: 37-69.

Zweers, G. A. 1982a. Pecking of the pigeon (*Columba livia* L.).Behaviour 81: 173-230.

Zweers, G. A. 1982b. Drinking of the pigeon (*Columba livia* L.).Behaviour 80: 274-317.

Zweers, G. A. 1982c. The feeding system of the pigeon (*Columba livia* L.).Adv. Anat. Embryol. Cell Biol., 73: 1-108.

Zweers, G. A. 1991a. Pathways and space for evolution of feeding mechanisms in birds. In: E. C. Dudly (ed.). The unity of Evolutionary Biology. Dioscorides Press, Portland, pp. 530-547.

Zweers, G. A. 1991b. Transformation of avian feeding mechanisms: a deductive approach. Acta Biotheor 39: 15-36.

Zweers, G. A., H. Berkhoudt., and J. C. Vanden Berge. 1994a. Behavioural mechanisms of avian feeding. Adv. Comp. Envir. Physiol 18: 241-279.

Zweers, G. A., J. C. Vanden Berge., and H. Berkhoudt. 1997. Evolutionary patterns of avian trophic diversification. Zoology 100: 25-57.

Zweers, G. A., and J. C. Vanden Berge. 1997/98. Birds at geological boundaries. Zoology 100: 183-202; Gustav Fischer Verlag.

Index

D

E

N

O

P

Q

R